CISM COURSES AND LECTURES

The series presents lecture notes, monographs, edited works and proceedings in the field of Mechanics, Engineering, Computer Science and Applied Mathematics.
Purpose of the series is to make known in the international scientific and technical community results obtained in some of the activities organized by CISM, the International Centre for Mechanical Sciences.

INTERNATIONAL CENTRE FOR MECHANICAL SCIENCES

COURSES AND LECTURES - No. 444

ADVANCED DYNAMICS AND CONTROL OF STRUCTURES AND MACHINES

EDITED BY

HANS IRSCHIK
JOHANNES KEPLER UNIVERSITY OF LINZ

KURT SCHLACHER
JOHANNES KEPLER UNIVERSITY OF LINZ

Springer-Verlag Wien GmbH

This volume contains 63 illustrations

Originally published by Springer-Verlag Wien New York in 2004
SPIN 11313434

In order to make this volume available as economically and as rapidly as possible the authors' typescripts have been reproduced in their original forms. This method unfortunately has its typographical limitations but it is hoped that they in no way distract the reader.

ISBN 978-3-211-22867-8 ISBN 978-3-7091-2774-2 (eBook)
DOI 10.1007/978-3-7091-2774-2

PREFACE

Advanced dynamics and control of structures and machines represents an up-to-date, multi-disciplinary field in engineering. It brings together researchers in mechanical, civil and control engineering, and in applied mathematics. This book is a result of the Advanced School entitled "Advanced Dynamics of Structures and Machines", which took place in the International Centre for Mechanical Sciences (CISM), Udine, Italy, during April 15–19, 2002.

It was the aim of this school to present modern foundations and methods of modeling and control of flexible structures and machines, under operating as well as under severe loading conditions. The need to understand such advanced formulations for deformable mechanical systems is evident, especially in problems of automotive engineering, of robotics, in rotating machinery and for civil engineering structures. The collection of lecture notes given in this book is an attempt to introduce the basic ideas of dynamics and control, to provide scientific knowledge for these fast expanding fields, as well as to present up-to-date research.

The book starts with a concise introduction into the models and equations of continuum mechanics, including the laws of thermodynamics and a discussion on the form of the constitutive relations for materials. Afterwards, the balance equations and jump relations of continuum mechanics are reviewed with a special emphasis upon the validity and consistent form of these relations. The Rayleigh-Ritz method and the Lagrange equations then are applied to the field equations of continuum mechanics, where formulations for industrial problems are given with reference to material as well as non-material volumes.

The book proceeds with a concise presentation of the principles and methods of analytical mechanics, including the canonical Hamiltonian equations, the Hamilton variational principle and the Lagrange-Maxwell equations of electromechanics. A further chapter presents exact solutions for deformation compensation by shaped actuation in elastic bodies in the presence of rigid-body motions. As an example for non-linear constitutive behavior, elasto-plasticity in multi-body dynamics is discussed from a computational point of view.

For the analysis of structures and machines in the high frequency range, two approaches are proposed, namely high frequency structural analysis and the vibrational conductivity approach.

Basic notions of differential geometry are introduced followed by the discussion of certain geometric representations of dynamic systems. Algorithms for checking the accessibility and observability of explicit control systems, as well as two nonlinear control design techniques are presented. Also, the use of differential geometry in order to obtain a coordinate-free representation of the equations of nonlinear mechanics is discussed. The port-based modeling of lumped-parameter complex physical systems leads to the class of port-Hamiltonian systems. It is shown how passivity-based control results from interconnecting the plant port-Hamiltonian system with a controller port-Hamiltonian system. Extensions to the distributed-parameter case are provided by formulating boundary control systems as infinite-dimensional port-Hamiltonian systems.

A discussion of the periodic motion induced by the Painlevé paradox commences with the introduction of a frictional impact oscillator containing the basic mechanism for a hopping phenomenon observed in many practical applications. It is shown that this type of systems can exhibit the Painlevé paradox for physically realistic values of the friction coefficient.

The problem of coordinating rigid and flexible joint robot manipulators clearly being attached with particular industrial importance is solved by means of a master-slave scheme incorporating a feedback control law and nonlinear observers. The performance of the proposed control law is illustrated by experimental results.

This book also focuses on structural control for dynamic hazard mitigation providing a discussion of structural control implementations for the area of civil and architectural engineering. Special care is devoted to the interdisciplinary character of the achievable progress and to the role played by laboratory testing.

It is a great pleasure to acknowledge the significant contributions made to this book by Alexander K. Belyaev of St. Petersburg, Russia, Fabio Casciati of Pavia, Italy, Arjan van der Schaft of Twente, and Henk Nijmeijer of Eindhoven, The Netherlands. They delivered excellent lectures in Udine in April 2002 and they contributed chapters to this book making our joint project a truly international effort. We wish to thank R.I. Leine and H.A. van Essen of Eindhoven, The Netherlands, B. Brogliato of Saint Ismier, France, and A. Rodriguez-Angeles of the Mexican Petroleum Institute, Mexico, for their contributions.

The school brought together a number of 42 participants from 10 countries. We are grateful to all participants for their interest and the numerous discussions that took place during and after the courses. We are particularly thankful to the Scientific Council of CISM for supporting the advanced school and recognizing the importance of topics related to dynamics and control. Last but not least, we wish to thank our co-authors J. Gerstmayr, M. Dibold, F. Hammelmüller, H.J. Holl, M. Nader, U. Pichler, Ch. Zehetner, K. Zehetleitner, J. Holl, G. Grabmair, S. Fuchshumer, H. Ennsbrunner and R. Stadlmayr for the fruitful discussions and their innovative contributions.

H. Irschik
K. Schlacher

CONTENTS

Basics of Continuum Mechanics

Alexander K. Belyaev

Department of Mechanics and Control, Polytechnic University of St. Petersburg, Russia

Basically we will follow the books by Truesdell (1977) and Palmov (1998).

1 Kinematics of a continuous medium

The motion of the continuous medium is considered in a certain fixed system of rectangular Cartesian coordinates $x_1x_2x_3$. The position of a generic material point M of the continuum at time t is given by the position vector $\mathbf{r}$. It allows us to individualize the material points of the continuum by means of their position at any instant of time.

The individualization of the material points in terms of their position at some fixed instant $t = t_0$ of time (e.g. at the initial instant of time) introduces the Lagrangian approach to continuum mechanics whereas individualization of the material points in terms of their position at the actual instant of time t introduces the Eulerian approach.

The position of a generic material point M at the initial instant $t = t_0$ of time is given by the vector $\mathbf{R}$. In the framework of the Lagrangian approach the material point is completely specified by the given value of $\mathbf{R}$. The motion of the point is given by a single-valued function of the arguments $\mathbf{R}$ and t

$$\mathbf{r} = \mathbf{r}(\mathbf{R}, t). \tag{1.1}$$

Provided that dependence (1.1) is given for each point $\mathbf{R}$ of the initial volume, this dependence is said to determine the motion of the material volume. The Lagrangian coordinate $\mathbf{R}$ of any material point remains constant during the motion of the point. Hence the velocity and the acceleration of the material points are as follows

$$\mathbf{v} = \dot{\mathbf{r}} = \frac{\partial \mathbf{r}}{\partial t} = \mathbf{v}(\mathbf{R}, t), \quad \mathbf{w} = \ddot{\mathbf{r}} = \frac{\partial^2 \mathbf{r}}{\partial t^2} = \mathbf{w}(\mathbf{R}, t). \tag{1.2}$$

In the framework of the Eulerian approach the material points are specified by their position $\mathbf{r}$ at the current instant of time t. This means that a single-valued dependence $\mathbf{R} = \mathbf{R}(\mathbf{r}, t)$ is assumed to be known and thus $\mathbf{v} = \mathbf{v}(\mathbf{r}, t)$.

The displacement of a generic material point is determined by the difference

$$\mathbf{u} = \mathbf{r} - \mathbf{R}. \tag{1.3}$$

The displacement $\mathbf{u}$ can be expressed either in terms of the Lagrangian variables $\mathbf{R}$ and t or in terms of the Eulerian variables $\mathbf{r}$ and t. This leads to two different representations for the displacement

$$\mathbf{u} = \mathbf{u}(\mathbf{R}, t), \quad \mathbf{u} = \mathbf{u}(\mathbf{r}, t). \tag{1.4}$$

In what follows we assume that functions $\mathbf{u}$, $\mathbf{r}$, $\mathbf{v}$ and $\mathbf{w}$ are continuous functions of time and spatial variables $\mathbf{R}$ and $\mathbf{r}$.

2 Law of conservation of mass

The mass density ρ is usually given in Eulerian variables $\rho = \rho(\mathbf{r}, t)$ and yields the mass of the material volume V

$$m = \int_V \rho dV. \tag{2.1}$$

An integral form of the mass conservation law is as follows

$$\dot{m} = \int_V \frac{\partial \rho}{\partial t} dV + \int_S \mathbf{n} \cdot \mathbf{v} \rho dS = 0. \tag{2.2}$$

While estimating the time derivative of the volume integral (2.1) we took into account that change of the mass m is due to the explicit dependence $\rho(t)$ and the volume change caused by the motion of the surface S of the material volume V. The volume change depends only upon the normal component of the velocity $v_n = \mathbf{n} \cdot \mathbf{v}$, with $\mathbf{n}$ being the exterior unit normal to the surface S. Applying the Gauss-Ostrogradsky integral theorem to the surface integral and taking into account that the material volume V is arbitrary yields a differential form of this law

$$\frac{\partial \rho}{\partial t} + \nabla \cdot (\mathbf{v} \rho) = 0, \tag{2.3}$$

where ∇ is the Hamilton operator. Introducing the material derivative of ρ, that is $\dot{\rho} = \dfrac{\partial \rho}{\partial t} + \mathbf{v} \cdot (\nabla \rho)$, one arrives at another differential form of the mass conservation law

$$\dot{\rho} + \rho (\nabla \cdot \mathbf{v}) = 0. \tag{2.4}$$

3 Strain and rate of strain

The instantaneous motion of an infinitesimally small vicinity of a material point $\mathbf{r}$ is characterized by the tensor $\nabla \mathbf{v}$ where $\mathbf{v} = \mathbf{v}(\mathbf{r}, t)$. The first invariant of this tensor, i.e. $\nabla \cdot \mathbf{v}$, determines the rate of change of the mass density, as it follows from the mass conservation law (2.4). The deviator of this tensor is

$$\mathbf{D} = \nabla \mathbf{v} - \frac{1}{3} \nabla \cdot \mathbf{v} \mathbf{E}, \tag{3.1}$$

where $\mathbf{E}$ is the unity tensor. The symmetric part of the deviator determines the rate of distortion of a small vicinity of a material point whereas the skew symmetric part (the spin tensor) represents the velocity of the rigid-body rotation of the above vicinity.

In continuum mechanics, finite deformation is characterized by various strain measures and strain tensors. In the framework of the Eulerian approach one usually uses the Almansi strain measure

$$\mathbf{g} = (\nabla \mathbf{R}) \cdot (\mathbf{R} \nabla). \tag{3.2}$$

The third invariant of tensor $\mathbf{\Lambda} = (\rho/\rho_0)^{-2/3}\,\mathbf{g}$ is equal to unity since

$$I_3(\mathbf{\Lambda}) = \left(\frac{\rho}{\rho_0}\right)^{-2} I_3(\mathbf{g}) = 1. \tag{3.3}$$

Therefore, tensor $\mathbf{\Lambda}$ characterises deformation without change of the mass density and for this reason is referred to as strain of shape change.

Linearised equations of continuum mechanics are utilised for analysis of engineering structures since a routine functioning of these structures is feasible only for small structural displacement $\mathbf{u} = \mathbf{r} - \mathbf{R}$. The Almansi strain measure $\mathbf{g}$ is then

$$\mathbf{g} = \mathbf{E} - 2\varepsilon + (\nabla \mathbf{u}) \cdot (\mathbf{u} \nabla), \tag{3.4}$$

where ε denotes the linear strain tensor

$$\varepsilon = (\nabla \mathbf{u})^s. \tag{3.5}$$

4 Equations governing dynamics of a continuum

The Cauchy true stress tensor τ is introduced for describing the stresses in a continuum. The traction vector on a surface with exterior unit normal $\mathbf{n}$ is given by the Cauchy formula

$$\tau_n = \mathbf{n} \cdot \tau. \tag{4.1}$$

The equations of dynamics of an arbitrary material volume V are as follows

$$\frac{d}{dt}\int_V \mathbf{v}\rho dV = \int_V \mathbf{K}\rho dV + \int_S \mathbf{n} \cdot \tau dS, \tag{4.2}$$

$$\frac{d}{dt}\int_V \mathbf{r} \times \mathbf{v}\rho dV = \int_V \mathbf{r} \times \mathbf{K}\rho dV - \int_S \mathbf{n} \cdot \tau \times \mathbf{r} dS, \tag{4.3}$$

where S is the boundary surface of volume V and $\mathbf{K}$ is a body force vector. These equations represent the law of change of linear momentum and the law of change of moment of momentum, respectively. Equations (4.2) and (4.3) represent an integral form of the equations of dynamics. The Gauss-Ostrogradsky theorem yields the following differential form

$$\nabla \cdot \tau + \rho (\mathbf{K} - \dot{\mathbf{v}}) = 0, \quad \tau^T = \tau. \tag{4.4}$$

The latter equation expresses the symmetry of the stress tensor.

The stress tensor can be decomposed into a spherical part and a deviator, such that

$$\boldsymbol{\tau} = \sigma \mathbf{E} + \mathbf{s}. \tag{4.5}$$

Here σ denotes the mean normal stress $\sigma = \frac{1}{3}\mathbf{E} : \boldsymbol{\tau}$ and $\mathbf{s}$ is a stress deviator. As follows from (4.5) the first invariant of a deviator vanishes, i.e. $\mathbf{E} : \mathbf{s} = \mathbf{0}$.

5 The laws of thermodynamics

The first law of thermodynamics for an arbitrary material volume is set in the following form

$$\dot{T} + \dot{u} = p + q. \tag{5.1}$$

Here T and u denote respectively the kinetic and internal energy of the material volume

$$T = \frac{1}{2}\int_V \mathbf{v} \cdot \mathbf{v} \rho dV, \quad u = \int_V U \rho dV, \tag{5.2}$$

U being the specific internal energy. Further, p is the power of the body forces and surface tractions of the material volume

$$p = \int_V \mathbf{K} \cdot \mathbf{v} \rho dV + \int_S \boldsymbol{\tau}_n \cdot \mathbf{v} dS \tag{5.3}$$

and q is the rate of heat supply

$$q = \int_V b dV - \int_S \mathbf{n} \cdot \mathbf{h} dS. \tag{5.4}$$

Here b denotes the heating supply whilst the surface integral describes the heat supplied to the volume through its surface, $\mathbf{h}$ being the heat flux.

The first law (5.1) is given in an integral form. In order to obtain a differential form we apply the rule of differentiation of integrals over the material volume, the Cauchy formula for traction and the Gauss-Ostrogradsky theorem to reduce eq. (5.1) to a volume integral. Since eq. (5.1) holds true for any material volume, the integrand of the volume integral must vanish, which yields the differential form of the first law of thermodynamics

$$\rho \dot{U} = \boldsymbol{\tau} : (\nabla \mathbf{v}) + b - \nabla \cdot \mathbf{h}. \tag{5.5}$$

The second law of thermodynamics is put in the form of the Clausius-Duhem inequality. This inequality says that the rate of change of the internal entropy of the volume exceeds or equals the rate of the entropy supplied to the volume by external heat sources, i.e.

$$\dot{s} \geq \int_V \frac{b}{\theta} dV - \int_S \frac{\mathbf{n} \cdot \mathbf{h}}{\theta} dS, \tag{5.6}$$

where s and θ are the internal entropy and the absolute temperature of the material volume, respectively. The first term on the right hand side of eq. (5.6) is the entropy supplied to the volume while the second term describes the entropy supply through the surface.

Introducing the specific internal entropy S

$$s = \int_V S\rho dV, \tag{5.7}$$

applying the Gauss integral theorem and taking into account the arbitrariness of the material volume yields a differential form of the second law

$$\rho \dot{S} \geq \frac{1}{\theta}(b - \nabla \cdot \mathbf{h}) + \frac{1}{\theta^2}\mathbf{h} \cdot (\nabla \theta). \tag{5.8}$$

6 Constitutive equations for materials

We suppose here and in what follows that the thermodynamic state of a continuum is determined by the temperature θ, its gradient $\nabla\theta$ and the tensor $\nabla\mathbf{R}$ in a small vicinity of each point. These quantities are referred to as the defining variables.

Dealing with the Eulerian approach one expresses the above variables in terms of the actual coordinate $\mathbf{r}$ and the time t. Given the time-history of the defining variables for a material point with a Lagrangian coordinate $\mathbf{R}$ the thermodynamic process at this material point is said to be prescribed. The time-history is given by the following functions of time

$$\begin{aligned} \theta &= \theta[\mathbf{r}(\mathbf{R}, \tau), \tau], \\ \nabla\theta &= \nabla\theta[\mathbf{r}(\mathbf{R}, \tau), \tau], \\ \nabla\mathbf{R} &= \nabla\mathbf{R}[\mathbf{r}(\mathbf{R}, \tau), \tau], \quad t_0 < \tau < t, \end{aligned} \tag{6.1}$$

where the Eulerian coordinate $\mathbf{r}$ is replaced by its expression (1.1).

The definition of a material implies that the thermodynamic behaviour of a continuum at a point $\mathbf{r}$ is completely determined provided that the stress tensor τ, the heat flux $\mathbf{h}$, the specific free energy $F = U - \theta S$ and the specific internal entropy S are prescribed as operators over the time-history (6.1) at this point, i.e.

$$\begin{aligned} \tau(\mathbf{R}, t) &= \tau\{\mathbf{R}, \theta, \nabla\theta, \nabla\mathbf{R}\}, \\ \mathbf{h}(\mathbf{R}, t) &= \mathbf{h}\{\mathbf{R}, \theta, \nabla\theta, \nabla\mathbf{R}\}, \\ F(\mathbf{R}, t) &= F\{\mathbf{R}, \theta, \nabla\theta, \nabla\mathbf{R}\}, \\ S(\mathbf{R}, t) &= S\{\mathbf{R}, \theta, \nabla\theta, \nabla\mathbf{R}\}, \quad t_0 < \tau < t. \end{aligned} \tag{6.2}$$

Equations (6.2) describe a material and are conventionally referred to as the constitutive equations for the material.

A process in the material described by the time-history (6.1) is called reversible if condition (5.8) holds with the equality sign, and is called irreversible if (5.8) holds with the inequality sign.

A material is said to be homogeneous if the operators (6.2) do not contain the Lagrangian coordinate $\mathbf{R}$ explicitly, i.e. the operators (6.2) have the same form for any material point of the medium.

A material is called isotropic if its operators τ, $\mathbf{h}$, F and S at point $\mathbf{r}$ at time instant t do not change their values when an arbitrary initial rigid rotation is superimposed onto the body elements. Mathematically, this is equivalent to replacement of the motion $\mathbf{R} = \mathbf{R}(\mathbf{r}, t)$ by $\mathbf{R}_H = \mathbf{R}(\mathbf{r}, t) \cdot \mathbf{H}$, where $\mathbf{H}$ is an arbitrary constant orthogonal tensor.

Materials are known to behave differently at the volume change than at the shape change. The dilatation is determined by the mass density and is related to the third invariant of $\mathbf{g}$ as follows

$$\rho = \rho_0 \sqrt{I_3(\mathbf{g})}, \tag{6.3}$$

where ρ_0 is the initial mass density and $I_3(\mathbf{g})$ is the third invariant of tensor $\mathbf{g}$.

As an example, we consider a compressible heat-conducting material with elastic dilatation. The mean normal stress is assumed to be a function (not an operator over time) of relative mass density ρ_0/ρ and temperature θ

$$\sigma = \sigma\left(\frac{\rho_0}{\rho}, \theta\right) \tag{6.4}$$

and the heat flux is a linear vector function of the temperature gradient

$$\mathbf{h} = -\kappa \nabla \theta. \tag{6.5}$$

Here κ denotes the coefficient of thermal conductivity which is a function of temperature, absolute value of its gradient and mass density $\kappa = \kappa\left(\rho_0/\rho, \theta, |\nabla\theta|\right)$.

The second law of thermodynamics play a special role since its equation contains only operators (6.2) and no external parameter. Hence, only the second law of thermodynamics imposes restrictions on the choice of operators (6.2). As shown by Truesdell 1977 and Palmov 1998 the imposed restriction yields

$$\sigma = \rho_0 \frac{\partial F_0}{\partial\left(\rho_0/\rho\right)}, \quad S_0 = -\frac{\partial F_0}{\partial \theta}, \quad \kappa\left(\nabla\theta\right)^2 \geq 0. \tag{6.6}$$

From this equation one can see that the specific free energy is a generating function both for mean normal stress and specific internal entropy. The thermal conductivity coefficient κ must be non-negative. If $\kappa > 0$ and $\nabla\theta \neq 0$, the second law of thermodynamics holds with inequality sign and the thermodynamic processes in the material is irreversible. Thermodynamic processes can be reversible if either $\kappa = 0$ or $\nabla\theta = 0$, i.e. when the heat conduction is absent as the heat flux $\mathbf{h}$ vanishes.

It is worth mentioning that it is rheology, i.e. a study on the behaviour of loaded materials, which is concerned with the prescription of the operators in constitutive equations (6.2) for various materials. The method of rheological models allows one to derive the constitutive equations with a broad variety of properties. An attractive side of the method of rheological models is that the obtained equations *a priori* satisfy the second law of thermodynamics, the property of material isotropy and the material frame indifference principle provided that the basic materials meet these requirements. The three classical rheological materials referred to as elastic, viscous and plastic elements describe respectively the elastic, viscous and plastic properties of solids. The basic concept of rheology is a rheological model which is composed of the above elements in series and in parallel, see Palmov 1998, for detail. In what follows, only elastic materials are studied.

7 Elastic material

Elastic materials are known to have no memory, that is all their variables are functions of the defining variables rather than operators over them. The stress deviator, free energy and entropy are assumed to be functions of the strain of shape change $\mathbf{\Lambda}(t)$, temperature $\theta(t)$ and mass density $\rho(t)$

$$\begin{aligned} \mathbf{s} &= \mathbf{s}(\rho, \mathbf{\Lambda}, \theta), \\ F &= F_0\left(\frac{\rho_0}{\rho}, \theta\right) + F_*(\mathbf{\Lambda}, \theta), \quad S = S_0\left(\frac{\rho_0}{\rho}, \theta\right) + S_*(\mathbf{\Lambda}, \theta). \end{aligned} \tag{7.1}$$

Tensor $\mathbf{\Lambda}$ is assumed to be prescribed in a fixed basis. Then the second law of thermodynamics yields the following constitutive equations

$$\mathbf{s} = -2\rho \mathrm{Dev}\left(\frac{\partial F_*}{\partial \mathbf{\Lambda}} \cdot \mathbf{\Lambda}\right)^s, \quad \left(\frac{\partial F_*}{\partial \mathbf{\Lambda}} \cdot \mathbf{\Lambda}\right)^a = 0, \; S_* = -\frac{\partial F_*}{\partial \theta}. \tag{7.2}$$

The free energy is seen to be a generating function for both the stress deviator and the specific entropy.

The isotropy of the material imposes some restrictions on the form of dependences (7.1). In particular, the free energy must be insensitive to a rotational transformation of the coordinate system. As a scalar, it must be a function of the first ($I = \mathbf{E} : \mathbf{\Lambda}$) and second $\left(II = \frac{1}{2}\mathbf{\Lambda} : \mathbf{\Lambda}\right)$ invariants of tensor $\mathbf{\Lambda}$. A good approximation to free energy in the case of finite deformations is the Mooney potential

$$F_* = \frac{A}{2\rho_0}(I-3) + \frac{B}{2\rho_0}\left(II - \frac{3}{2}\right), \tag{7.3}$$

where A and B may be some functions of temperature. This potential says that for $A > 0$ and $B > 0$ the free energy (7.3) turns out to be positive for any shape deformation, except the case of $\mathbf{\Lambda} = \mathbf{E}$ in which it is identically zero.

When the free energy is given by (7.3) expression (7.2) takes the especially simple form

$$\mathbf{s} = -\frac{\rho}{\rho_0}\mathrm{Dev}\left(A\mathbf{\Lambda} + B\mathbf{\Lambda}^2\right). \tag{7.4}$$

The second law of thermodynamics takes the form of the non-stationary heat conduction equation for elastic material

$$\rho\theta(\dot{S}_0 + \dot{S}_*) = b + \nabla \cdot (\kappa\nabla\theta), \tag{7.5}$$

where, by virtue of (7.2) and (7.3), the entropy S_* is given by

$$S_* = -\frac{1}{2\rho_0}\left[(I-3)\frac{\partial A}{\partial \theta} + \left(II - \frac{3}{2}\right)\frac{\partial B}{\partial \theta}\right]. \tag{7.6}$$

Bibliography

V.A. Palmov. *Vibrations of Elastoplastic Bodies.* Springer, 1998.

C.A. Truesdell. *A First Course in Rational Continuum Mechanics.* Academic Press, 1977.

A Treatise on the Equations of Balance and on the Jump Relations in Continuum Mechanics.

Hans Irschik[1]

1 Johannes Kepler University of Linz and
Linz Center of Excellence in Mechatronics (LCM), Linz, Austria

Abstract. In this Lecture, we are concerned with the range of applicability of the balance relations in continuum mechanics. These relations have been already introduced in Lecture 1 of the present book as an important foundation of the dynamics and control of structures and machines, Belyaev (2004). In the present Lecture, we start with the general form of the relations of balance, which we afterwards specialise to the equations of balance of mass, momentum and energy (the first law of thermodynamics) and to the balance relation of entropy (the second law of thermodynamics), the latter in the form of the Clausius-Duhem inequality. We also discuss the frequently used equations of balance of kinetic energy (the law of power), and we point out the consequences of inserting the latter into the first and second law of thermodynamics, leading to the balance relations for the internal energy and to the Clausius-Planck inequality, respectively. We lay special emphasis on reviewing the continuity conditions that must be satisfied in order that the global and local forms of the relations of balance do hold. When a surface of discontinuity, a so-called singular surface, is present, across which some entity shows different values when approaching from the two sides of the surface, jump relations are needed in order to connect the local forms of the relations of balance at the two sides. We summarise and extend a recent formulation, which allows to connect the local equations of balance of mass, momentum, energy, kinetic energy and internal energy in a consistent manner, Irschik (2003). In the latter reference, it has been shown that surface growth terms must be introduced for the sake of consistency, and relations between these surface growth terms have been derived. In the present Lecture, these results are extended with respect to the second law of thermodynamics, and with respect to the resolution of a contradictory result in the literature on the jump relations of energy and of internal energy.

1 On the Relations of Balance.

As the engineering background of the following treatise, we study the motion of some element of a structure or a machine, this element being virtually separated from the rest of the world by a closed surface, possibly together with some matter that surrounds the structural element. In the framework of *continuum mechanics*, the material body thus created is modelled as a dense, continuously distributed set of material particles. In order to describe the motion of this set of particles, in the following we refer our formulations to a *common inertial frame*, in which the

Newtonian dynamics of mass-points is assumed to hold. We furthermore use the *spatial or Eulerian formulation* of continuum mechanics throughout the Lecture, see Section 1 of Belyaev (2004). We thus express any mechanical entity as a function of its instantaneous place in the Euclidean space and of time, as observed in the common frame. The *material volume* that contains the set of material particles under consideration is denoted as $\mathcal{V}$. A material volume is a volume with a closed surface $\mathcal{S}$ moving together with the material particles located on it. In our background problem, $\mathcal{S}$ separates the structural element under consideration from the rest of the world. That the behaviour of the material particles of this structure or machine element is modelled in the framework of continuum mechanics does not mean that no *discontinuities* may occur in the fields to be studied. E.g., the element may consist of several solid parts made of different materials, the mass density attributed to the particles then showing a jump across the interface between the parts under consideration. Often, the material surface $\mathcal{S}$ will enclose not only solid parts, but also fluids. When the front of a wave is running through the material volume $\mathcal{V}$, various types of discontinuities may occur. Discontinuous fields also may arise when phase transformations have to be considered in an industrial production process, in which the machine element and its surrounding matter are involved. Moreover, a discontinuity is to be considered, when a contact between some surfaces takes place, etc. It is the scope of the present Lecture to discuss the basic *relations of balance of continuum mechanics*, with special emphasis laid upon the *continuity conditions* that must hold in order that these relations can be used, and upon the *jump relations* that must be set up in order to consistently relate discontinuous values of some entities under consideration.

In mechanics, basically three different equations of balance are contributed to the general theory, namely balance of mass, balance of momentum (also denoted as linear momentum) and balance of moment of momentum (or angular momentum). Formulations that are derived from the latter relations, such as the equation of balance of kinetic energy and Hamilton's principle, are frequently applied in connection with automatic control of structural vibrations. Additionally, balance equations from thermodynamics and, sometimes, from electrodynamics are needed. In the following, we deal with the mechanical equations of balance, as well as with the first and the second law of thermodynamics, the latter in the form of the Clausius-Duhem inequality. As has been shown in Lecture 1 of the present book, Belyaev (2004), these laws of thermodynamics are needed for a complete formulation of continuum mechanics. When the equation of balance of kinetic energy is substituted into the two fundamental laws of thermodynamics, relations known as balance of internal energy and the Clausius-Planck inequality are obtained, the latter two again representing derived formulations. In order to connect these various relations in a systematic manner, we first recall a general form of the laws of balance, which was stated in the article on the classical field theories by Truesdell and Toupin (1960), and which was extended in the book of Wang and Truesdell (1973) for inequalities of balance.

Let Ψ denote some mechanical or thermodynamic entity that is attached to the material particles of $\mathcal{V}$. We first formulate the instantaneous quantity of Ψ, namely the total $\mathcal{Q}_\Psi(t)$ of Ψ contained in the material volume $\mathcal{V}$ at the time t. In the spatial formulation, $\mathcal{Q}_\Psi$ follows from the integration over the spatial volume $\mathcal{v}_t$ that instantaneously coincides with the material volume $\mathcal{V}$:

$$\mathcal{Q}_\Psi(t) = \int_{\mathcal{v}_t} \Psi \, dv = \int_{\mathcal{V}} \Psi \, dv \,. \tag{1}$$

In general, the quantity $\mathcal{Q}_\Psi$ contained in the material volume $\mathcal{V}$ will change in the course of time. When $\mathcal{Q}_\Psi$ does not change, it is said to be *conserved.* A fundamental idea is *to balance* the change of $\mathcal{Q}_\Psi$ between two time instances, t and $t+\Delta t$, by the time-and space-wise resultant of *sources* acting upon the material particles in $\mathcal{V}$ during the time-interval $t \le \tau \le t+\Delta t$:

$$\mathcal{Q}_\Psi(t+\Delta t) - \mathcal{Q}_\Psi(t) - \int_t^{t+\Delta t} \mathcal{R}_\Psi(\tau)\, d\tau \ge 0\,. \tag{2}$$

This relation can be considered as the *fully general statement of any law of balance*, see Truesdell and Toupin (1960), Sect. 194 for the case of the equality sign in Eq.(2). In case of balance of mass, momentum and energy (the first law of thermodynamics), the equality sign holds, while balance of entropy (the second law of thermodynamics) requires the left hand side of Eq.(2) to be non-negative, compare Sects. 2, 4 and 5 of Belyaev (2004). The space-wise resultant $\mathcal{R}_\Psi$ can be decomposed into two parts:

$$\mathcal{R}_\Psi(\tau) = \int_{v_\tau} s[\Psi]\, dv - \oint_{s_\tau} \boldsymbol{da} \cdot \boldsymbol{i}[\Psi] = \int_{\mathcal{V}} s[\Psi]\, dv - \oint_{\mathcal{S}} \boldsymbol{da} \cdot \boldsymbol{i}[\Psi]\,. \tag{3}$$

The instantaneous bounding surface of v_τ at time τ is denoted as s_τ. In order to indicate that we refer to the change of $\mathcal{Q}_\Psi$ with respect to the material volume $\mathcal{V}$, we write the symbols $\mathcal{V}$ and $\mathcal{S}$ instead of v_τ and s_τ, respectively, see Eq.(3.2). The supply of Ψ within $\mathcal{V}$ is denoted by $s[\Psi]$. The influx of Ψ through $\mathcal{S}$ is written as $\boldsymbol{i}[\Psi]$. This latter entity represents surface sources of Ψ that are associated with the material particles located on $\mathcal{S}$. The oriented area element $\boldsymbol{da}$ of $\mathcal{S}$ is a vector pointing outwards of $\mathcal{V}$. In the balance equations of mechanics, Ψ is formed by a combination of kinematical entities such as mass density, position vector and velocity, and it remains restricted to scalars or vectors. In the laws of thermodynamics, we are concerned with scalar entities Ψ, namely with internal energy, kinetic energy and entropy. Hence, in our subsequent considerations the supply term $s[\Psi]$ will represent a scalar or a vector, while the influx $\boldsymbol{i}[\Psi]$ will be a vector or a tensor of second order, respectively. Throughout the present Lecture, we refer to the notation introduced in the article by Truesdell and Toupin (1960). Particularly, we make use of a generalised dot product introduced in the article on tensor fields by Ericksen (1960). Basically the same notation can be found, e.g., in Appendix A of the book by Slattery (1999). In the mechanical equations of balance, the generalised dot product in the surface integral of Eq.(3) denotes either the inner product of the two vectors $\boldsymbol{da}$ and $\boldsymbol{i}[\Psi]$, or a linear mapping of the vector $\boldsymbol{da}$ by means of the second order tensor $\boldsymbol{i}[\Psi]$, as defined in Sect.3 of Ericksen (1960). Of course, we have to require that the integrals in Eq.(3) do exist.

In the present Lecture, we assume that $\mathcal{Q}_\Psi(t)$ is continuously differentiable and $\mathcal{R}_\Psi(t)$ is bounded within $t \le \tau \le t+\Delta t$. We then can put Eq.(2) into the differential form

$$\frac{d}{dt}\mathcal{Q}_\Psi - \mathcal{R}_\Psi \ge 0\,, \tag{4}$$

which is known as the *general balance equation*, cf. Truesdell and Toupin (1960), Sect. 157 for the case of an equality sign, and see Chapter II.V of Wang and Truesdell (1973). The general balance equation attributes the time rate of change of $\mathcal{Q}_\Psi$ to the resultant $\mathcal{R}_\Psi$ of the sources that are instantaneously associated with the material particles of $\mathcal{V}$. (Recall that impulsive sources that tend to infinity during an impact time that tends to zero are approximately introduced in the mechanical theory of impacts. While Eq.(2) then remains still valid, the quantity $\mathcal{Q}_\Psi$ suffers a jump, see Chapter 12 of Ziegler (1998) for a comprehensive exposition on balance equations for impact problems.)

When it is possible to perform a proper continuation of the influx $\boldsymbol{i}[\Psi]$ into the interior of $\mathcal{V}$, the *divergence theorem* can be used to re-write the surface integral in Eq.(3) as a volume integral over $\mathcal{V}$. For the range of application of the divergence theorem, see Sect. 6 of Gurtin (1972). Performing the time-rate of change of $\mathcal{Q}_\Psi$ in Eq.(1), we may interchange differentiation with respect to time with the integration over the material volume $\mathcal{V}$. The material time derivative then has to be applied to the terms inside the volume integral of Eq.(1) according to the product rule of differentiation. The material time derivative $\dot{\Psi}$ of an entity Ψ splits into the local rate of change, $\partial\Psi / \partial t$, and into the convective part $\boldsymbol{v} \cdot grad\,\Psi$, the latter bringing into the play the gradient of the entity Ψ, see Sect. 72 of Truesdell and Toupin (1960). The material time derivative is indicated by a superposed dot, and $\boldsymbol{v}$ is the velocity of the material particles. The Euler expansion formula for the material time derivative of a differential volume dv asserts that

$$\dot{\overline{dv}} = (dv)\, div\, \boldsymbol{v} \ . \tag{5}$$

Using a well-known tensorial identity, the general balance relation, Eq.(4), eventually can be re-written in the form of an integral over $\mathcal{V}$. Since our formulation must hold for every part of the material body under consideration, this means that the integrand of the latter volume integral must be locally balanced. We thus arrive at the *local form of the general balance relation,*

$$\dot{\Psi} + \Psi\, div\, \boldsymbol{v} - s[\Psi] + div\, \boldsymbol{i}[\Psi] \geq 0 \ , \tag{6}$$

see Truesdell and Toupin (1960), Sect. 157 for the case of the equality sign, and cf. Section II.6 of Wang and Truesdell (1973). Note that Eq.(6) refers to a region in which the corresponding terms take on definite values. Hence, Ψ, $\dot{\Psi}$ and $s[\Psi]$ should be continuous, while $\boldsymbol{v}$ and $\boldsymbol{i}[\Psi]$ should be continuously differentiable.

It is important to note that, in principle, any quantity may be formally balanced as described by Eqs. (4) and (6). However, only the relations of balance of mass, momentum, angular momentum and energy (the first law of thermodynamics), as well as the relation of balance of entropy (the second law of thermodynamics), when written in the form of Eq.(4), represent the fundamental laws of balance of continuum mechanics. Their local expressions in the form of Eq.(6) refer to regions in which sufficient continuity conditions do hold. Other relations of balance, such as the relations of balance of kinetic and internal energy, which can be derived from the former fundamental laws of balance, are also of the form of Eq.(4) or (6), respectively. However, these latter relations should be applied with care, keeping in mind the conditions under which they have been derived.

Notwithstanding the different notation, by comparison to Lecture 1 of the present book, Belyaev (2004), we now are ready to identify Ψ, $s[\Psi]$ and $\boldsymbol{i}[\Psi]$ in the fundamental balance relations of continuum mechanics and thermodynamics. We start with *balance of mass*, see Section 2 of Belyaev (2004). Hence, we set $\Psi = \rho$ in the above equations, the scalar ρ denoting the instantaneous mass density, and we use the equality sign in Eqs.(4) and (6). We subsequently assume that sources of mass are absent, $s[\rho] = 0$ and $\boldsymbol{i}[\rho] = \boldsymbol{0}$, such that mass is conserved, $\dot{\rho} + \rho\, div\, \boldsymbol{v} = 0$. As a formula applied frequently, it follows for any φ carried by the particles in the material volume $\mathcal{V}$ that

$$\rho\,\dot{\varphi} - s[\rho\varphi] + div\,\boldsymbol{i}[\rho\varphi] \geq 0\,, \tag{7}$$

which follows from Eq.(6) by setting $\Psi = \rho\varphi$ and taking into account conservation of mass.

Balance of momentum, see Section 4 of Belyaev (2004), means to consider the *mass-flow* $\boldsymbol{\Psi} = \rho\,\boldsymbol{v}$ in Eqs.(1)-(6), together with the equality sign in Eqs.(4) and (6), the velocity of the material particles being denoted by $\boldsymbol{v}$ as above. The rate of change of momentum is balanced by a supply represented by body forces $\boldsymbol{b}$ per unit volume, $s[\rho\,\boldsymbol{v}] = \boldsymbol{b}$, as well as by an influx associated with the surface traction or stress vector $\boldsymbol{t}_{(\boldsymbol{n})}$ on S, such that $\boldsymbol{da}\cdot\boldsymbol{i}[\rho\,\boldsymbol{v}] = -\boldsymbol{t}_{(\boldsymbol{n})}\,da$. The index $(\boldsymbol{n})$ indicates that the stress vector belongs to a surface element with outer unit normal vector $\boldsymbol{n}$, and da denotes the scalar surface element, $\boldsymbol{da} = \boldsymbol{n}\,da$. The equation of balance of momentum (as observed in the common inertial frame) thus reads

$$\frac{d}{dt}\int_V \rho\,\boldsymbol{v}\,dv - \boldsymbol{\mathcal{F}} = \boldsymbol{0}\,, \tag{8}$$

where $\boldsymbol{\mathcal{F}}$ denotes the resultant of the external forces acting upon the material volume V,

$$\boldsymbol{\mathcal{F}} = \boldsymbol{\mathcal{R}}_{\rho\boldsymbol{v}} = \int_V \boldsymbol{b}\,dv + \oint_S \boldsymbol{t}_{(\boldsymbol{n})}\,da = \int_V \boldsymbol{b}\,dv + \oint_S \boldsymbol{da}\cdot\boldsymbol{T}\,. \tag{9}$$

The relation stated in Eq.(9.2) follows from the *Cauchy fundamental theorem on stresses*, which asserts that

$$\boldsymbol{t}_{(\boldsymbol{n})} = \boldsymbol{n}\cdot\boldsymbol{T} = \boldsymbol{T}^T\cdot\boldsymbol{n}\,, \tag{10}$$

where $\boldsymbol{T}$ is the Cauchy stress tensor, a tensor of second order. The influx of momentum thus may be written as $\boldsymbol{i}[\rho\,\boldsymbol{v}] = -\boldsymbol{T}$. The transpose of a second-order tensor is indicated by the superscript *T*. The surface integrals in Eq.(9) represent the resultant of the surface traction at S. The surface traction $\boldsymbol{t}_{(\boldsymbol{n})}$ and the body forces $\boldsymbol{b}$ form the system of *external forces* acting upon V. When mass is conserved, the local form of the equation of balance of momentum reads

$$\rho\,\dot{\boldsymbol{v}} = \rho\,\boldsymbol{a} = \boldsymbol{b} + div\,\boldsymbol{T}\,, \tag{11}$$

cf. Eq.(6). The acceleration of the material particle at the place under consideration is denoted as $\boldsymbol{a} = \dot{\boldsymbol{v}}$. The relation given in Eq.(11) represents the *fundamental equation of dynamics,* or *Cauchy's first law of motion*, Truesdell and Toupin (1960), Sect. 205, and Ziegler (1998), Chapter 7. Note that Eq.(11) refers to a region in which $\rho\,\boldsymbol{a}$ and $\boldsymbol{b}$ are continuous and $\boldsymbol{T}$ is continuously differentiable.

In order to establish *balance of moment of momentum* (or angular momentum), the entity $\boldsymbol{\Psi} = \rho\,\hat{\boldsymbol{p}}\times(\boldsymbol{v} - \boldsymbol{v}_A)$ is to be introduced into the general relations of Eqs.(1)-(4), together with the equality sign. Here, $\hat{\boldsymbol{p}}$ denotes the position vector of the material particle from an origin *A* that possibly moves at the velocity $\boldsymbol{v}_A$, the vector $(\boldsymbol{v} - \boldsymbol{v}_A)$ being the relative velocity of the material particle with respect to *A*. In the absence of assigned couples, and when there are no couple stresses, the local form of the equation of balance of moment of momentum is found to assert the *symmetry of the stress tensor* $\boldsymbol{T}$, see Truesdell and Toupin (1960), Sect. 205, and Section 4 of Belyaev (2004). Media with a symmetric stress tensor are called *non-polar media.* Typically, problems in the dynamics of structures and machines refer to the non-polar case, and we therefore consider $\boldsymbol{T}$ to be symmetric throughout the Lecture, $\boldsymbol{T} = \boldsymbol{T}^T$. For a comprehensive presentation of *global balance of moment of momentum* with various examples, we particularly refer to Chapter 7 of Ziegler (1998). In short, when A is either not accelerated, or when A represents the center of the mass in $\mathcal{V}$, the angular momentum with respect to A is balanced by the resultant torque of the external forces about A. Otherwise, an additional term containing the acceleration of A enters the equation of balance of angular momentum. When this term is absent, and when the resultant torque about A also vanishes, then the angular momentum is conserved.

When the continuity requirements for Cauchy's first law are met, Eq.(11) can be used to derive various useful, nevertheless delicate representations that may be applied instead of the equation of balance of momentum. In the following, we present a formulation often used in practice, namely the *equation of balance of kinetic energy*. We first perform a scalar multiplication of Eq.(11) by a time-dependent test vector field $\bar{\boldsymbol{u}}$. We then use the identities

$$\dot{\boldsymbol{v}} \cdot \bar{\boldsymbol{u}} = \frac{d}{dt}(\boldsymbol{v} \cdot \bar{\boldsymbol{u}}) - \boldsymbol{v} \cdot \dot{\bar{\boldsymbol{u}}} , \tag{12}$$

$$(div\, \boldsymbol{T}) \cdot \bar{\boldsymbol{u}} = -\; tr\,(\boldsymbol{T} \cdot grad\, \bar{\boldsymbol{u}}) + div\,(\bar{\boldsymbol{u}} \cdot \boldsymbol{T}) , \tag{13}$$

where the symmetry of $\boldsymbol{T}$ has been taken into account, and the trace of a second-order tensor is indicated by the abbreviation *tr*. Here, $\boldsymbol{T} \cdot grad\, \bar{\boldsymbol{u}}$ means the second-order tensor representing the product of $\boldsymbol{T}$ with the second-order tensor $grad\, \bar{\boldsymbol{u}}$. For the trace of the product of two second-order tensors, see e.g. Appendix A of Slattery (1999). Integrating over the material volume $\mathcal{V}$, and using Eq.(7) together with the divergence theorem and Cauchy's fundamental theorem on stresses, Eq.(10), we obtain a relation which is of the form of the general balance equation, Eq.(4):

$$\frac{d}{dt}\int_{\mathcal{V}} \frac{1}{2}(\boldsymbol{v} \cdot \boldsymbol{v})\, \rho\, dv = \int_{\mathcal{V}} \boldsymbol{b} \cdot \boldsymbol{v}\, dv + \oint_{\mathcal{S}} \boldsymbol{t}_{(\boldsymbol{n})} \cdot \boldsymbol{v}\, da - \int_{\mathcal{V}} tr\,(\boldsymbol{T} \cdot grad\, \boldsymbol{v})\, dv , \tag{14}$$

where we have used the actual velocity field as test field, $\bar{\boldsymbol{u}} = \boldsymbol{v}$. In Eq.(14), we identify the *kinetic energy* $\mathcal{K}$ of the particles contained in $\mathcal{V}$ as

$$\mathcal{K} = \frac{1}{2}\int_{\mathcal{V}} \boldsymbol{v} \cdot \boldsymbol{v}\, \rho\, dv . \tag{15}$$

The first two integrals on the right hand side of Eq.(14) are the *power of the external forces*,

$$\mathcal{P}_{ext} = \int_{\mathcal{V}} \boldsymbol{b} \cdot \boldsymbol{v}\, dv + \oint_{\mathcal{S}} \boldsymbol{t}_{(\boldsymbol{n})} \cdot \boldsymbol{v}\, da , \tag{16}$$

while the last integral is known as the *power of the internal forces*, or the *stress power*,

$$\mathcal{P}_{int} = -\int_{\mathcal{V}} tr\,(\boldsymbol{T} \cdot grad\, \boldsymbol{v})\, dv . \tag{17}$$

Eq.(15) thus represents an equation of balance of kinetic energy:

$$\frac{d}{dt}\mathcal{K} - \mathcal{P}_{ext} - \mathcal{P}_{int} = 0 . \tag{18}$$

When Eq.(18) is written in the form of Eq.(2), it is found that the difference of kinetic energy between two time instances is equal to the work of the internal and external forces during the respective time-interval. The kinetic energy $\mathcal{K}$ is conserved, when the total of the power of the external and of the internal forces vanishes. Various useful applications of the equation of balance of kinetic energy are presented in Chapter 8 of Ziegler (1998).

It becomes apparent from Eq.(13), that the equation of balance of kinetic energy, Eq.(18), rests upon continuity requirements that are enhanced with respect to the local form of the equation of balance of momentum, Eq.(11), from which it has been derived by further manipulation. Despite its usefulness, the equation of balance of kinetic energy thus must be applied with some care. In order to see this, we first note that Eq.(18) may be brought into coincidence with the general equation of balance, Eq.(4), by setting $\Psi = \mathscr{k} = \boldsymbol{v} \cdot \boldsymbol{v} \rho / 2$, the local kinetic energy per unit volume being denoted as $\mathscr{k}$. The corresponding sources in Eq.(4) follow from Eqs.(16) and (17). For the influx of kinetic energy, we have $\boldsymbol{n} \cdot \boldsymbol{i}[\mathscr{k}] = -\boldsymbol{t}_{(n)} \cdot \boldsymbol{v}$, such that we may set $\boldsymbol{i}[\mathscr{k}] = -\boldsymbol{T} \cdot \boldsymbol{v}$, cf. Eq.(10). Moreover, the supply of kinetic energy can be written as $s[\mathscr{k}] = \boldsymbol{b} \cdot \boldsymbol{v} - tr(\boldsymbol{T} \cdot grad\, \boldsymbol{v})$. The last term in this supply, the stress power per unit volume, however contains the velocity gradient, which will become unbounded, when the velocity suffers a jump discontinuity, cf. also Chapter 2 of Liu (2002). When the stress also shows a jump discontinuity, then the integration to be performed in Eq.(17) can not be performed directly.

The equation of balance of kinetic energy, Eq.(18), must not be confused with the *first law of thermodynamics* (also denoted as the *law of balance of energy*), compare Section 5 of Belyaev (2004). The equation of balance of kinetic energy is not concerned with the conversion of mechanical work into thermal energy, this conversion being the topic of the first law of thermodynamics. The equation of balance of kinetic energy, when it holds, however may be used for replacing some terms in the first and second law of thermodynamics, which will be shortly discussed subsequently. The first law of thermodynamics is obtained from Eq.(4) by considering the total energy per unit volume as the entity to be balanced, $\Psi = \mathscr{k} + \rho\, u$, where u denotes the internal energy per unit mass. The supply of the total energy is assumed to be provided by the power of the body forces per unit volume, and by external non-mechanical volume sources q per unit volume, such that $s[\mathscr{k} + \rho\, u] = \boldsymbol{b} \cdot \boldsymbol{v} + q$. The influx of the total energy is given by the negative power density of the surface traction, and by a non-mechanical influx vector $\boldsymbol{h}$, such that $\boldsymbol{i}[\mathscr{k} + \rho\, u] = -\boldsymbol{T} \cdot \boldsymbol{v} + \boldsymbol{h}$. Denoting the quantity of the internal energy contained in the material volume $\mathcal{V}$ by

$$\mathcal{U} = \int_{\mathcal{V}} u\, \rho\, dv\,, \tag{19}$$

the first law of thermodynamics reads,

$$\frac{d}{dt}(\mathcal{K} + \mathcal{U}) - \mathcal{P}_{ext} - \mathcal{P}_{non} = 0\,, \tag{20}$$

see Sect. 240 and 241 of Truesdell and Toupin (1960). The power of the non-mechanical sources is

$$\mathcal{P}_{non} = \int_{\mathcal{V}} q\, dv - \oint_{\mathcal{S}} d\boldsymbol{a} \cdot \boldsymbol{h}\,. \tag{21}$$

Note, furthermore, that the first law basically is stated with reference to the power of the external forces $\mathcal{P}_{ext}$, see Eq.(16), and not with reference to the stress power, Eq.(17). The stress power can be brought into the play by substituting the equation of balance of kinetic energy, Eq.(18), into Eq.(20). This leads to an alternative form of the first law of thermodynamics, appearing in the form of an *equation of balance of internal energy*:

$$\frac{d}{dt}\mathcal{U} + \mathcal{P}_{int} - \mathcal{P}_{non} = 0\,. \tag{22}$$

It is seen that Eq.(22) again is of the form of the general relation of balance, Eq.(4). The corresponding supply of internal energy is $s[\rho\ u] = q + tr\,(\boldsymbol{T} \cdot grad\ \boldsymbol{v})$, and the influx can be written as $\boldsymbol{i}[\rho\ u] = \boldsymbol{h}$. As it should be, we thus arrive at the relations

$$s[\mathscr{k} + \rho\ u] = s[\mathscr{k}] + s[\rho\ u]\,,\ \boldsymbol{i}[\mathscr{k} + \rho\ u] = \boldsymbol{i}[\mathscr{k}] + \boldsymbol{i}[\rho\ u]\,. \tag{23}$$

Nevertheless, the equation of balance of internal energy, Eq.(22), represents a derived relation, resting upon the applicability of the global form of the equation of balance of kinetic energy, Eq.(18), or upon a procedure that is equivalent to the derivation of the latter relation. Since the equation of balance of kinetic energy rests upon the local form of the equation of balance of momentum, Eq.(11), this can also be said with respect to balance of internal energy. Moreover, since in the equation of balance of kinetic energy special care has to taken with respect to the stress power, Eq.(17), this is also the case with respect to the global form of the equation of balance of internal energy, Eq.(22).

When Eq.(22) can be directly applied, however, it can be conveniently used to derive the following local form, which is of frequent use in the literature, and which again contains the stress power per unit volume, see Eq.(7):

$$\rho\ \dot{u} - q - tr\,(\boldsymbol{T} \cdot grad\ \boldsymbol{v}) + div\ \boldsymbol{h} = 0\,, \tag{24}$$

compare Eq.(5.5) of Belyaev (2004), and see Section 241 of Truesdell and Toupin (1960).

The *second law of thermodynamics*, see Section 5 of Belyaev (2004), or the relation of balance of entropy, represents an inequality, which is also shortly addressed. Denoting the entropy per unit mass as η, and setting $\Psi = \rho\eta$, we here assume with respect to Eq.(3) that the supply of entropy can be formulated as $s[\rho\ \eta] = q\ \theta^{-1}$, and that the influx is $\boldsymbol{i}[\rho\ \eta] = \boldsymbol{h}\ \theta^{-1}$, where θ denotes the absolute temperature. From Eq.(4), we thus arrive at the *Clasius-Duhem inequality*:

$$\frac{d}{dt}\int_{\mathcal{V}} \rho\ \eta\ dv - \int_{\mathcal{V}} q\ \theta^{-1}\ dv + \oint_{\mathcal{S}} d\boldsymbol{a} \cdot \boldsymbol{h}\ \theta^{-1} \geq 0\,. \tag{25}$$

The Clausius-Duhem inequality requires the left hand side of Eq.(25), the so-called *entropy production*, to be a non-negative quantity. When mass is conserved in the material volume $\mathcal{V}$, the local form of Eq.(25) reads, cf. Eq.(7),

$$\rho\ \dot{\eta} - q\ \theta^{-1} + div\,(\boldsymbol{h}\ \theta^{-1}) \geq 0\,, \tag{26}$$

which refers to a region with suitable continuity conditions, compare the remarks given above with respect to Eq.(6), see Section 258 of Truesdell and Toupin (1960). For the restrictions that are imposed upon the constitutive relations of continuum mechanics by the two fundamental laws of thermodynamics, Eqs. (20) and (25), see also Section 6 of Belyaev (2004).

The local form of the second law, Eq.(26), suggests to eliminate the sources of heat by means of the local form of the first law given in Eq.(24). Taking into account additional assumptions, the result may be eventually written in the form of the so-called *Clausius-Planck inequality*,

$$\rho\,(\theta\ \dot{\eta} - \dot{u}) + tr\,(\boldsymbol{T} \cdot grad\ \boldsymbol{v}) \geq 0\,, \tag{27}$$

a relation that is also often used in the literature. Here, we only note that this elimination process again brings into the play the stress power per unit volume, which, as has been indicated above, represents an entity to be handled with care when discontinuities are to be expected.

2 On the Relations of Jump.

As has been noted above, situations may arise in which, at certain points in the material volume $\mathcal{V}$, the necessary continuity conditions for the local form of the general balance relation, Eq.(6), fail to be valid, such that the latter does not hold. A situation to be frequently encountered, e.g. in wave propagation, is represented by a material volume $\mathcal{V}$ that is subdivided into two non-material sub-volumes by a persistent *singular surface* Σ. The latter represents a surface of discontinuity, across which some entity Ψ shows different values when approaching from the two sides of the surface. Note that the singular surface may move at a velocity $\boldsymbol{w}_\Sigma$ that is different from the velocity $\boldsymbol{v}$ of the material particles located on it. In order to connect the local forms of the balance relations at the two sides of the singular surface, jump relations are needed. The classical treatment of this situation is to apply the Reynolds transport theorem to the two sub-volumes, taking into account the motion of Σ relative to the material particles, and to add the results, see Section 192 of Truesdell and Toupin (1960). In a straightforward manner, this classical strategy brings into the play the jump of the entity Ψ across Σ. However, this procedure means that the particles instantaneously located "on the singular surface" may not be properly taken into account, but may be removed from the material volume. In order to circumvent this criticism, a singular surface can be modelled as a *thin shell-type non-material layer of transition* $v_\Sigma(t)$, within which Ψ and its sources are subjected to considerable changes in their spatial distribution, the thickness of the layer remaining thin or tending to zero, depending on the nature of the problem in hand. In extension of the classical treatment of singular surfaces, it was found to be necessary in various fields to introduce into the equations of balance the rate of change of certain quantities contained in the thin non-material layer $v_\Sigma(t)$, together with the corresponding supply terms, see the book on interfacial transport phenomena by Slattery (1990) for a comprehensive exposition. In the procedure presented in the latter reference, the thin non-material layer is replaced by a fictitious dividing surface, a notion that dates back to G. W. Gibbs. The dividing surface plays the role of an equivalent singular surface. This equivalent singular surface is attached with surface terms $\boldsymbol{i}_\Sigma[\Psi]$ that stand for the time rate of the respective quantity contained in the thin non-material layer, as well as for the source terms. This modelling eventually leads to the following *general jump relation*, connecting the two sides of the effective singular surface Σ:

$$\boldsymbol{n}_\Sigma \cdot \{\boldsymbol{i}_\Sigma[\Psi] + [\![\boldsymbol{i}[\Psi] + (\boldsymbol{v} - \boldsymbol{w}_\Sigma)\Psi]\!] \} \geq \boldsymbol{0} . \tag{28}$$

where the doubled bracket indicates a jump, $[\![\varphi]\!] = \varphi^+ - \varphi^-$ for any entity φ. Here, φ^+ denotes the limit of φ at the positive side of the singular surface Σ, the positive side having the outer unit normal vector $\boldsymbol{n}_\Sigma$, and φ^- denotes the limit of φ at the negative side. In writing down Eq.(28), we need to assume that the corresponding entities are continuous functions of the position at the respective sides of the singular surface, taking on definite values at the two sides. For a detailed derivation of Eq.(28) in the case of an equality sign, see Irschik (2003). Proceeding very similar to the formulation given in the latter reference, one approaches to the inequality presented in Eq.(28). In this relation, the equality sign holds for the jump relations of balance of mass, momentum and energy, while the jump relation of balance of entropy requires the left hand side of Eq.(28) to be non-negative.

With $\boldsymbol{i}_\Sigma[\Psi] = \mathbf{0}$, Eq.(28) yields the *Kotchine jump conditions* at a singular surface. For this latter classical form of the general jump relation, see Sect. 194 of Truesdell and Toupin (1960) in case of the equality sign, and Section II.6 of Wang and Truesdell (1973). The formulation given in Eq.(28) extends the Kotchine jump conditions with respect to the presence of the fictitious influx $\boldsymbol{i}_\Sigma[\Psi]$ defined on the equivalent singular surface Σ. This term $\boldsymbol{i}_\Sigma[\Psi]$ has been denoted as *surface growth term* in Irschik (2003), a notation that we use also in the following.

As has been mentioned above, there are physical motivations that have been given in the literature for introducing entities of the type of surface growth terms $\boldsymbol{i}_\Sigma[\Psi]$ in the relations of jump, see Irschik (2003) for a review. A mathematical motivation is the following. Note that supply terms do not appear explicitly in Eq.(28). As has been discussed in the previous Section, however, it may happen that certain supply terms may become unbounded when the thickness of the thin layer $v_\Sigma(t)$, which is equivalently replaced by $\boldsymbol{\Sigma}$, tends to zero. This is particularly the case for the stress power, Eq.(17), which contains the velocity gradient, the latter becoming unbounded, when the velocity suffers a jump discontinuity, the integral in Eq.(17) then possibly being ill-posed. In order to account for such a situation, surface growth terms $\boldsymbol{i}_\Sigma[\Psi]$ must be introduced. We thus note that, willy nilly, growth terms are needed in the jump equations for balance of kinetic and internal energy, since the stress power comes into the play in these relations. Recall that the stress power also appears in the Clausius-Planck inequality, Eq.(29). Surface growth terms are also needed, when the time rate of the quantity contained in the thin region of transition or other sources can not be neglected, see again Irschik (2003).

Corresponding to the preceding remarks, we now show that the surface growth terms $\boldsymbol{i}_\Sigma[\Psi]$ in Eq.(28) are necessary to assure that the various relations of balance presented in Section 1 above are associated by consistent jump conditions at a singular surface. First, we exemplary treat *consistency between the equations of balance of momentum and balance of kinetic energy*. In studying balance of momentum, we have $\Psi = \rho\,\boldsymbol{v}$ and $\boldsymbol{i}[\rho\,\boldsymbol{v}] = -\boldsymbol{T}$, see above. Due to the Cauchy fundamental theorem on stresses, Eq.(10), we may put Eq.(28) into the form

$$\boldsymbol{n}_\Sigma \cdot \boldsymbol{i}_\Sigma[\rho\,\boldsymbol{v}] - \left[\!\left[\boldsymbol{t}_{(\boldsymbol{n}_\Sigma)} - \boldsymbol{n}_\Sigma \cdot (\boldsymbol{v} - \boldsymbol{w}_\Sigma)\,\rho\,\boldsymbol{v} \right]\!\right] = \mathbf{0}\,. \tag{29}$$

Note that the fictitious influx or surface growth term $\boldsymbol{i}_\Sigma[\rho\,\boldsymbol{v}]$ associated with balance of momentum represents a tensor of second order. An enlightening physical example for this surface growth of momentum is represented by an interfacial tension in the dividing surface between two phases, see Section 2.4.3 of Slattery (1999). For balance of kinetic energy, we have $\Psi = k = \boldsymbol{v}\cdot\boldsymbol{v}\,\rho/2$ and there is $\boldsymbol{i}[k] = -\boldsymbol{T}\cdot\boldsymbol{v}$, such that

$$\boldsymbol{n}_\Sigma \cdot \boldsymbol{i}_\Sigma[k] - \left[\!\left[\boldsymbol{t}_{(\boldsymbol{n}_\Sigma)} \cdot \boldsymbol{v} - \frac{1}{2}\boldsymbol{n}_\Sigma \cdot (\boldsymbol{v} - \boldsymbol{w}_\Sigma)\,\rho\,\boldsymbol{v}\cdot\boldsymbol{v} \right]\!\right] = 0\,. \tag{30}$$

In Eq.(30), the surface growth term $\boldsymbol{i}_\Sigma[k]$ represents a vector. Now, for the generalised product of two entities there is

$$[\![\Psi\cdot\Phi]\!] = [\![\Psi]\!]\cdot\langle\Phi\rangle + \langle\Psi\rangle\cdot[\![\Phi]\!]\,, \tag{31}$$

the mean value of Ψ across Σ being denoted by $\langle\Psi\rangle$, see e.g. Irschik (2003). For the sake of brevity, we assume that the classical form of the Kotchine jump conditions holds with respect to balance of mass, such that $\boldsymbol{i}_\Sigma[\rho]=\mathbf{0}$ and $\boldsymbol{i}[\Psi]=\mathbf{0}$. Thus, Eq.(28) yields

$$[\![\boldsymbol{n}_\Sigma\cdot(\boldsymbol{v}-\boldsymbol{w}_\Sigma)\,\rho\,]\!]=0\,. \tag{32}$$

From Eqs.(29)-(32), we obtain a relation between the respective surface growth terms,

$$\boldsymbol{n}_\Sigma\cdot\boldsymbol{i}_\Sigma[k]-(\boldsymbol{n}_\Sigma\cdot\boldsymbol{i}_\Sigma[\rho\,\boldsymbol{v}])\cdot\langle\boldsymbol{v}\rangle=\langle\boldsymbol{t}_{(\boldsymbol{n}_\Sigma)}\rangle\cdot[\![\,\boldsymbol{v}\,]\!]\,. \tag{33}$$

Hence, even when we assume $\boldsymbol{n}_\Sigma\cdot\boldsymbol{i}_\Sigma[\rho\,\boldsymbol{v}]=\mathbf{0}$ in Eq.(33), a non-vanishing surface growth term of kinetic energy, $\boldsymbol{n}_\Sigma\cdot\boldsymbol{i}_\Sigma[k]=\langle\boldsymbol{t}_{(\boldsymbol{n}_\Sigma)}\rangle\cdot[\![\,\boldsymbol{v}\,]\!]\neq 0$, is generally required in order to formulate a consistent jump condition in the framework of balance of kinetic energy, Eq.(30). An enhanced formula, not assuming that the surface growth terms for balance of mass vanish, has been presented in Irschik (2003). A further result proved in Irschik (2003) concerns balance of internal energy, namely that

$$\boldsymbol{n}_\Sigma\cdot\boldsymbol{i}_\Sigma[k+\rho\,u]=\boldsymbol{n}_\Sigma\cdot\boldsymbol{i}_\Sigma[k]+\boldsymbol{n}_\Sigma\cdot\boldsymbol{i}_\Sigma[\rho\,u]\,. \tag{34}$$

Hence, when we assume that the surface growth terms of momentum and energy are absent in Eqs. (33) and (34), $\boldsymbol{n}_\Sigma\cdot\boldsymbol{i}_\Sigma[\rho\,\boldsymbol{v}]=\mathbf{0}$ and $\boldsymbol{n}_\Sigma\cdot\boldsymbol{i}_\Sigma[k+\rho\,u]=0$, the surface growth term of internal energy becomes

$$\boldsymbol{n}_\Sigma\cdot\boldsymbol{i}_\Sigma[\rho\,u]=-\boldsymbol{n}_\Sigma\cdot\boldsymbol{i}_\Sigma[k]=-\langle\boldsymbol{t}_{(\boldsymbol{n}_\Sigma)}\rangle\cdot[\![\,\boldsymbol{v}\,]\!]\,. \tag{35}$$

The jump relation suitable for the local form of the first law of thermodynamics given in Eq.(24) thus follows from Eq.(28) as

$$\begin{aligned}&\boldsymbol{n}_\Sigma\cdot\boldsymbol{i}_\Sigma[\rho\,u]+\boldsymbol{n}_\Sigma\cdot[\![\boldsymbol{h}+(\boldsymbol{v}-\boldsymbol{w}_\Sigma)\,\rho\,u]\!]=\\&-\langle\boldsymbol{t}_{(\boldsymbol{n}_\Sigma)}\rangle\cdot[\![\,\boldsymbol{v}\,]\!]+\boldsymbol{n}_\Sigma\cdot[\![\boldsymbol{h}+(\boldsymbol{v}-\boldsymbol{w}_\Sigma)\,\rho\,u]\!]=0\,,\end{aligned} \tag{36}$$

where we have used $\boldsymbol{i}[\rho\,u]=\boldsymbol{h}$ in Eq.(28). Substituting Eqs.(35) and (36) into Eq.(30) we obtain the jump relation for the first law of thermodynamics:

$$[\![\boldsymbol{t}_{(\boldsymbol{n}_\Sigma)}\cdot\boldsymbol{v}-\boldsymbol{n}_\Sigma\cdot[\boldsymbol{h}+(\boldsymbol{v}-\boldsymbol{w}_\Sigma)\,(k+\rho\,u)]]\!]=0\,. \tag{37}$$

The same result can be obtained from the general form of Eq.(28) by substituting $\boldsymbol{i}[k+\rho\,u]=-\boldsymbol{T}\cdot\boldsymbol{v}+\boldsymbol{h}$ and setting $\boldsymbol{n}_\Sigma\cdot\boldsymbol{i}_\Sigma[k+\rho\,u]=0$. This result is in coincidence with the literature, see Eq.(5.1.3-9) of Slattery (1999). We note, however, that a contradictory formulation for the jump relation of internal energy was also presented by Slattery (1999), in his Exercise 5.1.3-6. This latter formulation is obtained by replacing the term $\langle\boldsymbol{t}_{(\boldsymbol{n}_\Sigma)}\rangle\cdot[\![\,\boldsymbol{v}\,]\!]$ in Eq.(36) by $[\![\boldsymbol{t}_{(\boldsymbol{n}_\Sigma)}\cdot\boldsymbol{v}]\!]$, the result then being in contradiction to Eq.(37). Slattery remarked that there is not a sufficient experimental evidence so far in order to distinguish between the two formulations. Our above derivations should give theoretical evidence for ruling out the formulation presented in Exercise 5.1.3-6 of Slattery (1999). Instead of the latter, we suggest to use Eq.(36) for the jump relation of internal energy.

We now turn to the second law thermodynamics, the balance of entropy, Eq.(25). With $\boldsymbol{i}[\rho\,\eta]=\boldsymbol{h}\,\theta^{-1}$, the following jump relation is obtained from Eq.(28):

$$\boldsymbol{n}_\Sigma \cdot \{\boldsymbol{i}_\Sigma [\rho\,\eta] + [\![\boldsymbol{h}\,\theta^{-1} + (\boldsymbol{v} - \boldsymbol{w}_\Sigma)\,\rho\,\eta]\!] \} \geq 0\,. \tag{38}$$

This is the jump relation suitable for the local form of the second law given in Eq.(26). For the case of $\boldsymbol{i}_\Sigma [\rho\,\eta] = \boldsymbol{0}$, see Section 258 of Truesdell and Toupin (1960) and Chapter 5.2 of Slattery (1999).

Now assume that θ is continuous across the singular surface. Eq.(36.$_1$) then can be used in to eliminate the flux vector $\boldsymbol{h}$, in Eq.(38), leading to

$$\boldsymbol{n}_\Sigma \cdot \{\boldsymbol{i}_\Sigma [\rho\,\eta]\,\theta - \boldsymbol{i}_\Sigma [\rho\,u] + [\![(\boldsymbol{v} - \boldsymbol{w}_\Sigma)\,\rho\,(\theta\,\eta - u)]\!] \} \geq 0\,, \tag{39}$$

with $\boldsymbol{i}_\Sigma [\rho\,u]$ of Eq.(35.$_2$). The relation given in Eq.(39) represents a jump relation that is suitable for the Clausius-Planck inequality, Eq.(27). The entity $F = u - \theta\,\eta$ represents the *free energy density*, compare Section 6 of Belyaev (2004). In Chapter 5.6 of the book of Liu (2002), an relation similar to Eq.(39) has been presented in the material (Lagrangian) formulation of continuum mechanics, not taking into account a surface growth of entropy, $\boldsymbol{i}_\Sigma [\rho\,\eta] = \boldsymbol{0}$. We note that we would not have arrived at Eq.(38), when the surface growth term for the internal energy, $\boldsymbol{i}_\Sigma [\rho\,u]$ of Eq.(35.$_2$), had been neglected in Eqs.(36) and (39). This gives further evidence for the appropriateness of the our formulation with surface growth terms. The surface growth terms of mass, momentum, energy and entropy may or may not be needed, depending on the specific problem. The surface growth terms for internal and kinetic energy, however, must be introduced, when jumps of the latter two quantities are given.

References:

Belyaev, A.K. (2004). Basics of Continuum Mechanics. In: Irschik, H. and Schlacher, K., eds., *Advanced Dynamics and Control of Structures and Machines*. CISM Courses and Lectures No. 444, Wien–New York: Springer-Verlag.

Gurtin, M. E. (1972). The Linear Theory of Elasticity. In: *Handbuch der Physik*, Vol. VIa/2. Berlin: Springer-Verlag.

Ericksen, J.L. (1960). Tensor Fields. In: *Handbuch der Physik*, Vol. III/1, Springer-Verlag.

Irschik, H. (2003). On the Necessity of Surface Growth Terms for the Consistency of Jump Relations at a Singular Surface. *Acta Mechanica*, 162: 195 – 211.

Liu, I-S. (2002). *Continuum Mechanics*. New York: Springer-Verlag.

Truesdell, C., Toupin, R. (1960). The Classical Field Theories. In: *Handbuch der Physik*, Vol. III/1, Springer-Verlag.

Slattery, J.C. (1990). *Interfacial Transport Phenomena*. New York: Springer-Verlag.

Slattery, J.C.(1999). *Advanced Transport Phenomena*. Cambridge: Cambridge Univ. Press.

Wang, C.C., Truesdell, C. (1973). *Introduction to Rational Elasticity*. Nordhoff Int. Publ.

Ziegler, F., (1998). *Mechanics of Solids and Fluids*, 2nd Engl. Edition, corrected 2nd printing. New York: Springer-Verlag.

Acknowledgement:

The present Lecture Note is a contribution to the K+ Linz Center of Competence in Mechatronics (LCM), Strategic Project 4.4. Support of LCM by the Austrian K+ Fund and the Government of Upper Austria is gratefully acknowledged.

The Rayleigh-Ritz Technique and the Lagrange Equations in Continuum Mechanics: Formulations for Material and Non-Material Volumes

Hans Irschik[1], Helmut J. Holl[1], Franz Hammelmüller[2]

1 Johannes Kepler University of Linz and LCM, Linz, Austria
2 Linz Center of Excellence in Mechatronics (LCM), Linz, Austria

Abstract. In the present Lecture, we use the Rayleigh-Ritz technique in connection with the Lagrange equations in order to approximate the partial differential equations of continuum mechanics by means of a system of ordinary differential equations in time. We start from the local form of the equation of balance of momentum, from which we proceed to an extended Hamilton principle, eventually ending up with the Lagrange equations. We lay special emphasis upon the functional dependencies to be considered in the Rayleigh-Ritz technique with respect to the spatial and the material descriptions of continuum mechanics. In analogy to the notion of a material-time derivative, we define material variations and material partial derivatives of the quantities and entities that enter the Lagrange equations, and we relate these derivatives to local partial derivatives, the latter being particularly feasible with respect to the spatial formulation of continuum mechanics. Having formulated the Lagrange equations for the case of a material volume, we present a re-formulation for problems that are posed with respect to non-material volumes. As an example for such a problem, we shortly treat the coiling of a strip. The presented re-formulation of the Lagrange equations for non-material volumes represents an alternative derivation of recent results by Irschik and Holl (2002).

1 Introduction

From the continuum mechanics point of view, the equations of motion of structures and machines basically are given as *partial differential equations* that are imposed by the *local form of the balance equations*. (When a surface of discontinuity is present, *jump conditions* are to be introduced in order to connect the values of the entities at the two sides of that surface, see Lecture 2 of the present book, Irschik (2004). In the present Lecture, discontinuities however will not be taken into account.) Additional to the local equations of balance, the *kinematic differential relations* between displacement and strain are to be considered, and *constitutive relations* between stress and strain must be formulated in order to describe the material behaviour, see Lecture 1 of the present book, Belyaev (2004a). The resulting system of equations is accompanied by the conditions given at the beginning of the motion, the *initial conditions*, and by the conditions given at the surface of the respective body, the *boundary conditions*. In case the displacements are prescribed at the boundary, one talks about *kinematic boundary conditions* (also denoted as boundary conditions of place, or of the first kind). When the traction is prescribed at the boundary, the boundary conditions are called *dynamic boundary conditions* (the boundary conditions of traction, or of the second kind). Due to the complex geometry of usual structural elements, analytical solutions of this *initial-boundary-value problem* can be seldom found. It therefore has become a common computational strategy to put the problem into the approximate form of a *system of ordinary differential equations in time (ODEs)*. It is often necessary, or at least computationally advantageous, to accompany this system of ODEs by algebraic constraint equations, resulting in a *system of differential-algebraic equations (DAEs)*. This computational strategy is also appropriate with respect to *automatic control*, since powerful control algorithms have been developed, which are applicable to DAE systems stemming from the dynamics of structures and machines, see e.g. Schlacher and Kugi (2002). Moreover, the cited computational strategy connects the formulations of continuum mechanics with the *analytical mechanics of systems of mass-points and rigid bodies*, which represents a branch of mechanics of its own right, see Lecture 4 of the present book, Belyaev (2004b). Unfortunately, the elegant and useful forms that analytical mechanics provides for systems of mass-points do not carry over in full generality to continuum mechanics. However, the fact that both, continuum as well as analytical mechanics can be reduced to DAE systems has given raise to an intense and successful research in *numerical and computational mathematics*. In the present Lecture, we give an exemplary discussion of how ideas of analytical mechanics can be utilised in the derivation of approximate solution schemes for problems of continuum mechanics. We thereby also intend to provide a basis for Lecture 4 of the present book on analytical mechanics, Belyaev (2004b).

In the following, we use the *Rayleigh-Ritz technique* in order to approximate the partial differential equations of continuum mechanics by means of a system of ODEs. In the Rayleigh-Ritz technique, the motion of a deformable body is approximated by expressing the displacement field by *admissible spatial functions of time-dependent generalised coordinates*. The corresponding representation is called a *Ritz-Ansatz*, which in its original form is represented as a series of functions that are separable in space and time. The Rayleigh-Ritz technique nowadays represents a widely recognised method of mathematical analysis for finding approximate solutions to partial differential equations, see Chapter 8 of the book by Gershenfeld (1999). When the Ritz-Ansatz is separately applied to a set of finite sub-bodies, this set being assembled to the whole body afterwards, the resulting computational scheme is called a *finite element method (FEM)*. For an exposition on the use of the Rayleigh-Ritz technique in solid mechanics with reference to FEM, we refer to Chapter 11 of Ziegler (1998). Recently, a Rayleigh-Ritz approximation has been reported even for the complex case of large deformations of elastic *multi-body dynamic (MBD) systems*, see Chapters 6 and 7 of the book of Shabana (1998). In *elastic MBD* systems, the generalised coordinates represent rigid-body degrees-of-freedom as well as flexible elastic coordinates. For the motion of *elasto-plastic bodies* with rigid-body degrees-of-freedom and small deformations, see Lecture 6 of the present book, Gerstmayr, Irschik and Dibold (2004).

In any case, it is necessary to relate the respective Ritz-Ansatz with the initial-boundary value problem under consideration, in order to obtain ODEs for the generalised coordinates. This should not be done directly, but a spatial integration technique is required, such that the errors produced by the Ritz-Ansatz can be studied in a well-defined setting. There are several techniques to reach this goal, for instance the *Ritz-Galerkin technique*, see again Chapter 11 of

Ziegler (1998). As an alternative strategy that turned out to be equally fruitful in practice, we subsequently present the method of *Lagrange equations*, also discussed in Ziegler (1998). For the use of the Lagrange equations in analytical mechanics, see Lecture 4 of the present book, Belyaev (2004b). As will be demonstrated below, the Lagrange equations can be derived from the local form of the equation of balance of momentum of continuum mechanics by mathematical manipulation and subsequent spatial integration, which in an intermediate step provides an extended *Hamilton principle*. Inserting the Ritz-Ansatz and its *variations* into the spatial integrals of the latter principle eventually leads to the Lagrange equations, providing a connection to the *kinetic energy* of the body and to the *virtual work* or *virials* of the forces under consideration, as will be demonstrated below in more detail. For the required continuity conditions to be observed in such formulations, compare Lecture 2 of the present book, Irschik (2004). The notions of kinetic energy and work represent a framework, in which the Rayleigh-Ritz approximation of continuum mechanics can be performed on a firm ground. A further advantage of the Lagrange equations is the fact that the admissible functions need only to satisfy the kinematic boundary conditions of the continuum mechanics problem in hand.

In the present Lecture, we lay special emphasis upon the functional dependencies to be observed in the Rayleigh-Ritz technique, when formulated in the *spatial* and in the *material descriptions of continuum mechanics*, respectively. In analogy to the notion of the material-time derivative, we define *material variations* and *material partial derivatives* of the quantities and entities that occur in the Lagrange equations, and we relate these derivatives to *local partial derivatives*, the latter being particularly feasible with respect to a formulation in the spatial formulation of continuum mechanics. Having formulated the Lagrange equations for the case of a material volume, we re-formulate them for problems that are posed with respect to a *non-material volume*. A non-material volume is a volume with a closed surface that moves with a velocity different from the material particles located on it. In general, the mass contained in a non-material volume is not conserved. Problems posed with respect to a non-material volume are of a practical interest not only in fluid dynamics, but also in the dynamics of solid structures and machines. An example for such a problem is represented by the case of the coiling of a strip, which is shortly treated at the end of the present Lecture. The presented re-formulation of the Lagrange equations for a non-material volume represents an alternative derivation of results by Irschik and Holl (2002).

1 The Equations of Lagrange Written for a Material Volume

In the *spatial version of the Rayleigh-Ritz technique*, the instantaneous displacement vector $\boldsymbol{u}$ of a material particle is approximated in the form of an *admissible function* of its instantaneous place in the Euclidean space, of a number of n time-dependent *generalised coordinates* $q_i(t)$, $i = 1, \ldots, n$, and of time t:

$$\boldsymbol{u} = \boldsymbol{u}(\boldsymbol{p}, \ldots, q_i(t), \ldots, t), \tag{1}$$

which is called a *spatial Ritz-Ansatz* for the displacement. The position vector of the instantaneous place of the particle from some fixed origin is denoted as $\boldsymbol{p}$. For details of the spatial description of continuum mechanics, see Lectures 1 and 2 of the present book, Belyaev (2004a) and Irschik (2004), respectively. In order to be admissible, a Ritz-Ansatz in the following is required to satisfy the *kinematic boundary conditions* at the corresponding part of the boundary S of the material volume V under consideration. In case that Eq.(1) would be the exact solution of the problem in hand, the dynamic boundary conditions were to be satisfied, too. In case of an approximate statement, we will not require this.

The generalized coordinates q_i in the following are assumed to be independent of one another, such that no constraint conditions between them are to be considered. In our approximation of the initial value problem under consideration, the q_i are assumed to describe completely the desired solution. The problem thus is discretised to a system with n degrees-of-freedom, the generalized coordinates $q_i(t)$, $i = 1, ..., n$ being also denoted as *system coordinates*. Note that we allow an explicit dependence of $\boldsymbol{u}$ on the time t in Eq.(1).

For an element of a structure or machine, it is often more natural to formulate a Ritz-Ansatz with respect to the place $\boldsymbol{P}$ of the particle in an undeformed reference configuration of this element, i.e. to work in the material description of continuum mechanics, see Lecture 1 of the present book, Belyaev (2004a). The *material Ritz-Ansatz* for the displacement $\boldsymbol{u}$ reads

$$\boldsymbol{u} = \boldsymbol{U}(\boldsymbol{P}, ..., q_i(t), ..., t) \,, \tag{2}$$

where we use a capital letter in order to indicate the functional dependency on $\boldsymbol{P}$, as opposed to Eq.(1). From Eq.(2), the functional dependence

$$\boldsymbol{p} = \boldsymbol{P} + \boldsymbol{U}(\boldsymbol{P}, ..., q_i(t), ..., t) \tag{3}$$

is at our disposal. Assuming, as usual, the latter relation to be invertible with respect to $\boldsymbol{P}$, and re-inserting the inverted relation

$$\boldsymbol{P} = \boldsymbol{P}(\boldsymbol{p}, ..., q_i(t), ..., t) \tag{4}$$

into the material Ritz-Ansatz for the displacement, Eq.(2), we arrive at the Ritz-Ansatz in the spatial description, Eq.(1). In the first part of the following derivations, we need to work not only in the spatial, but also in the material description, in order to have a sound basis for our formulations. Later on, we will focus on the spatial description, since there occur circumstances in practice in which it is not feasible to formulate a Ritz-Ansatz with respect to an undeformed reference configuration, see the example of the coiling of a band presented at the end of the present Lecture. To provide the computational apparatus for such situations is a main topic of the present derivations.

With respect to the material formulations given in Eqs.(2)-(4), we now define a *material variation* $\delta\boldsymbol{u}$ of the displacement vector $\boldsymbol{u}$, the symbol δ denoting the material variation. In exemplary defining this latter operation for the case of $\boldsymbol{u}$, we consider a virtually displaced neighbouring position of the particle by adding the vector $\lambda\,\delta\boldsymbol{u}$ to the material formulation for $\boldsymbol{u}$, where we take the place $\boldsymbol{P}$ of the particle in the undeformed reference configuration as well as the time t to be fixed in Eq.(2), and where we let the scalar λ then tend to zero. The material variation $\delta\boldsymbol{u}$ thus can be expressed as

$$\begin{aligned}\delta\boldsymbol{u} &= \lim_{\lambda\to 0}\frac{1}{\lambda}\left(\boldsymbol{U}(\boldsymbol{P}, ..., q_i + \lambda\,\delta q_i, ..., t) - \boldsymbol{U}(\boldsymbol{P}, ..., q_i, ..., t)\right) \\ &= \frac{d}{d\lambda}\,\boldsymbol{U}(\boldsymbol{P}, ..., q_i + \lambda\,\delta q_i, ..., t)\Big|_{\lambda=0} \\ &= \frac{\partial \boldsymbol{U}(\boldsymbol{P}, ..., q_i + \lambda\,\delta q_i, ..., t)}{\partial(q_i + \lambda\,\delta q_i)}\,\frac{\partial(q_i + \lambda\,\delta q_i)}{\partial\lambda}\Big|_{\lambda=0} = \frac{\partial_P \boldsymbol{U}}{\partial q_i}\,\delta q_i\,,\end{aligned} \tag{5}$$

where Einstein's convention of summation about repeated indices is understood. The variation of the i-th generalized co-ordinate is denoted as δq_i, $i = 1, ..., n$. In Eq.(5.1), the virtually displaced neighbouring position is given by $\boldsymbol{U}(\boldsymbol{P}, ..., q_i + \lambda\, \delta q_i, ..., t)$. In the present Lecture, the variations δq_i are assumed to be independent, the differential calculus eventually leading to Eqs.(5.3) and (5.4). As is seen from the latter relation, the variation $\delta \boldsymbol{u}$ emerges as a linear mapping of the n independent variations δq_i. The functional dependencies in Eq. (2) of course must be such that the partial derivatives $\partial_P \boldsymbol{U} / \partial q_i$ in Eq.(5.4) remain bounded and take on definite values, which we assume throughout the Lecture for all of the derivatives coming into further consideration. Recall that the notion of a material variation introduced in Eq.(5) means that both, time t and the place $\boldsymbol{P}$ are fixed, which is indicated by the notation $\partial_P / \partial q_i$ in Eq.(5.4), this operation being denoted as a *material partial derivative* in the following. We note that $\delta \boldsymbol{u}$ in the literature is also denoted as the first variation of the displacement, since the first derivatives appear in Eq.(5), or, more precisely as the *first Gateaux variation* of $\boldsymbol{u}$, the latter notion indicating the linear dependence on the variations δq_i, see e.g. Sagan (1992) for the foundations of the calculus of variations. In the course of the motion, different variations δq_i are assigned at the respective instants of time, where we assume $\delta q_i(t)$ to represent sufficiently smooth functions of time. In the literature, $\delta \boldsymbol{u}$ is called a *virtual displacement*.

Analogous to the displacement, see Eqs.(1)-(5), we now formulate the Ritz-Ansatz for the velocity in both, the spatial and the material formulation. The material formulation of the velocity is found by differentiating Eq.(2) with respect to time, holding the place $\boldsymbol{P}$ in the reference configuration to be fixed:

$$\boldsymbol{v} = \dot{\boldsymbol{u}} = \frac{\partial_P}{\partial t} \boldsymbol{U}(\boldsymbol{P}, ..., q_i(t), ..., t) = \frac{\partial_P \boldsymbol{U}}{\partial q_i} \dot{q}_i + \frac{\partial_{P, q_i} \boldsymbol{U}}{\partial t} . \tag{6}$$

In Eq.(6.1), the superimposed dot indicates the material time derivative, revealed by the symbol $\partial_P / \partial t$ in Eq.(6.2). In Eq.(6.3), the notation $\partial_{P, q_i} / \partial t$ indicates that $\boldsymbol{P}$ as well as q_i are to be fixed. The generalized velocities are $\dot{q}_i = dq_i / dt$. From Eq.(6.3), the following functional dependency is seen to exist in the material description:

$$\boldsymbol{v} = \boldsymbol{V}(\boldsymbol{P}, ..., q_i(t), \dot{q}_i(t), ..., t) , \tag{7}$$

which, after inserting Eq.(4), leads to the spatial formulation

$$\boldsymbol{v} = \boldsymbol{v}(\boldsymbol{p}, ..., q_i(t), \dot{q}_i(t), ..., t) . \tag{8}$$

By analogy to Eq.(5), the material variation of the velocity, or the *virtual velocity*, becomes

$$\delta \boldsymbol{v} = \frac{d}{d\lambda} \boldsymbol{V}(\boldsymbol{P}, ..., q_i + \lambda\, \delta q_i, \dot{q}_i + \lambda\, \dot{\overline{\delta q}}_i, ..., t) \Big|_{\lambda = 0} = \frac{\partial_P \boldsymbol{V}}{\partial q_i} \delta q_i + \frac{\partial_P \boldsymbol{V}}{\partial \dot{q}_i} \dot{\overline{\delta q}}_i , \tag{9}$$

where $\dot{\overline{\delta q}}_i = d(\delta q_i) / dt$. In the present Lecture, we assume that non-integrable constraints between the generalized velocities $\dot{q}_i$ or their variations $\dot{\overline{\delta q}}_i$ are absent, such that these entities can be taken as independent. Applying the variation to Eq.(6.2) directly, we obtain

$$\delta \boldsymbol{v} = \delta \dot{\boldsymbol{u}} = \frac{d}{d\lambda} \frac{\partial_P}{\partial t} \boldsymbol{U}(\boldsymbol{P}, ..., q_i + \lambda\, \delta q_i, ..., t) \Big|_{\lambda = 0}$$

$$= \frac{\partial_P}{\partial t} \Big(\frac{\partial_P \boldsymbol{U}}{\partial q_i} \delta q_i \Big) = \frac{\partial_P \boldsymbol{V}}{\partial q_i} \delta q_i + \frac{\partial_P \boldsymbol{U}}{\partial q_i} \dot{\overline{\delta q}}_i , \tag{10}$$

where we have exchanged the derivative with respect to λ and the material time derivative in Eq.(10.3), as well as the material time derivative and the material partial derivative in Eq. (10.4). Accordingly, we set $\dot{\overline{\delta q_i}} = \delta\dot{q}_i = \delta\,(dq_i\,/\,dt)$ from now on. Comparing Eq.(10.4) with Eq.(9.2), the following useful relation is obtained:

$$\frac{\partial_P \boldsymbol{V}}{\partial \dot{q}_i} = \frac{\partial_P \boldsymbol{U}}{\partial q_i}. \tag{11}$$

Again in the material description, we now consider the kinetic energy per unit mass

$$\mathcal{K}'' = \frac{1}{2}\boldsymbol{v}\cdot\boldsymbol{v} = K''(\boldsymbol{P}, \ldots, q_i(t), \dot{q}_i(t), \ldots, t)\,, \tag{12}$$

the functional dependency given in Eq.(12.2) following by inserting Eq.(7) into (12.1). The symbol K indicates that $\mathcal{K}$ is expressed as a function of $\boldsymbol{P}$. In the spatial description, we have

$$\mathcal{K}'' = \frac{1}{2}\boldsymbol{v}\cdot\boldsymbol{v} = \mathcal{K}''(\boldsymbol{p}, \ldots, q_i(t), \dot{q}_i(t), \ldots, t)\,. \tag{13}$$

The material variation of Eq.(12.2) reads, compare Eqs. (5) and (9.2),

$$\delta\mathcal{K}'' = \frac{d}{d\lambda}K''(\boldsymbol{P},\ldots, q_i + \lambda\,\delta q_i, \dot{q}_i + \lambda\,\delta\dot{q}_i, \ldots, t)\,\Big|_{\lambda=0} = \frac{\partial_P K''}{\partial q_i}\,\delta q_i + \frac{\partial_P K''}{\partial \dot{q}_i}\,\delta\dot{q}_i. \tag{14}$$

From (12.1), we obtain the following expressions for the material partial derivatives in Eq.(14.2):

$$\frac{\partial_P K''}{\partial q_i} = \boldsymbol{V}\cdot\frac{\partial_P \boldsymbol{V}}{\partial q_i}, \tag{15}$$

$$\frac{\partial_P K''}{\partial \dot{q}_i} = \boldsymbol{V}\cdot\frac{\partial_P \boldsymbol{V}}{\partial \dot{q}_i} = \boldsymbol{V}\cdot\frac{\partial_P \boldsymbol{U}}{\partial q_i}, \tag{16}$$

see Eq.(11) for Eq.(16.2). Inserting Eqs.(15) and (16.1) into the material variation of the kinetic energy per unit mass, Eq.(14.2), it is seen from Eq.(9.2) that

$$\delta\mathcal{K}'' = \delta(\tfrac{1}{2}\boldsymbol{v}\cdot\boldsymbol{v}) = \boldsymbol{v}\cdot\delta\boldsymbol{v} = \boldsymbol{v}\cdot\delta\dot{\boldsymbol{u}}\,. \tag{17}$$

The latter relation demonstrates that the operation of variation follows the chain rule of differential calculus. This fact has given raise to a vast literature in which the virtual displacements and velocities are taken as infinitesimal from the onset. Note, however, that we have not required the variations of the generalized coordinates and velocities to be small in the above derivations. For variational formulations in elasticity without taking the variations to be infinitesimal, see also Sections 38 and 65 of Gurtin (1972) and Chapter 9.2 of Gummert and Reckling (1987).

From now on, we completely turn to the spatial description of continuum mechanics by substituting Eq. (4) into the above material formulations for the material variations and the material partial derivatives. For example, the variation $\delta\boldsymbol{u}$ in Eq.(5) now is formulated as a function of the place $\boldsymbol{p}$ of the particle in the instantaneous configuration as

$$\delta\boldsymbol{u} = \frac{\partial_P \boldsymbol{U}}{\partial q_i}\,\Big|_{\boldsymbol{P}=\,\boldsymbol{P}(\boldsymbol{p},\,\ldots,\,q_i(t),\,\ldots,t)}\,\delta q_i = \frac{\partial_p \boldsymbol{u}}{\partial q_i}\,\delta q_i\,. \tag{18}$$

The substitution of Eq.(4) indicated in Eq.(18.1) can be denoted as a *push-forward*. It turns out to be essential that this substitution is made after the partial derivative has been performed, since the substitution and the material partial derivative in general are not interchangeable. There are important exceptions, however, namely that

$$\frac{\partial_P V(\boldsymbol{P}, \ldots, q_i(t), \dot{q}_i(t), \ldots, t)}{\partial \dot{q}_i}\Big|_{\boldsymbol{P}=\boldsymbol{P}(\boldsymbol{p}, \ldots, q_i(t), \ldots, t)} = \frac{\partial \boldsymbol{v}(\boldsymbol{p}, \ldots, q_i(t), \dot{q}_i(t), \ldots, t)}{\partial \dot{q}_i} = \frac{\delta \boldsymbol{v}}{\delta \dot{q}_i}, \tag{19}$$

since the substitution of Eq.(4) does not affect the functional dependency with respect to $\dot{q}_i$. Hence the material partial derivative of V with respect to $\dot{q}_i$ is equal to the partial derivative of the spatial representation of $\boldsymbol{v}$ with respect to $\dot{q}_i$, Eq.(19.1). In the latter spatial partial derivative, the place $\boldsymbol{p}$ is fixed, which we indicate by the symbol $\delta / \delta \dot{q}_i$, see Eq.(19.2). This operation is called a *local partial derivative* in the following. From now on, in order to retain a compact notation, the push-forward operation is understood for the material partial and material time derivatives. In order to indicate this push forward more clearly, we use the abbreviated notation indicated in Eq.(18.2). Analogous to Eqs.(18) and (19), we have

$$\frac{\partial_P K''}{\partial \dot{q}_i}\Big|_{\boldsymbol{P}=\boldsymbol{P}(\boldsymbol{p}, \ldots, q_i(t), \ldots, t)} = \frac{\partial_P \mathcal{K}''}{\partial \dot{q}_i} = \frac{\delta \mathcal{K}''}{\delta \dot{q}_i}. \tag{20}$$

However, there is

$$\frac{\partial_P K''}{\partial q_i}\Big|_{\boldsymbol{P}=\boldsymbol{P}(\boldsymbol{p}, \ldots, q_i(t), \ldots, t)} = \frac{\partial_P \mathcal{K}''}{\partial q_i} = \frac{\delta \mathcal{K}''}{\delta q_i} + \frac{\partial_P \boldsymbol{u}}{\partial q_i} \cdot grad\, \mathcal{K}''. \tag{21}$$

The relation given in Eq.(21) holds in analogy to the well known expression for the material time derivative of some entity Ψ:

$$\frac{d\Psi}{dt} = \dot{\Psi} = \frac{\partial \Psi}{\partial t} + \boldsymbol{v} \cdot grad\, \Psi = \frac{\delta \Psi}{\delta t} + \frac{\partial_P \boldsymbol{u}}{\partial t} \cdot grad\, \Psi. \tag{22}$$

For Eq.(22.2), see Sect. 72 of Truesdell and Toupin (1960), and compare Lecture 2 of the present book, Irschik (2004). We have used the above introduced notation in order to obtain Eq.(22.3). For a more detailed proof of Eqs.(20) and (21), see Irschik and Holl (2002).

In the next step, we multiply Eq.(17) by the elementary mass of a material particle, $\rho\, dv$, and we integrate over the material volume $\mathcal{V}$ under consideration:

$$\int_{\mathcal{V}} \delta \mathcal{K}'' \rho\, dv = \int_{\mathcal{V}} \boldsymbol{v} \cdot \delta \boldsymbol{v}\, \rho\, dv = \int_{\mathcal{V}} \boldsymbol{v} \cdot \delta \dot{\boldsymbol{u}}\, \rho\, dv. \tag{23}$$

Now recall Eqs.(12) and (13) of Lecture 2 of the present book, Irschik (2004). In analogy to the equation of balance of kinetic energy, see Eq.(14) of the latter reference, but identifying the test vector field $\bar{\boldsymbol{u}}$ with the variation of the displacement, $\bar{\boldsymbol{u}} = \delta \boldsymbol{u}$ and $\dot{\bar{\boldsymbol{u}}} = \delta \dot{\boldsymbol{u}}$ in Eqs.(12) and (13) of Irschik (2004), we obtain the following relation:

$$\frac{d}{dt} \int_{\mathcal{V}} \boldsymbol{v} \cdot \delta \boldsymbol{u}\, \rho\, dv - \int_{\mathcal{V}} \boldsymbol{v} \cdot \delta \dot{\boldsymbol{u}}\, \rho\, dv$$

$$= \int_{\mathcal{V}} \boldsymbol{b} \cdot \delta \boldsymbol{u}\, dv + \oint_{S} \boldsymbol{t}_{(\boldsymbol{n})} \cdot \delta \boldsymbol{u}\, da - \int_{\mathcal{V}} tr\,(\boldsymbol{T} \cdot grad\, \delta \boldsymbol{u})\, dv, \tag{24}$$

which however is not of the form of the general balance law. The surface of the material volume $\mathcal{V}$ is denoted as $\mathcal{S}$. Integrating Eq.(24) over the time-interval $t \leq \tau \leq t + \Delta t$, and substituting Eq.(23.2), it is found that Eq.(24) can be put into the form

$$\int_t^{t+\Delta t} \{\int_{\mathcal{V}} \delta\mathcal{K}'' \rho \, dv + \int_{\mathcal{V}} \boldsymbol{b} \cdot \delta\boldsymbol{u} \, dv$$

$$+\oint_{\mathcal{S}} \boldsymbol{t}_{(\boldsymbol{n})} \cdot \delta\boldsymbol{u} \, da - \int_{\mathcal{V}} tr(\boldsymbol{T} \cdot grad\, \delta\boldsymbol{u}) \, dv\} \, d\tau = 0 \,. \tag{25}$$

The latter relation may be considered as an extension of the celebrated *Hamilton principle of analytical mechanics*. For the useful classical form of Hamilton's principle, in which it assumed that all of the forces are derivable from potential functions, compare Lecture 4 of the present book, Belyaev (2004a). For the extended form of Eq.(25), see also Section 236 of Truesdell and Toupin (1960). The relation given in Eq.(25) remains valid in case of an arbitrary constitutive behaviour of the body under consideration. In deriving Eq.(25), the variations at the beginning and at the end of the respective time interval have been assumed to be such that

$$\int_{\mathcal{V}} \boldsymbol{v} \cdot \delta\boldsymbol{u} \, \rho \, dv \,|_{\tau = t,\, t+\Delta t} = \int_{\mathcal{V}} \boldsymbol{v} \cdot \frac{\partial_P \boldsymbol{u}}{\partial q_i} \delta q_i \, \rho \, dv \,|_{\tau = t,\, t+\Delta t} = 0 \,, \tag{26}$$

see Eq. (18). Eq.(26) is satisfied by requiring that $\delta q_i |_{\tau = t, t+\Delta t} = 0, i = 1, ..., n$.

Of course, Eqs.(24) and (25) can be applied also in the static limit, then representing the *principles of virtual work* or *virials*, respectively. When the virtual displacements $\delta\boldsymbol{u}$ are infinitesimal, then the respective volume and surface integrals in Eq.(25) represent the *virtual work of the external and internal forces*, a notion, which is related to the definition of elementary work, see Chapter 5 of Ziegler (1998). An advantage of taking $\delta\boldsymbol{u}$ to be infinitesimal is that formulas of the kinematics of rigid bodies may be used by analogy in practical applications. When the variations are not infinitesimal, the notion of a *virtual displacement virial* of the external and internal forces appears to be more appropriate, see Irschik (2001). In the latter reference, it has been demonstrated that the use of finite virtual displacements $\delta\boldsymbol{u}$ represents a computational appealing alternative to the principle of virtual work. In the following derivations, the question of the finiteness of the variations $\delta\boldsymbol{u}$ however does not come into the play.

We now insert Eq.(14) into the first volume integral in Eq.(25), which shows that

$$\int_{\mathcal{V}} \delta\mathcal{K}'' \rho \, dv = \int_{\mathcal{V}} (\frac{\partial_P \mathcal{K}''}{\partial q_i} \delta q_i + \frac{\partial_P \mathcal{K}''}{\partial \dot{q}_i} \delta\dot{q}_i) \rho \, dv \,. \tag{27}$$

But in the absence of sources of mass, mass is conserved in material volume $\mathcal{V}$ under consideration, such that the mass of a material particle, $\rho \, dv$, must not change with the generalized coordinates. We are therefore allowed to interchange the material partial derivatives in Eq.(27) with the integration over the material volume $\mathcal{V}$:

$$\int_{\mathcal{V}} (\frac{\partial_P \mathcal{K}''}{\partial q_i} \delta q_i + \frac{\partial_P \mathcal{K}''}{\partial \dot{q}_i} \delta\dot{q}_i) \rho \, dv = \frac{\partial_P}{\partial q_i} \int_{\mathcal{V}} \mathcal{K}'' \rho \, dv \, \delta q_i + \frac{\partial_P}{\partial \dot{q}} \int_{\mathcal{V}} \mathcal{K}'' \rho \, dv \, \delta\dot{q}_i$$

$$= \frac{\partial \mathcal{K}}{\partial q_i} \delta q_i + \frac{\partial \mathcal{K}}{\partial \dot{q}} \delta \dot{q}_i , \tag{28}$$

where $\mathcal{K} = \int_{\mathcal{V}} K'' \rho \, dv$ is the total kinetic energy in $\mathcal{V}$, see Eq.(13) above, and recall Eq.(15) of Lecture 2 of the present book, Irschik (2004). We have suppressed the Index P in the last expression of Eq.(28) in order to be consistent with the usual notation. Substituting Eqs.(27) and (28.2) into Eq.(25) and integrating by parts it is found that

$$\int_t^{t+\Delta t} \int_{\mathcal{V}} \delta \mathcal{K}'' \rho \, dv \, d\tau$$

$$= \int_t^{t+\Delta t} \left(\frac{\partial \mathcal{K}}{\partial q_i} - \frac{d}{d\tau} \frac{\partial \mathcal{K}}{\partial \dot{q}_i} \right) \delta q_i \, d\tau + \frac{\partial \mathcal{K}}{\partial \dot{q}_i} \delta q_i \big|_{\tau = t+\Delta t} - \frac{\partial \mathcal{K}}{\partial \dot{q}_i} \delta q_i \big|_{\tau = t} . \tag{29}$$

The last two terms in Eq.(29) can be re-formulated by means of Eq.(16), which tells us that

$$\frac{\partial \mathcal{K}}{\partial \dot{q}_i} = \int_{\mathcal{V}} \frac{\partial_P \mathcal{K}''}{\partial \dot{q}_i} \rho \, dv = \int_{\mathcal{V}} \boldsymbol{v} \cdot \frac{\partial_P \boldsymbol{u}}{\partial q_i} \rho \, dv . \tag{30}$$

Hence, from Eq.(26), the last two terms in Eq.(29) do vanish, such that

$$\int_t^{t+\Delta t} \int_{\mathcal{V}} \delta \mathcal{K}'' \rho \, dv \, d\tau = \int_t^{t+\Delta t} \left(\frac{\partial \mathcal{K}}{\partial q_i} - \frac{d}{d\tau} \frac{\partial \mathcal{K}}{\partial \dot{q}_i} \right) \delta q_i \, d\tau . \tag{31}$$

The generalized Hamilton's principle, Eq.(25), thus can be written as

$$\int_t^{t+\Delta t} \left(\frac{\partial \mathcal{K}}{\partial q_i} - \frac{d}{d\tau} \frac{\partial \mathcal{K}}{\partial \dot{q}_i} + Q_i \right) \delta q_i \, d\tau = 0 , \tag{32}$$

where the *generalised forces* Q_i in Eq.(32) are given by

$$Q_i = \int_{\mathcal{V}} \boldsymbol{b} \cdot \frac{\partial_P \boldsymbol{u}}{\partial q_i} dv + \oint_{\mathcal{S}} \boldsymbol{t}_{(\boldsymbol{n})} \cdot \frac{\partial_P \boldsymbol{u}}{\partial q_i} da - \int_{\mathcal{V}} tr \left(\boldsymbol{T} \cdot grad \frac{\partial_P \boldsymbol{u}}{\partial q_i} \right) dv , \tag{33}$$

compare Eq.(18), and recall that the push-forward is understood for the material partial derivatives. Substituting Eqs.(11) and (19), this can be expressed by local partial derivatives of the velocity in the form:

$$Q_i = \int_{\mathcal{V}} \boldsymbol{b} \cdot \frac{\delta \boldsymbol{v}}{\delta \dot{q}_i} dv + \oint_{\mathcal{S}} \boldsymbol{t}_{(\boldsymbol{n})} \cdot \frac{\delta \boldsymbol{v}}{\delta \dot{q}_i} da - \int_{\mathcal{V}} tr \left(\boldsymbol{T} \cdot grad \frac{\delta \boldsymbol{v}}{\delta \dot{q}_i} \right) dv , \tag{34}$$

which is particularly feasible for the spatial description. Assuming that the variations δq_i can be taken independent from one another and to be arbitrary, with exception of the conditions formulated in Eq.(26), the following set of n ordinary differential equations in time for the generalized coordinates q_i, $i = 1, ..., n$, is provided by Eq.(32):

$$\frac{d}{dt} \frac{\partial \mathcal{K}}{\partial \dot{q}_i} - \frac{\partial \mathcal{K}}{\partial q_i} - Q_i = 0, \quad i = 1, ..., n . \tag{35}$$

The latter set represents the desired *Lagrange equations*, also denoted as Lagrange equations of the second kind. Note that the partial derivatives in Eq.(35) are taken to be material with respect to the motion of the material volume $\mathcal{V}$ under consideration, likewise to the material time derivative. In order to present a secure formula for considering this fact in the spatial description, we use Eqs.(20) and (21), which show that:

$$\frac{\partial \mathcal{K}}{\partial \dot{q}_i} = \int_{\mathcal{V}} \frac{\partial_P \mathcal{K}''}{\partial \dot{q}_i} \rho \, dv = \int_{\mathcal{V}} \frac{\delta \mathcal{K}''}{\delta \dot{q}_i} \rho \, dv , \tag{36}$$

$$\frac{\partial \mathcal{K}}{\partial q_i} = \int_{\mathcal{V}} \frac{\partial_P \mathcal{K}''}{\partial q_i} \rho \, dv = \int_{\mathcal{V}} \left(\frac{\delta \mathcal{K}''}{\delta q_i} + \frac{\delta \boldsymbol{v}}{\delta \dot{q}_i} \cdot grad \, \mathcal{K}''\right) \rho \, dv , \tag{37}$$

such that it is possible again to work solely with local partial derivatives. The kinetic energy per unit mass, $\mathcal{K}''$, here is computed according to the spatial formulation of Eq.(13). The corresponding spatial Ritz-Ansatz for the velocity, Eq.(8), of course should be such, that the volume integrals in Eqs.(34), (36) and (37.2) do exist. In order to evaluate the generalised forces in Eq.(34), the Cauchy stress $\boldsymbol{T}$ furthermore must be expressed by *constitutive equations*, resulting in an expression, which is related to the Ritz-Ansatz of Eq.(1) through certain algebraic or differential operations. We furthermore need to introduce the *dynamic boundary conditions*, concerning the surface traction $\boldsymbol{t}_{(n)}$ in the surface integral of Eq.(12). If $\boldsymbol{t}_{(n)}$ is not given, since a kinematic boundary condition is prescribed, the Cauchy fundamental theorem on stresses, Eq.(10) of Irschik (2004), must be used in order to relate $\boldsymbol{t}_{(n)}$ to the stress $\boldsymbol{T}$. Should the boundary be required to remain fixed in space, then $\delta \boldsymbol{v} / \delta \dot{q}_i$ must vanish, and the surface integral in Eq.(34) needs not to be computed.

In deriving Eq.(35), we have assumed that the generalized coordinates can be taken independent of one another. However, sometimes it may be convenient to use a set of dependent coordinates, or there may be non-integrable constraint conditions between the generalised coordinates and the generalised velocities, the latter case being said to be *non-holonomic.* For a discussion on *non-holonomic constraint conditions* in analytical mechanics, see Eqs.(1.2) and (4.3) of Lecture 4 of the present book, Belyaev (2004b). The Lagrange equations, Eq.(35), must be extended by additional constraint forces in order that the formulation does hold also in the non-holonomic case. These additional constraint forces can be conveniently introduced into the Lagrange equations using the method of Lagrange multipliers, see Chapter (3.4) of the book of Shabana (1998) and Lecture 4 of the present book, Belyaev (2004b). The extended Lagrange equations and the constraint equations form a set of *differential algebraic equations (DAE)* for the generalised coordinates and the Lagrange multipliers. For a classical treatment of non-holonomic conditions we refer to Hamel (1967). See also the expositions by Bremer and Pfeiffer (1992) on elastic robots.

3 The Equations of Lagrange for a Non-Material Volume

We now proceed to re-write the Lagrange equations, Eqs. (35), such that they become applicable for problems posed with respect to a non-material volume. A non-material volume $\mathcal{v}(t)$ is a volume with a closed surface $\mathcal{s}(t)$ that moves with a velocity $\boldsymbol{w}$, which is possibly different from the velocity $\boldsymbol{v}$ of the material particles instantaneously located on $\mathcal{s}(t)$. The mass contained in a non-volume thus is not conserved in general. As has been pointed out, e.g., in Chapter 7 of the book by Parkus (1988), such situations are of interest also in the mechanics of solid bodies.

When we assume that the non-material volume $\nu(t)$ is instantaneously coinciding with the material volume $\mathcal{V}$ considered in the Lagrange equations above, Eq.(35), these latter relations can be used as a starting point. The value of the kinetic energy contained in the material volume of course coincides with the kinetic energy instantaneously contained in the non-material volume, however the rates of change and the partial derivatives are not equal in general. Now, in order to transform the material time derivative of some quantity that is contained in the material volume $\mathcal{V}$ to the time rate of change of this quantity contained in the respective non-material volume $\nu(t)$, the *transport theorem of Reynolds* is at our disposal. For a review including some historical remarks on the Reynolds transport theorem, see Irschik and Holl (2002). Here, we start to derive the Reynolds theorem by first considering the material time rate of the quantity $\mathcal{Q}$ of some scalar entity $\Psi = \Psi(\boldsymbol{p}, t)$ that is contained in the material volume $\mathcal{V}$ having the surface $\mathcal{S}$, see Lecture 2 of the present book, Irschik (2004). Interchanging the differentiation and integration, it is found that

$$\frac{d}{dt}\mathcal{Q} = \frac{d}{dt}\int_{\mathcal{V}} \Psi \, dv = \int_{\mathcal{V}} \dot{\Psi} \, dv + \int_{\mathcal{V}} \Psi \, div \, \boldsymbol{v} \, dv \,, \tag{38}$$

where the material time derivative of Ψ is given by Eq.(22.$_1$), and the Euler expansion formula for the material time derivative of a differential volume has been used, see Eq.(5) of Irschik (2004). Using the formulation of Eq.(22.$_2$) as well as the divergence theorem, Eq.(38) can be eventually written as

$$\frac{d}{dt}\mathcal{Q} = \int_{\mathcal{V}} \frac{\delta\Psi}{\delta t} \, dv + \oint_{\mathcal{S}} d\boldsymbol{a} \cdot \boldsymbol{v} \Psi \,. \tag{39}$$

We now consider the non-material volume $\nu(t)$. Following Sect. 81 of Truesdell and Noll (1960), in forming the time rate of change of the quantity contained in $\nu(t)$ one only needs to consider fictitious particles moving at the velocity $\boldsymbol{w}$ of the non-material surface $s(t)$, and to formulate the material time derivative by analogy to Eq.(39):

$$\frac{d_w}{dt}\mathcal{Q} = \int_{\nu(t)} \frac{\delta\Psi}{\delta t} \, dv + \oint_{s(t)} d\boldsymbol{a} \cdot \boldsymbol{w} \Psi \,. \tag{40}$$

The notation d_w / dt indicates that the volume of integration is taken material with respect to the velocity $\boldsymbol{w}$. Recalling that, instantaneously, $\mathcal{V} = \nu(t)$ and $\mathcal{S} = s(t)$, and subtracting Eq.(40) from (39), the following useful form of the Reynolds transport theorem is obtained:

$$\frac{d}{dt}\mathcal{Q} = \frac{d_w}{dt}\mathcal{Q} + \oint_{\mathcal{S}} d\boldsymbol{a} \cdot (\boldsymbol{v} - \boldsymbol{w}) \, \Psi = \frac{d_w}{dt}\mathcal{Q} + \oint_{\mathcal{S}} d\boldsymbol{a} \cdot \left(\frac{\partial_{\boldsymbol{p}} \boldsymbol{u}}{\partial t} - \frac{\partial_{\boldsymbol{p}} \boldsymbol{u}_w}{\partial t}\right) \Psi \,. \tag{41}$$

For the application of Eq.(41.$_1$) in fluid dynamics, see Sect. 1.12 of Warsi (1999). The symbol $\boldsymbol{u}_w$ in Eq.(41.$_2$) stands for a fictitious displacement field corresponding to the velocity $\boldsymbol{w}$ of the non-material surface $s(t)$, the latter instantaneously coinciding with the material surface $\mathcal{S}$.

By analogy to the material time derivative, the formulation of Eq.(41) can be used in order to relate the material partial derivatives of some quantity contained in $\mathcal{V}$ to the partial derivative of this quantity contained in $\nu(t)$. In connecting Eq.(41) to the Lagrange equations, Eq. (35), it turns out to be convenient to work with the kinetic energy per unit volume,

$$\mathcal{K}' = \frac{1}{2}\boldsymbol{v} \cdot \boldsymbol{v}\,\rho = \mathcal{K}''\,\rho\,, \tag{42}$$

instead of $\mathcal{K}''$ in Eq.(13). The total kinetic energy in $\mathcal{V}$ is $\mathcal{K} = \int_{\mathcal{V}} \mathcal{K}'\,dv$. Setting $\Psi = \mathcal{K}'$, the formulations in Eqs.(41.$_2$) and (38.$_1$) thus suggest to write

$$\frac{\partial}{\partial q_i}\mathcal{K} = \frac{\partial_w}{\partial q_i}\mathcal{K} + \oint_S d\boldsymbol{a} \cdot \left(\frac{\partial\,\boldsymbol{p}\boldsymbol{u}}{\partial q_i} - \frac{\partial\,\boldsymbol{p}\boldsymbol{u}_w}{\partial q_i}\right)\mathcal{K}' = \frac{\partial_w}{\partial q_i}\mathcal{K} + \oint_S d\boldsymbol{a} \cdot \left(\frac{\delta \boldsymbol{v}}{\partial \dot{q}_i} - \frac{\delta \boldsymbol{w}}{\partial \dot{q}_i}\right)\mathcal{K}'\,, \tag{43}$$

where we have used Eqs.(11) and (19) in order to bring into the play the local partial derivatives in Eq.(43.$_2$). Analogously, we have

$$\frac{\partial}{\partial \dot{q}_i}\mathcal{K} = \frac{\partial_w}{\partial \dot{q}_i}\mathcal{K} + \oint_S d\boldsymbol{a} \cdot \left(\frac{\partial\,\boldsymbol{p}\boldsymbol{u}}{\partial \dot{q}_i} - \frac{\partial\,\boldsymbol{p}\boldsymbol{u}_w}{\partial \dot{q}_i}\right)\mathcal{K}' = \frac{\partial_w}{\partial \dot{q}_i}\mathcal{K}\,, \tag{44}$$

since the displacements do not depend on the generalized velocities $\dot{q}_i$, see Eq.(2). Furthermore, from Eq.(30.$_1$), there is

$$\frac{\partial \mathcal{K}}{\partial \dot{q}_i} = \int_{\mathcal{V}} \frac{\delta \mathcal{K}''}{\delta \dot{q}_i}\,\rho\,dv = \int_{\mathcal{V}} \frac{\delta \mathcal{K}'}{\delta \dot{q}_i}\,dv\,. \tag{45}$$

The formulation in Eq.(45.$_1$) follows by substituting Eq.(20) into Eq.(30.$_1$), while Eq.(45.$_2$) is due to the fact that the mass density must not depend on the generalized velocities,

$$\rho = \rho(\boldsymbol{p}, \ldots, q_i(t), \ldots, t),\; i = 1, \ldots, n\,. \tag{46}$$

In short, the functional dependency in Eq.(46) holds because the instantaneous mass density is related to the mass density in the reference configuration by means of the displacement gradient, see Sect. 156 of Truesdell and Toupin (1960), and compare Eq.(1). The Reynolds transport theorem, Eq.(41) thus can be applied by analogy to state the following relation:

$$\frac{d}{dt}\frac{\partial \mathcal{K}}{\partial \dot{q}_i} = \frac{d_w}{dt}\frac{\partial_w \mathcal{K}}{\partial \dot{q}_i} + \oint_S d\boldsymbol{a} \cdot (\boldsymbol{v} - \boldsymbol{w})\,\frac{\delta \mathcal{K}'}{\delta \dot{q}_i}\,. \tag{47}$$

For a detailed proof of Eqs.(44) and (47), see Irschik and Holl (2002). The Lagrange equations for a non-material volume follow by inserting Eqs.(44) and (47) into Eq.(35):

$$\frac{d_w}{dt}\frac{\partial_w \mathcal{K}}{\partial \dot{q}_i} - \frac{\partial_w}{\partial q_i}\mathcal{K} + \oint_S d\boldsymbol{a} \cdot (\boldsymbol{v} - \boldsymbol{w})\,\frac{\delta \mathcal{K}'}{\delta \dot{q}_i} - \oint_S d\boldsymbol{a} \cdot \left(\frac{\delta \boldsymbol{v}}{\partial \dot{q}_i} - \frac{\delta \boldsymbol{w}}{\partial \dot{q}_i}\right)\mathcal{K}' - Q_i = 0\,, \tag{48}$$

for $i = 1, \ldots, n$.

A typical process involving a non-material volume is the coiling process, when material in the form of a band, sheet or rope is stored on a rotating coiling drum. For the case of a coiling drum, which is mounted on a shaft, a plane mechanical model perpendicular to the axis of rotation is shown in Fig. 1. The non-material control volume is indicated by a dashed line surrounding the coiled material, the radius of the drum growing with the speed $\dot{r}$. There is only one entrance surface area in this problem, A_e, through which the material is flowing into the control volume. The velocity of the band at this entrance surface is

$$\boldsymbol{v} = \left(\dot{x} + v_{rel}\right)\boldsymbol{e}_x + \left(\dot{y} - \dot{r}\right)\boldsymbol{e}_y\,, \tag{49}$$

where $\dot{x}$ and $\dot{y}$ are the speeds of the shaft in the directions of the horizontal unit normal vector $\boldsymbol{e}_x$ and of the vertical unit normal vector $\boldsymbol{e}_y$, respectively. The velocity of the non-material entrance surface A_e is

$$\boldsymbol{w} = \dot{x}\,\boldsymbol{e}_x + \left(\dot{y} - \dot{r}\right)\boldsymbol{e}_y . \tag{50}$$

Hence, the relative velocity of the inflowing material is taken as horizontal, $\boldsymbol{v} - \boldsymbol{w} = v_{rel}\,\boldsymbol{e}_x$.

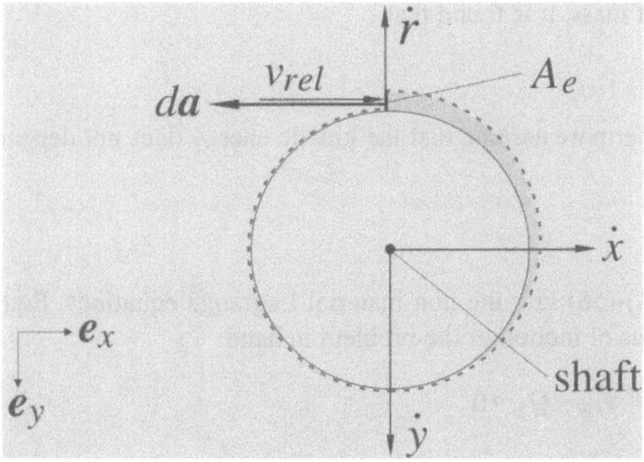

Fig.1: Rotating drum coiling up a sheet or band.

At the remaining parts of the surface of the non-material control volume, there is $\boldsymbol{v} - \boldsymbol{w} = \boldsymbol{0}$, such that these surface parts do not contribute to the surface integrals in Eq.(48). We now assume that the relative velocity v_{rel} is prescribed, and that $\dot{r}$ can be computed from v_{rel} alone, thus representing a given quantity. As generalized coordinates of the problem, we choose the shaft speeds, $\dot{q}_1 = \dot{x}$ and $\dot{q}_2 = \dot{y}$. The surface integrals in Eq.(48) then become

$$\int_{A_e} \boldsymbol{da} \cdot (\boldsymbol{v} - \boldsymbol{w}) \frac{\delta \mathcal{K}'}{\delta \dot{x}} = -\rho\, A_e\, v_{rel} \left(\dot{x} + v_{rel}\right) , \tag{51}$$

$$\int_{A_e} \boldsymbol{da} \cdot (\boldsymbol{v} - \boldsymbol{w}) \frac{\delta \mathcal{K}'}{\delta \dot{y}} = -\rho\, A_e\, v_{rel} \left(\dot{y} - \dot{r}\right) , \tag{52}$$

$$\int_{A_e} \boldsymbol{da} \cdot \left(\frac{\delta \boldsymbol{v}}{\partial \dot{x}} - \frac{\delta \boldsymbol{w}}{\partial \dot{x}}\right) \mathcal{K}' = \int_{A_e} \boldsymbol{da} \cdot \left(\frac{\delta \boldsymbol{v}}{\partial \dot{y}} - \frac{\delta \boldsymbol{w}}{\partial \dot{y}}\right) \mathcal{K}' = 0 . \tag{53}$$

We furthermore assume that the partial derivatives of the kinetic energy contained in the non-material volume can be approximately written as

$$\frac{d_w}{dt}\frac{\partial_w \mathcal{K}}{\partial \dot{x}} = m_w\, \ddot{x} + \frac{d_w m_w}{dt}\,\dot{x}, \quad \frac{d_w}{dt}\frac{\partial_w \mathcal{K}}{\partial \dot{y}} = m_w\, \ddot{y} + \frac{d_w m_w}{dt}\,\dot{y} , \tag{54}$$

where m_w denotes the mass contained in the non-material control volume, and $d_w m_w / dt$ represents the time rate of change of that mass. Corresponding to the assumption of a prescribed entrance velocity v_{rel}, we assume m_w to be given as a function of time. The relations given in Eq.(54) are well-known from the problem of a mass-point with a time-variable mass. For a review on systems with variable mass, see Irschik and Holl (2004). More complex expressions than those in Eq.(54) would emerge, if the center of the mass contained in the control volume could not be approximated by the location of the shaft. From elementary considerations with respect to balance of mass, it is found that

$$\frac{d_w m_w}{dt} = \rho A_e v_{rel} \tag{55}$$

in Eq.(54). We furthermore assume that the kinetic energy does not depend on the position of the shaft, such that

$$\frac{\partial_w}{\partial q_i} \mathcal{K} = 0 . \tag{56}$$

Substituting Eqs.(51)-(56) into the non-material Lagrange equations, Eq.(48), we eventually arrive at the equations of motion of the problem in hand:

$$m_w \ddot{x} - \frac{d_w m_w}{dt} v_{rel} - Q_x = 0 , \tag{57}$$

$$m_w \ddot{y} + \frac{d_w m_w}{dt} \dot{r} - Q_y = 0 . \tag{58}$$

The case in which v_{rel} is not prescribed will be treated in some detail elsewhere.

References

Belyaev, A.K. (2004). Basics of Continuum Mechanics. In: Irschik, H. and Schlacher, K., eds., *Advanced Dynamics and Control of Structures and Machines*. CISM Courses and Lectures No. 444. Wien–New York: Springer-Verlag.

Belyaev, A.K. (2004). Basics of Analytical Mechanics. In: Irschik, H. and Schlacher, K., eds., *Advanced Dynamics and Control of Structures and Machines*. CISM Courses and Lectures No. 444. Wien–New York: Springer-Verlag.

Bremer, H.. and Pfeiffer, F. (1992). Elastische Mehrkörpersysteme. Stuttgart: B.G. Teubner.

Gershenfeld, N. (1999). *The Nature of Mathematical Modelling*. Cambridge: University Press.

Gerstmayr, J., Irschik, H. and Dibold, M. (2004). Computational Dynamics of an Elasto-Plastic Structural Element With Rigid-Body Degrees-of-Freedom. In: Irschik, H. and Schlacher, K., eds., *Advanced Dynamics and Control of Structures and Machines*. CISM Courses and Lectures No. 444, Wien–New York: Springer-Verlag.

Gurtin, M. E. (1972). The linear theory of elasticity. In: Flügge, S., ed., *Handbuch der Physik, Vol. VIa/2*, Berlin: Springer-Verlag.

Gummert, P., Reckling, K.-A. (1987). *Mechanik*, 2nd Edition. Braunschweig: F. Vieweg & Sohn.

Hamel, G. (1967). *Theoretische Mechanik*. Berlin: Springer-Verlag.

Irschik, H. (2001). Zum Fingerschen Virial verformbarer Körper in der nichtlinearen Statik. *Sitzungsberichte der Öster. Akademie der Wiss.*, Math.-Nat. Klasse Abt. II, 209: 47-66.

Irschik, H. (2004). A Treatise on the Equations of Balance and on the Jump Relations in Continuum Mechanics. In: Irschik, H. and Schlacher, K., eds., *Advanced Dynamics and Control of Structures and Machines*. CISM Courses and Lectures No. 444, Wien–New York: Springer-Verlag.

Irschik, H.. and Holl, H. J. (2002). The Equations of Lagrange Written for a Non-Material Volume, *Acta Mechanica* 153: .231-248.

Irsc hik H., Holl H.J. (2004): Mechanics of Variable-Mass Systems: Part 1: Balance of Mass and Linear Momentum. ASME-Applied Mechanics Reviews, to appear.

Parkus, H. (1988). *Mechanik der festen Körper*, Nachdruck der 2. Auflage. Wien-New York: Springer-Verlag.

Schlacher K. and Kugi A. (2002): Symbolic Methods for Systems of Implicit Ordinary Differential Equations. *Mechanics of Structures and Machines*, 30: 411 - 429.

Truesdell, C. and Toupin, R. (1960). The Classical Field Theories. In: Flügge, S., ed., *Handbuch der Physik*, Vol. III/1, Berlin: Springer-Verlag.

Sagan, H. (1992). *Introduction to the Calculus of Variations*, 2nd edition. New York: Dover Publ.

Shabana, A.A. (1998). *Dynamics of Multibody Systems*, Second Ed. Cambridge: University Press.

Ziegler, F. (1998). *Mechanics of Solids and Fluids*, 2nd Engl. Edition, corrected 2nd printing. Berlin: Springer-Verlag.

Warsi, Z. U. A. (1999). *Fluid Dynamics, Theoretical and Computational Approaches*, 2nd Edition. CRC Press.

Acknowledgement:

The present Lecture Note is a contribution to the K+ Linz Center of Excellence in Mechatronics (LCM), Strategic Project 4.4. Support of LCM by the Austrian K+ Fund and the Government of Upper Austria is gratefully acknowledged.

Basics of Analytical Mechanics

Alexander K. Belyaev
Department of Mechanics and Control, Polytechnic University of St. Petersburg, Russia

Basically we will follow the books by Whittaker (1937) and Lurie (2002).

1 Constraints and generalised coordinates

From a dynamical perspective any material system can be regarded as a collection of material particles. One refers to the relationships between the quantities determining the position and the velocity of the system of particles as constraints. These relationships must be satisfied regardless of the initial conditions and the forces acting on the system. A standard example of a system subject to constraints is a rigid body which is a collection of material particles kept at invariable distances from each other.

The position of a particle M_i of a system is determined by its coordinates in an inertial Cartesian coordinate system $Oxyz$. The position vector $\mathbf{r}_i = \overrightarrow{OM_i}$ when the subscript i is $1, 2, ...N$, and N denotes the number of particles in the system.

A simple and the most important class of constraints are holonomic constraints. These are expressed analytically in terms of the constraint equations

$$f_i\left(x_1, y_1, z_1, ..., x_N, y_N, z_N; t\right) = 0 \quad (i = 1, ..., r), \tag{1.1}$$

which include only coordinates of the system's points (and not any of their velocities!). It is clear that the number of constraint equations $r \leq 3N$, where the equality condition corresponds to an *a priori* prescribed motion. A holonomic constraint is named stationary if time t does not appear explicitly in constraint equation (1.1).

A system is said to be holonomic when all of the constraints acting on it are holonomic. Non-holonomic constraints express the relations existing between the velocities of the particles, provided that these relations are not reducible to dependences between the coordinates. A classical example of a system subjected to non-holonomic constraints is a rigid body which rolls on a surface without slip. The non-integrable constraint equations take the form

$$\sum_{k=1}^{N}\left(a_{ik}\dot{x}_k + b_{ik}\dot{y}_k + c_{ik}\dot{z}_k\right) + g_i = 0 \quad (i = 1, ..., r'), \tag{1.2}$$

where factors a_{ik}, b_{ik} and c_{ik} depend on the coordinates of the particles and time.

In order to shorten the equations we adopt the abridged notation

$$x_\nu = \xi_{3\nu-2}, \quad y_\nu = \xi_{3\nu-1}, \quad z_\nu = \xi_{3\nu} \quad (\nu = 1, ..., N), \tag{1.3}$$

which allows us to rewrite the holonomic constraint equations in the form

$$f_i(\xi_1, ..., \xi_{3N}; t) = 0 \quad (i = 1, ..., r). \tag{1.4}$$

These equations are assumed to be independent. In this case, the latter system of equations may be resolved for $\xi_1, ..., \xi_r$, and the latter may be expressed in terms of the remaining $3N - r$ variables $\xi_{r+1}, ..., \xi_{3N}$ and time t. Thus we obtain the relationships

$$\xi_k = \xi_k(\xi_{r+1}, ..., \xi_{3N}; t) \quad (k = 1, ..., r), \tag{1.5}$$

in which coordinates $\xi_{r+1}, ..., \xi_{3N}$ are independent of each other. If a system is holonomic, i.e. there are no non-holonomic constraints, the number of degrees of freedom of the system is $n = 3N - r$. The remaining coordinates $\xi_1, ..., \xi_r$ are determined in terms of the independent coordinates by means of eq. (1.5). Given non-holonomic constraints, the number of independent parameters determining the system configuration minus the number of equations of non-holonomic constraints (i.e. $n - r'$) denotes the number of degrees of freedom.

It is not necessary to take solely Cartesian coordinates $\xi_{r+1}, ..., \xi_{3N}$. Instead of these one can use any other independent quantities $q_1, ..., q_n = q_{3N-r}$ which together determine the system configuration. These may be distances, angles, Gaussian coordinates of the point on a surface, areas etc. It is only significant that these introduced quantities allow unique determination of independent Cartesian coordinates

$$\xi_{r+k} = \xi_{r+k}(q_1, ..., q_n; t) \quad (k = 1, ..., n). \tag{1.6}$$

Substituting eq. (1.6) into the right hand side of eq. (1.5), the Cartesian coordinates of the system particles can be then expressed in terms of quantities $q_1, ..., q_n$ referred to as the generalised coordinates and time t. Returning to the original notation we can set

$$\mathbf{r}_i = \mathbf{r}_i(q_1, ..., q_n; t) \quad (i = 1, ..., N), \tag{1.7}$$

where $\mathbf{r}_i$ denotes the position vector of point M_i in an inertial Cartesian coordinate system. As mentioned above, time t does not appear in this equation explicitly if the constraints are stationary.

2 Kinetic energy

The kinetic energy of a system of particles is given by

$$T = \frac{1}{2}\sum_{i=1}^{N} m_i v_i^2 = \frac{1}{2}\sum_{i=1}^{N} m_i \mathbf{v}_i \cdot \mathbf{v}_i. \tag{2.1}$$

where m_i and $\mathbf{v}_i$ denote the mass and velocity of the $i-th$ particle, respectively.

Replacing velocities $\mathbf{v}_i$ in terms of the generalised velocities $\dot{q}_s$ we obtain that the kinetic energy can be represented as the sum of three terms

$$T = T_2 + T_1 + T_0, \quad T_2 = \frac{1}{2}\sum_{s=1}^{n}\sum_{k=1}^{n} A_{sk}\dot{q}_s\dot{q}_k, \; T_1 = \sum_{s=1}^{n} B_s\dot{q}_s, \tag{2.2}$$

where

$$A_{sk} = A_{ks} = \sum_{i=1}^{N} m_i \frac{\partial \mathbf{r}_i}{\partial q_s} \cdot \frac{\partial \mathbf{r}_i}{\partial q_k}, \; B_s = \sum_{i=1}^{N} m_i \frac{\partial \mathbf{r}_i}{\partial q_s} \cdot \frac{\partial \mathbf{r}_i}{\partial t}, \; T_0 = \frac{1}{2} \sum_{i=1}^{N} m_i \left| \frac{\partial \mathbf{r}_i}{\partial t} \right|^2 .$$

The kinetic energy of the system with stationary constraints is seen to be a quadratic form of the generalised velocities

$$T = \frac{1}{2} \sum_{s=1}^{n} \sum_{k=1}^{n} A_{sk} \dot{q}_s \dot{q}_k. \tag{2.3}$$

According to definition, the kinetic energy is a non-negative value which is equal to zero only if the velocities $\mathbf{v}_i$ of all points of the system vanish. For this reason the quadratic form (2.3) of the generalised coordinates is positive definite.

The kinetic energy of a rigid body is given by

$$T = \frac{1}{2} \left[M v_0^2 + 2M \left(\mathbf{v}_0 \times \omega \right) \cdot \mathbf{r}'_C + \omega \cdot \mathbf{\Theta}^O \cdot \omega \right]. \tag{2.4}$$

Here $\mathbf{v}_0$ is the velocity of the pole O, $\mathbf{r}'_C$ denotes the position vector of the centre of inertia of the body relative to the pole O, and $\mathbf{\Theta}^O$ denotes the inertia tensor at this point.

It is known that the vector $m_i \mathbf{v}_i$ determines the momentum of the particle. The principal momentum of the system $\mathbf{Q}$ is, by definition, equal to

$$\mathbf{Q} = \sum_{i=1}^{N} m_i \mathbf{v}_i = \sum_{i=1}^{N} m_i \dot{\mathbf{r}}_i = M \mathbf{v}_C, \tag{2.5}$$

that is the principal momentum of the system is equal to product of the total mass and the velocity of the centre of inertia, since the velocity of the centre of inertia is $\dot{\mathbf{r}}_C = \mathbf{v}_C = \frac{1}{M} \sum_{i=1}^{N} m_i \dot{\mathbf{r}}_i$.

3 Work and potential energy

The sum of the elementary work due to forces $\mathbf{F}_i$ applied at points M_i of a system undergoing virtual displacements $\delta \mathbf{r}_i$ of these points from their positions at a fixed time instant is given by

$$\delta' W = \sum_{i=1}^{N} \mathbf{F}_i \cdot \delta \mathbf{r}_i. \tag{3.1}$$

The notation δ' is used to indicate that we are dealing with an infinitesimal quantity which is not necessarily a variation of a quantity W.

Replacing in this equation the virtual displacements by expressions in terms of variations of the generalised coordinates δq_s one obtains

$$\delta' W = \sum_{i=1}^{N} \mathbf{F}_i \cdot \sum_{s=1}^{n} \frac{\partial \mathbf{r}_i}{\partial q_s} \delta q_s = \sum_{s=1}^{n} \delta q_s \sum_{i=1}^{N} \mathbf{F}_i \cdot \frac{\partial \mathbf{r}_i}{\partial q_s} = \sum_{s=1}^{n} Q_s \delta q_s. \tag{3.2}$$

The quantity $Q_s = \sum_{i=1}^{N} \mathbf{F}_i \cdot \frac{\partial \mathbf{r}_i}{\partial q_s}$ is referred to here as the generalised force corresponding to the generalised coordinate q_s. Hence, the generalised force Q_s is equal to the coefficient of variation δq_s of the corresponding generalised coordinate in the expression for the elementary work.

By analogy we define the elementary work done by the actual displacements as

$$d'W = \sum_{i=1}^{N} \mathbf{F}_i \cdot d\mathbf{r}_i. \tag{3.3}$$

Using this quantity we define the work of the forces during a finite displacement of each point in the system as the following integral

$$W_{12} = \int_{(0)}^{(1)} d'W = \int_{(0)}^{(1)} \sum_{i=1}^{N} \mathbf{F}_i \cdot d\mathbf{r}_i, \tag{3.4}$$

whose limits are given by the coordinates of the system at the initial (0) and final (1) positions.

Let us consider the case where the forces depend only on the positions of the points referred to an inertial coordinate system. These forces are called potential forces if there exists a single-valued and twice-differentiable function $\Pi = \Pi\,(x_1, y_1, z_1, \ldots, x_N, y_N, z_N)$ such that

$$F_{ix} = -\frac{\partial \Pi}{\partial x_i}, \; F_{iy} = -\frac{\partial \Pi}{\partial y_i}, \; F_{iz} = -\frac{\partial \Pi}{\partial z_i}. \tag{3.5}$$

Function Π is assumed to depends only upon the point coordinates. In addition to this, it is assumed that time does not appear explicitly in the latter expression. This function is referred to as the potential energy, and as follows from the latter equation this function is determined up to an additive constant.

By virtue of eqs. (3.3) and (3.5), the elementary work of the potential forces due to actual displacements is given by

$$d'W = -\sum_{i=1}^{N} \left(\frac{\partial \Pi}{\partial x_i} dx_i + \frac{\partial \Pi}{\partial y_i} dy_i + \frac{\partial \Pi}{\partial z_i} dz_i \right) = -d\Pi, \tag{3.6}$$

where $d\Pi$ is the total differential of the potential energy.

Let us consider the case of stationary constraints in which the potential energy becomes a function of the generalised coordinates only $\Pi = \Pi\,(q_1, \ldots, q_n)$. The latter equation suggests that the elementary work of the potential forces due to a virtual displacement is

$$\delta' W = -\delta \Pi = \sum_{s=1}^{n} \frac{\partial \Pi}{\partial q_s} \delta q_s. \tag{3.7}$$

Comparing this with eq. (3.2) and taking into account that the variations of the generalised coordinates are independent yields the following expression for the generalised forces in terms of the potential energy

$$Q_s = -\frac{\partial \Pi}{\partial q_s} \quad (s = 1, \ldots, n). \tag{3.8}$$

4 Lagrange's equations of motion

Formal transformation of the fundamental equation of dynamics yields

$$\sum_{s=1}^{n} \left(\frac{d}{dt}\frac{\partial T}{\partial \dot{q}_s} - \frac{\partial T}{\partial q_s} - Q_s \right) \delta q_s = 0. \tag{4.1}$$

This equality is as general as the fundamental equation of dynamics and, for this reason, it is applicable for both holonomic and non-holonomic systems. In the case of holonomic constraints and independent generalised coordinates the variations δq_s are independent and thus the coefficients in front of each δq_s in this equation must be zero independently. We obtain the system of differential equations of motion expressed in terms of the generalised coordinates in the form of Lagrange's equations

$$\frac{d}{dt}\frac{\partial T}{\partial \dot{q}_s} - \frac{\partial T}{\partial q_s} = Q_s \quad (s = 1, \ldots, n). \tag{4.2}$$

Their number coincides with the number of the generalised coordinates, i.e. the number of degrees of freedom of the holonomic system.

Provided that there exist non-holonomic constraints described by the relationships

$$\sum_{s=1}^{n} a_{ks}\dot{q}_s + a_k = 0 \quad (k = 1, \ldots, l), \tag{4.3}$$

then, by means of Lagrange's multipliers one obtains

$$\frac{d}{dt}\frac{\partial T}{\partial \dot{q}_s} - \frac{\partial T}{\partial q_s} = Q_s + \sum_{k=1}^{l} \lambda_k a_{ks} \quad (s = 1, \ldots, l). \tag{4.4}$$

These are Lagrange's equations in the case of non-holonomic constraints. The total number of equations (4.3) and (4.4) is $n+l$ and these equations have $n+l$ unknown variables which are n generalised coordinates $q_1, ..., q_n$ and l constraint multipliers $\lambda_1, ..., \lambda_l$.

Introducing Euler's operator over f

$$E_s(f) = \frac{d}{dt}\frac{\partial f}{\partial \dot{q}_s} - \frac{\partial f}{\partial q_s} \quad (s = 1, \ldots, n) \tag{4.5}$$

and the kinetic potential $L = T - \Pi$ we cast Lagrange's equations (4.2) in the general form

$$E_s(T) = Q_s \quad (s = 1, \ldots, n), \tag{4.6}$$

and in the case of potential forces

$$E_s(L) = 0 \quad (s = 1, \ldots, n). \tag{4.7}$$

When the non-potential forces are present (along with the potential forces) and some of them are described by the dissipation function Φ, Lagrange's equations take the form

$$E_s(L) = Q_s - \frac{\partial \Phi}{\partial \dot{q}_s} \quad (s = 1, \ldots, n). \tag{4.8}$$

In the case of non-holonomic constraints (4.3) Lagrange's equations are written as follows

$$E_s(L) = Q_s + \sum_{k=1}^{l} \lambda_k a_{ks} - \frac{\partial \Phi}{\partial \dot{q}_s} \quad (s = 1, \ldots, n). \tag{4.9}$$

The terms $\sum_{k=1}^{l} \lambda_k a_{ks}$ in eq. (4.9) present the generalised constraint forces of the non-holonomic constraints.

5 Canonical Hamiltonian equations

Let us consider a continuous function $\Phi(x_1, ..., x_n)$ of variables $x_1, ..., x_n$ which has continuous derivatives of the first and second order with respect to all variables. The transformation from the "old" variables $x_1, ..., x_n$ to the "new" ones

$$y_s = \frac{\partial \Phi}{\partial x_s} \quad (s = 1, \ldots, n). \tag{5.1}$$

is carried out by means of function Φ referred to as the generating function. The Jacobian of the variables y is the Hessian $H(\Phi)$ of function Φ. Its elements are the second partial derivatives of Φ with respect to x_s. The system of equations (1) can be resolved for the old variables if the Hessian is non-singular. Then one can obtain the formulae for the inverse transformation by expressing the old variables in terms of the new variables, i.e. $x_s = x_s(y_1, \ldots, y_n)$. This transformation can be performed with the help of the generating functions of the new variables Ψ

$$\Psi(y_1, \ldots, y_n) = \sum_{k=1}^{n} x_k y_k - \Phi(x_1, \ldots, x_n), \tag{5.2}$$

where all old variables are expressed in terms of the new ones. Indeed,

$$\frac{\partial \Psi}{\partial y_s} = x_s + \sum_{k=1}^{n} \frac{\partial x_k}{\partial y_s} y_k - \sum_{k=1}^{n} \frac{\partial \Phi}{\partial x_k} \frac{\partial x_k}{\partial y_s} = x_s, \tag{5.3}$$

since the sums cancel out by virtue of equalities (5.1). Hence

$$x_s = \frac{\partial \Psi}{\partial y_s} \quad (s = 1, \ldots, n). \tag{5.4}$$

The formulae (5.1) and (5.4) for the direct and the inverse transformations are referred to as Legendre's transformation. Clearly, applying inverse transformation twice results in the recovering of the original transformation.

If, in particular, the generating function Φ is a quadratic form

$$\Phi = \frac{1}{2}\sum_{i=1}^{n}\sum_{k=1}^{n} a_{ik}x_i x_k, \tag{5.5}$$

then its Hessian is the determinant $|a|$ of the matrix a of the coefficients of this form, i.e. $H(\Phi) = |a|$. Provided that a is a non-singular matrix, i.e. $|a| \neq 0$, then the system of linear equations

$$y_s = \frac{\partial \Phi}{\partial x_s} = \sum_{k=1}^{n} a_{sk}x_k \quad (s = 1, \ldots, n) \tag{5.6}$$

can be solved for the old variables. Denoting the inverse matrix by $b = a^{-1}$ we find

$$x_s = \sum_{k=1}^{n} b_{sk}y_k \quad (s = 1, \ldots, n). \tag{5.7}$$

It follows from Euler's theorem on homogeneous functions that the generating function $\Psi(y_1, \ldots, y_n)$ of the inverse transformation is obtained by direct substitution of eq. (5.7) into eq. (5.5). Indeed, by virtue of eqs. (5.1) and (5.2) we have

$$\Psi = \sum_{k=1}^{n} x_k y_k - \Phi = \sum_{k=1}^{n} x_k \frac{\partial \Phi}{\partial x_k} - \Phi = 2\Phi - \Phi = \Phi, \tag{5.8}$$

where Φ should be expressed in terms of the new variables. Hence,

$$\Psi = \Phi = \frac{1}{2}\sum_{s=1}^{n} x_s y_s = \frac{1}{2}\sum_{s=1}^{n}\sum_{k=1}^{n} b_{sk}y_s y_k, \tag{5.9}$$

i.e. the quadratic form Ψ obtained by means of matrix $b = a^{-1}$ is the associate expression for the quadratic form Φ.

Let us consider a material system with n degrees of freedom which is subject to holonomic constraints. Lagrange's equations of this system is a system of n second order differential equations for the generalised coordinates resolved for the second derivatives $\ddot{q}_s$. In order to reduce this system of equations to a system of $2n$ first order equations we introduce $2n$ variables which are the generalised coordinates $q_1, \ldots, q_n$ and the generalised momenta $p_1, \ldots, p_n$

$$p_s = \frac{\partial T}{\partial \dot{q}_s} \quad (s = 1, \ldots, n). \tag{5.10}$$

The generalised momenta are linear functions of the generalised velocities $\dot{q}_1, \ldots, \dot{q}_n$. For these variables and the potential active forces, the equations of motion are set in a very compact and symmetrical form referred to as the canonical form. This simplifies the analysis of the general properties of the motion and reduces the problem of integration of the canonical equations to a search for the complete integral of the equation with first order partial derivatives (Jacobi's theorem). The variations q_s, p_s are independent and are symmetric in the forthcoming equations and transformations.

The kinetic energy, regarded as a function of the generalised velocities, plays the part of the generating function transforming the old variables $\dot{q}_s$ to the new ones p_s. The Hessian of the transformation is the determinant of matrix A composed of the coefficients of the quadratic form T_2 which is the part of T. This determinant does not vanish since it is positive and thus equations (5.10), which are linear in the generalised velocities $\dot{q}_i$, are solvable. This results in the relationships

$$\dot{q}_s = \dot{q}_s \left(p_1, \ldots, p_n, q_1, \ldots, q_n; t\right), \tag{5.11}$$

which are linear in the generalised momenta. They comprise the first set of the canonical equations, namely a system of n first order differential equations defining the time-derivatives of the generalised coordinates in terms of the generalised momenta, generalised coordinates and time.

Legendre's transformation enables the latter equation to be represented in another form. By virtue of eq. (5.2), we construct the generating function of the inverse transformation

$$\tilde{T} = \sum_{s=1}^{n} p_s \dot{q}_s - T = \tilde{T} \left(q_1, \ldots, q_n, p_1, \ldots, p_n; t\right), \tag{5.12}$$

in which the generalised velocities are replaced by their expressions (5.11) in terms of the generalised momenta. Then, by means of eq. (5.4), we obtain

$$\dot{q}_s = \frac{\partial \tilde{T}}{\partial p_s} \quad (s = 1, \ldots, n). \tag{5.13}$$

It is important that the second set of equations of motion is obtained by means of the same function $\tilde{T}$. In contrast to equations (5.13), which simply express the transformation, the equations of the second set represent the general equation of dynamics. Lagrange's equations can be lead to the form

$$\dot{p}_s = -\frac{\partial \tilde{T}}{\partial q_s} + Q_s \quad (s = 1, \ldots, n). \tag{5.14}$$

This is the second set of the canonical equations for the system of variables q_s, p_s. It will coincide with the second set of Hamiltonian equations provided that the forces are potential. Similar to Lagrange's equations, which can be obtained by means of a single function, namely the kinetic potential L, a single function H referred to as the Hamiltonian function suffices to construct the canonical system. This function is defined as follows

$$H \left(q_1, \ldots, q_n, p_1, \ldots, p_n; t\right) = \tilde{T} + \Pi = \sum_{s=1}^{n} p_s \dot{q}_s - L. \tag{5.15}$$

Function H depends on the generalised coordinates, generalised momenta and time, whilst the generalised velocities are removed by means of relationship (5.11).

Taking into account that the potential energy does not depend on the generalised momenta we obtain, instead of eqs. (5.13) and (5.14),

$$\dot{q}_s = \frac{\partial \tilde{T}}{\partial p_s} = \frac{\partial H}{\partial p_s}, \quad \dot{p}_s = -\frac{\partial \tilde{T}}{\partial q_s} - \frac{\partial \Pi}{\partial q_s} = -\frac{\partial H}{\partial q_s}. \tag{5.16}$$

Thus we derived the system of canonical Hamiltonian equations

$$\dot{q}_s = \frac{\partial H}{\partial p_s}, \quad \dot{p}_s = -\frac{\partial H}{\partial q_s} \quad (s = 1, \ldots, n). \tag{5.17}$$

Variables q_s and p_s satisfying the system of canonical Hamiltonian equations are referred to as canonical.

By using the canonical equations (5.17) we obtain the total time-derivative of the Hamiltonian function

$$\begin{aligned} \frac{dH}{dt} &= \frac{\partial H}{\partial t} + \sum_{s=1}^{n} \left(\frac{\partial H}{\partial q_s} \dot{q}_s + \frac{\partial H}{\partial p_s} \dot{p}_s \right) \\ &= \frac{\partial H}{\partial t} + \sum_{s=1}^{n} \left(\frac{\partial H}{\partial q_s} \frac{\partial H}{\partial p_s} - \frac{\partial H}{\partial p_s} \frac{\partial H}{\partial q_s} \right) = \frac{\partial H}{\partial t}. \end{aligned} \tag{5.18}$$

Hence the Hamiltonian function H retains a constant value throughout the motion provided that H does not contain time explicitly. Therefore we obtain the energy integral

$$H(q_1, \ldots, q_n, p_1, \ldots, p_n) = h. \tag{5.19}$$

For a system with stationary constraints it states that the sum of the kinetic and potential energies is constant.

Let us now consider two systems of variables, the old variables $q_1, \ldots, q_n;\ p_1, \ldots, p_n$ and the new variables $Q_1, \ldots, Q_n;\ P_1, \ldots, P_n$, which are taken as functions of the old variables and time t

$$\left. \begin{array}{c} Q_s = Q_s(q_1, \ldots, q_n; p_1, \ldots, p_n; t) = Q_s(q|p;t), \\ P_s = P_s(q_1, \ldots, q_n; p_1, \ldots, p_n; t) = P_s(q|p;t) \\ (s = 1, \ldots, n). \end{array} \right\} \tag{5.20}$$

These equations are assumed to be solvable for the old variables, i.e.

$$q_k = q_k(Q|P;t), \quad p_k = p_k(Q|P;t) \quad (k = 1, \ldots, n). \tag{5.21}$$

The transformation (5.20) is called canonical provided that the following expression

$$\sum_{k=1}^{n} p_k \delta q_k - \sum_{i=1}^{n} P_i \delta Q_i \tag{5.22}$$

is a variation of a certain function of the old and new variables and time t.

The fundamental importance of the canonical transformations in analytical mechanics is due to the theorem that variables Q_s, P_s related to the canonical variables q_s, p_s by a canonical transformation are also canonical, that is, they satisfy the system of canonical Hamilton's equations

$$\dot{Q}_s = \frac{\partial K}{\partial P_s}, \quad \dot{P}_s = -\frac{\partial K}{\partial Q_s}. \tag{5.23}$$

Here K denotes the Hamiltonian function depending on the new canonical variables Q_s, P_s and time t which is related to the Hamiltonian function of the old canonical variables q_s, p_s as follows

$$K = H + \frac{\partial V}{\partial t}, \tag{5.24}$$

where V denotes the generating function of the canonical transformation. In particular, if the latter does not depend explicitly on t, then

$$K = H. \tag{5.25}$$

Thus the canonical equations are retained under the canonical transformation of variables.

6 The Hamilton variational principle

Let the system be subject to holonomic ideal constraints and the external forces be prescribed potential forces. Let the generalised coordinates $q_1(t), \ldots, q_n(t)$ describe the real motion of the material system. This set of functions is said to determine the true path of the system whilst any of the configurations

$$q_1^*(t) = q_1(t) + \delta q_1, \ldots, q_n^* = q_n(t) + \delta q_n, \tag{6.1}$$

admitted by the constraints and infinitesimally close the true path defines the varied path. In eq. (6.1) variations δq_s imply arbitrary, infinitesimal, differentiable functions of time. From all possible varied paths we select those which coincide with the true path at two fixed (however arbitrarily taken) time instants t_0 and t_1, so that

$$\delta q_s(t_0) = 0, \quad \delta q_s(t_1) = 0 \quad (s = 1, \ldots, n). \tag{6.2}$$

Let us now introduce quantity S referred to as Hamilton's action as follows

$$S = \int_{t_0}^{t_1} L dt, \tag{6.3}$$

where $L = L(q_1, \ldots, q_n, \dot{q}_1, \ldots, \dot{q}_n, t)$ denotes the kinetic potential.

We proceed now to calculate the increment in S when we take one of the varied paths instead of the true one. The value of the kinetic potential along this varied path should be inserted into eq. (6.3). Restricting our consideration to first order values in δq_s and

$\delta\dot{q}_s$ we replace L by $L+\delta L$. Then, the increment in Hamilton's action calculated with the mentioned accuracy is given by the equality

$$\delta S=\int_{t_0}^{t_1}(L+\delta L)\,dt-\int_{t_0}^{t_1}Ldt=\int_{t_0}^{t_1}\delta Ldt \tag{6.4}$$

and is called the variation in Hamilton's action

$$\delta S=\int_{t_0}^{t_1}\sum_{s=1}^{n}\left[\frac{\partial L}{\partial q_s}\delta q_s+\frac{\partial L}{\partial\dot{q}_s}\left(\delta q_s\right)^{\bullet}\right]dt. \tag{6.5}$$

Integrating by parts and taking into account the additional condition (6.2) imposed on the varied paths we have

$$\delta S=\sum_{s=1}^{n}\int_{t_0}^{t_1}\left(\frac{\partial L}{\partial q_s}-\frac{d}{dt}\frac{\partial L}{\partial\dot{q}_s}\right)\delta q_s dt=-\sum_{s=1}^{n}\int_{t_0}^{t_1}E_s\left(L\right)\delta q_s dt, \tag{6.6}$$

where $E_s\left(L\right)$ denotes the Euler's operator (4.5).

We consider the following equality

$$\delta S=0. \tag{6.7}$$

as an independent requirement. This leads to the statement of the Hamilton principle which states that Hamilton's action S has a stationary value for the true path as compared with all arbitrary neighbouring paths coinciding with the true path at the initial t_0 and final t_1 time instants.

In what follows we prove that Lagrange's equations (4.7) are a consequence of equality (6.7) then we can state that the Hamilton principle contains the basic laws of dynamics. The proof is as follows. As variations δq_s are independent of each other we can choose the following neighbouring path

$$\delta q_1=0,\ldots,\delta q_{k-1}=0,\quad \delta q_{k+1}=0,\ldots,\delta q_n=0, \tag{6.8}$$

whereas $\delta q_k\neq 0$. Equation (6.7) takes the form

$$-\int_{t_0}^{t_1}E_k\left(L\right)\delta q_k dt=\int_{t_0}^{t_1}\left(\frac{\partial L}{\partial q_k}-\frac{d}{dt}\frac{\partial L}{\partial\dot{q}_k}\right)\delta q_k dt=0. \tag{6.9}$$

Assume now that

$$E_k\left(L\right)=\frac{d}{dt}\frac{\partial L}{\partial\dot{q}_k}-\frac{\partial L}{\partial q_k}\neq 0 \tag{6.10}$$

at $t=t_*$. Due to continuity of $E_k\left(L\right)$ we can find such a time interval (t_{*0},t_{*1}) that it contains t_* and $E_k\left(L\right)$ retains its sign. But δq_k is an arbitrary function of time. Let us choose it so that it retains its sign within the above interval and is identically equal to zero at $t_0 < t_{*0}$ and $t_1 < t_{*1}$. Then,

$$\int_{t_0}^{t_1} E_k(L)\, \delta q_k dt = \int_{t_{*0}}^{t_{*1}} E_k(L)\, \delta q_k dt. \tag{6.11}$$

The integrand retains its sign and thus the integral does not vanish. This contradiction means that the assumed inequality (4) does not hold. Hence,

$$E_k(L) = 0. \tag{6.12}$$

This reasoning is valid for any $k = 1, \ldots, n$ which completes the proof.

Let us also demonstrate derivation of the canonical equations of motion in terms of the Hamilton principle. We begin with equality (5.15)

$$L = \sum_{s=1}^{n} \frac{\partial L}{\partial \dot{q}_s} \dot{q}_s - H = \sum_{s=1}^{n} p_s \dot{q}_s - H. \tag{6.13}$$

Then

$$\delta L = \sum_{s=1}^{n} \left(\dot{q}_s - \frac{\partial H}{\partial p_s} \right) \delta p_s + \sum_{s=1}^{n} \left[p_s (\delta q_s)^{\bullet} - \frac{\partial H}{\partial q_s} \delta q_s \right]. \tag{6.14}$$

Equalities

$$\dot{q}_s = \frac{\partial H}{\partial p_s} \quad (s = 1, \ldots, n) \tag{6.15}$$

are the first set of the canonical equations and express the relationship between the generalised coordinates and momenta. Thus

$$\delta L = \sum_{s=1}^{n} \left[p_s (\delta q_s)^{\bullet} - \frac{\partial H}{\partial q_s} \delta q_s \right]. \tag{6.16}$$

Taking into account eq. (6.2) we have

$$\int_{t_0}^{t_1} p_s (\delta q_s)^{\bullet} dt = p_s \delta q_s \big|_{t_0}^{t_1} - \int_{t_0}^{t_1} \dot{p}_s \delta q_s dt = - \int_{t_0}^{t_1} \dot{p}_s \delta q_s dt \tag{6.17}$$

and therefore

$$\delta S = \int_{t_0}^{t_1} \delta L dt = - \int_{t_0}^{t_1} \sum_{s=1}^{n} \left(\dot{p}_s + \frac{\partial H}{\partial q_s} \right) \delta q_s dt = 0. \tag{6.18}$$

By repeating the reasoning which yields Lagrange's equations from eq. (6.7), we arrive at the second set of the Hamiltonian canonical equations

$$\dot{p}_s = - \frac{\partial H}{\partial q_s} \quad (s = 1, \ldots, n), \tag{6.19}$$

which completes the proof.

In the case of non-potential forces the Hamilton principle is cast in the form

$$\delta S = \int_{t_0}^{t_1} \delta' W dt = \delta S + \sum_{s=1}^{n} \int_{t_0}^{t_1} Q_s \delta q_s dt = 0 \tag{6.20}$$

and reduces to the form

$$\sum_{s=1}^{n} \int_{t_0}^{t_1} \left(\frac{\partial L}{\partial q_s} - \frac{d}{dt} \frac{\partial L}{\partial \dot{q}_s} + Q_s \right) \delta q_s dt = 0, \tag{6.21}$$

which yields Lagrange's equations.

Attention should be paid to the principal difference of equality (6.20) from the previous relationships (6.7) and (6.18). The latter require a functional S and a search for the necessary conditions of stationarity of this functional, that is, the mechanical problem was reduced to the problem of calculus of variation. In contrast to this, eq. (6.20) contains only the statement that the value

$$\delta' R = \delta S + \int_{t_0}^{t_1} \delta' W dt = \int_{t_0}^{t_1} (\delta T - \delta \Pi + \delta' W) \, dt \tag{6.22}$$

turns to zero, however no functional R exists since there exists no quantity whose variation equals $\delta' R$. Hence, eq. (6.20) implies no variational statement of the problem.

The Hamilton principle is also applicable to systems subject to non-holonomic constraints and yields eq. (4.4). However the equations for non-holonomic constraints should hold also for varied generalised coordinates along the neighbouring path. This restriction results in additional conditions for the kinematic feasibility of adjacent motion selecting a narrower class of variations δq_k admitting motions adjacent to the true one.

7 The Lagrange-Maxwell equations of electromechanics

We refer to systems in which mechanical and electromagnetic processes are intimately linked with each other as electromechanical systems. In mechanics, the state of the system is considered to be given if its generalised coordinates and velocities are prescribed. However, for electromechanical systems they define only part of its character and the quantities describing the electromagnetic processes comprise another part. In order to characterise the electromagnetic field it is necessary to prescribe the vector of magnetic induction $\mathbf{B}$ and the vector of electric intensity $\mathbf{E}$. These quantities are known to satisfy the system of partial differential equations of Maxwell which differ drastically from the equations of mechanics. However, while considering a wide class of applied electromechanical problems, the field variables $\mathbf{B}$ and $\mathbf{E}$ can be expressed in terms of a finite number of other variables which are equivalent to the generalised coordinates and momenta to some extent. To meet this requirement, the conditions of quasi-stationarity enabling electromagnetic waves to be neglected must hold. Besides, the transverse dimensions of conductors are assumed to be much smaller than their longitudinal dimensions, with the exception of capacitor plates. Such conductors and currents in them are referred to as being linear. Linear conductors connected to each other, capacitor plates and external power sources form an electric circuit. Let an electromechanical system under consideration consist of l parallel circuits connected with each other by means of inductors. Each parallel circuit has nodes where more than two conductors and branches, i.e. non-parallel subcircuits between the nodes consisting of linear conductors, capacitors and power sources, are connected. The current in any cross-section of the linear conductor in the branch remains constant. It is assumed that the circuit does not change topologically under mechanical motion.

Let z_c and y_c denote respectively the number of branches and nodes in the $s-th$ circuit. Let us arbitrarily choose the positive direction of the current in any branch and designate these currents as $i_1, \ldots, i_N$ $(N = z_1 + \ldots + z_l)$. According to Kirchhoff's first law these currents are related as follows

$$\sum_{j=1}^{N} \gamma_{jk} i_j = 0, \quad k = 1, \ldots, \sum_{s=1}^{l} y_s. \tag{7.1}$$

Here $\gamma_{ik} = 0$ if the $j-th$ branch is not adjacent to the $k-th$ node and $\gamma_{ik} = \pm 1$ if this branch is adjacent to the $k-th$ node, with a positive sign corresponding to the current entering the branch and a negative sign for the circuit leaving the branch.

We see that the number of independent equations in (7.1) is equal to $\sum\limits_{s=1}^{l} y_s - l$, thus we can introduce m independent currents, $m = N + l - \sum\limits_{s=1}^{l} y_s$, in such a way that the other currents can be expressed in terms of these independent circuits.

As will be explained below, the currents play the role of the generalised velocities, that is $i_j = \dot{g}_j$ where g_j denotes the charge transferred by current i_j in the corresponding branch. This means that relationships (7.1) after integration play the role of the holonomic stationary constraints

$$\sum_{j=1}^{N} \gamma_{jk} \left(g_j - g_{j0} \right) = 0. \tag{7.2}$$

Currents $i_1, \ldots, i_m$ cause a magnetic field to develop in the surrounding. Vector $\mathbf{B}$ describing this field can be viewed as a function of the corresponding point in the space and these currents. Let the space be filled by a medium in which $\mathbf{B}$ is a linear function of currents $i_1, \ldots, i_m$. Then

$$\mathbf{B} = \sum_{s=1}^{m} \mathbf{B}_s \left(x, y, z \right) i_s, \tag{7.3}$$

where x, y, z are Cartesian coordinates of the point at which vector $\mathbf{B}$ is determined. Electrical field is considered only in the space between the capacitor plates, $\mathbf{E}$ being a linear function of the capacitor charge g_j

$$\mathbf{E} = g_j \mathbf{E}_{js} \left(x, y, z \right). \tag{7.4}$$

Here charge g_j is related to current i_j which flows through the branch with the capacitor.

The energy of the magnetic field is given by

$$W = \frac{1}{2}\int \frac{\mathbf{B}\cdot\mathbf{B}}{\mu}dV, \tag{7.5}$$

where the integral is evaluated over the whole space and $\mu = \mu(x,y,z)$ denotes the magnetic permeability of the medium. In accordance with eq. (7.3) we have

$$W = \frac{1}{2}\sum_{s,r=1}^{m} L_{rs} i_r i_s, \quad L_{rs} = \int \frac{\mathbf{B}_r\cdot\mathbf{B}_s}{\mu}dV, \tag{7.6}$$

that is, W is a homogeneous quadratic form of the currents $i_1,\ldots,i_m$ which are understood as being analogous to the generalised coordinates. With this in view, W is an analogue of the kinetic energy for the electromagnetic field.

The energy of the electric field is given by

$$V = \frac{1}{2}\sum_{j=1}^{N}\int_{\Omega_j} \varepsilon\mathbf{E}\cdot\mathbf{E}dV, \tag{7.7}$$

where $\varepsilon = \varepsilon(x,y,z)$ denotes dielectric permittivity of the medium and the integral is evaluated over the space between the capacitor plates. Substituting eq. (7.4) in eq. (7.7) we obtain

$$V = \frac{1}{2}\sum_{j=1}^{N}\frac{g_j^2}{c_j}, \quad c_j^{-1} = \int_{\Omega_j} \varepsilon\mathbf{E}\cdot\mathbf{E}dV. \tag{7.8}$$

Here c_j^{-1} denotes the inverse capacitance of the capacitors in the branch corresponding to i_j. For simplicity, we assume that each branch has not more than one capacitor. In the case of no capacitor in the branch it is necessary to take $c_j = \infty$. By virtue of relationships (7.2), function V can be expressed in terms of m independent charges $g_1,\ldots,g_m$ and their initial values $g_{10},\ldots,g_{m0}$

$$V = \frac{1}{2}\sum_{s,r=1}^{m}\frac{g_r g_s}{c_{rs}} + \sum_{r=1}^{m} b_r g_r + V_0. \tag{7.9}$$

The values c_{rs}^{-1} for $r = s$ and for $r \neq s$ are called the inverse capacitance and mutual inverse capacitance, respectively, and are expressed in terms of the capacitance of the capacitors. Values b_r and V_0 are expressed in terms of the initial values of the charges and the capacitance of the capacitors. As follows from eq. (7.9), V is a function of charges $g_1,\ldots,g_m$ treated as the generalised coordinates, therefore, V can be understood to be the potential energy of the field.

Another characteristic of circuits is active resistances R_j of the branches. Then we can enter an analogue of Rayleigh's dissipation function

$$\Psi = \frac{1}{2}\sum_{j=1}^{N} R_j i_j^2. \tag{7.10}$$

Function Ψ, referred to as the electric dissipation function, is related to the power P of the generated heat by the formula $P = 2\Psi$. Expressing, as before, N currents in terms of $i_1, \ldots, i_m$ we obtain

$$\Psi = \frac{1}{2} \sum_{r,s=1}^{m} R_{rs} i_r i_s. \tag{7.11}$$

The mechanical motion of the system can be described by the generalised coordinates $q_1, \ldots, q_n$. The induction coefficients L_{rs} and the inverse capacitances c_{rs}^{-1} depend on these coordinates. If $T(q, \dot{q})$ and $\Pi(q)$ denote the kinetic and potential energies of the system, respectively, then the system of equations governing mechanical and electrical processes, referred to as the Lagrange-Maxwell equations, have the following general form

$$\begin{aligned} \left(\frac{d}{dt}\frac{\partial}{\partial i_r} - \frac{\partial}{\partial g_r} \right)(W - V) &= G_r, \quad r = 1, \ldots, m, \\ E_i (T + W - \Pi - V) &= Q_i, \quad i = 1, \ldots, n, \end{aligned} \tag{7.12}$$

where E_i denotes Euler's operator. Lagrange's electric generalised forces G_r take into account the influence of the external power sources and the energy dissipation in the active resistances. Let us notice that, in general, the values of the inductances L_{rs} and the capacitances c_{rs}, see eqs. (7.6) and (7.9), change under mechanical displacements. Therefore, magnetic W and electric V energies of the electromechanical system of the general form are functions of the mechanical coordinates. At the same time, mechanical energies T and Π do not explicitly depend on the electrical coordinates $g_1, \ldots, g_m$ and their velocities.

When the generalised forces on the right hand side of eq. (7.12) are absent, the system is conservative and admits the first integral

$$T + W + \Pi + V = const, \tag{7.13}$$

which can be considered as integral of the total energy. It is needless to say that the validity of eq. (7.13) requires that system (7.12), for $G_r = Q_i = 0$, is autonomous and the constraints imposed on the mechanical part are stationary.

Bibliography

A.I. Lurie. *Analytical Mechanics.* Springer, 2002.

E.T. Whittaker. *A Treatise on the Analytical Dynamics of Particles and Rigid Bodies.* Cambridge University Press, 1937.

Compensation of Deformations in Elastic Solids and Structures in the Presence of Rigid-Body Motions

Hans Irschik[1], Uwe Pichler[1], Manfred Nader[2] and Christian Zehetner[2]

1 Johannes Kepler University of Linz and LCM, Linz, Austria
2 Linz Center of Excellence in Mechatronics (LCM), Linz, Austria

Abstract. The present Lecture is concerned with vibrations of linear elastic solids and structures. Some part of the boundary of the structure is suffering a prescribed large rigid-body motion, while an imposed external traction is acting at the remaining part of the boundary, together with given body forces in the interior. Due to this combined loading, vibrations take place. The latter are assumed to remain small, such that the linear theory of elasticity can be applied. As an illustrative example for the type of problems in hand, we mention the flexible wing of an aircraft in flight. In this example, the rigid-body motion is defined through the motion of the comparatively stiff fuselage to which a part of the boundary of the wing is attached. The goal of the present paper is to derive a time-dependent distribution of actuating stresses produced by additional eigenstrains, such that the deformations produced by the imposed forces and the rigid-body motion are exactly compensated. This is called a shape control problem, or a deformation compensation problem. We show that the distribution of the actuating stresses for shape control must be equal to a quasi-static stress distribution that is in temporal equilibrium with the imposed forces and the inertia forces due to the rigid-body motion. Our solution thus explicitly reflects the non-uniqueness of the inverse problem under consideration. The present Lecture extends previous results by Irschik and Pichler (2001, 2004) for problems without rigid-body degrees of freedom. As a computational example, we present results for a rectangular domain in a state of plane strain under the action of a translatory support motion.

1 Introduction

The present Lecture is concerned with the compensation of deformations in elastic solids and structures. One of the first successful applications in the literature was reported by Haftka and Adelmann (1985), who developed a procedure to determine the temperature in control elements so as to minimise the static deformation of a large space-antenna from its original shape. The latter authors used the notion of *shape control*. The notions of *displacement compensation* or *deformation compensation* appear to be equally suitable. In the following we do not use the notion of shape control in the sense of *automatic* or *closed-loop control*, but we assume the disturbances responsible for unwanted distortions to be given, such that we deal with an *open-loop control*.

An important practical topic in shape control concerns the physical nature of the actuation that is used for compensating deformations. Since the deformations are intended to be compensated everywhere in the body under consideration, a *distributed actuation* is required. Such a distributed actuation can be conveniently realised by various physical effects, which are denoted as *eigenstrains* or *sources of selfstress* in the mechanics literature. Examples for eigenstrain-type actuation are the *temperature* in a thermoelastic element, or the *electric field* in a *piezoelastic* element. In the last decades, piezoelectricity has been extensively applied for the sake of actuation and sensing of structural vibrations. Frequently, the piezoelectric effects are realised in practice by means of piezoelastic stack actuators, patches, layers or nets, which are bonded or otherwise integrated into the bodies to be actuated or sensed. It has become customary to summarise structures with such integrated actuators and sensors under the notion of *smart materials and structures.* Various other physical effects, such as magneto- or electrostriction, magneto-rheologic behaviour or shape memory effects, are also utilised in smart structures. These effects are to be classified as eigenstrains in most cases, too. In large scale applications pre-stress by tendons or hydraulic actuators may represent the methods of choice. The important questions of thresholds or delay of the actuation of course have to be treated from the point of view of the physical nature of the respective distributed actuation. Here, we only note that also temperature may represent a convenient tool of suppressing force-induced vibrations, as has been recently demonstrated in a study by Nader and Irschik (2003).

A literature survey on static and dynamic shape control was presented by Irschik (2002). In the following, we mention only some contributions of our own group. Exact solutions for static shape control of linear elastic solids and structures by means of eigenstrains were presented by Irschik and Ziegler (2001). Exact solutions for dynamic shape control of vibrations of force-loaded three-dimensional bodies about an equilibrium position were derived by Irschik and Pichler (2001, 2004). For recent applications to beam vibrations, see Irschik, Krommer and Pichler (2001, 2003); for a recent application to plate vibrations see Nader et al. (2003). The present Lecture is concerned with the extension of these contributions with respect to a deformation compensation in the presence of rigid-body motions. We first present a three-dimensional formulation in the framework of the theory of elastodynamics in the small-strain regime. Vibrations of a rectangular domain in a prescribed rigid-body motion then serve as a computational example. A first account of the present work has been presented in Irschik, Nader and Zehetner (2003).

2 Formulation of the Elastodynamic Shape Control Problem

Consider a solid body B in motion. Some part ∂B_1 of the boundary of B is assumed to perform a prescribed, possibly large, rigid-body motion with respect to an inertial frame. The remaining part of the boundary, ∂B_2, is assumed to be loaded by an imposed surface traction, while imposed body forces are considered to act upon the particles in the interior of B. Due to the rigid-body motion of ∂B_1, and due to the system of imposed forces, the body is deformed from its undistorted shape. The rigid-body motion of ∂B_1 is used to define a rigid reference configuration, relative to which this deformation of B can be conveniently measured. This rigid co-moving configuration is called the *floating reference configuration.* As an example for the class of problems under consideration, consider the wing of an aircraft during flight. Taking the fuselage to which the wing is attached as rigid, the motion of the rigid floating reference configuration of the flexible wing is defined through the rigid-body motion of the fuselage.

In order to describe the displacement and deformation of B with respect to the floating reference configuration, we use a frame which is fixed in the latter configuration, denoted as *co-moving frame* in the following. Note that this frame in general is moving relative to the inertial frame. Seen from the co-moving frame, however, the position vector $\boldsymbol{P}$ of the place of a particle of B in the floating reference configuration does not change with time, which is the reason for introducing the co-moving frame. The position vector $\boldsymbol{P}$ is drawn from the origin A of the spatial coordinate system used in this latter frame. In the following, we use the *material description of continuum mechanics*, in which all of the mechanical entities associated with a particle are described as a function of the position vector $\boldsymbol{P}$ of its place in the floating reference configuration and of time t, the latter being taken as synchronised between the inertial and the co-moving frame. E.g., the displacement vector of a particle, connecting its place in the floating reference configuration with its actual place, is written as $\boldsymbol{u} = \boldsymbol{u}(\boldsymbol{P}, t)$, see Lectures 1 and 3 of the present book, Belyaev (2004) and Irschik et.al.(2004), for the material description of continuum mechanics. Throughout the paper we however assume that both, $\boldsymbol{u}$ and its gradient with respect to the place $\boldsymbol{P}$ in the floating reference configuration are small, see Sections 12 and 16 of Gurtin (1972) for a precise discussion of the consequences of these assumptions. The corresponding theory is called the *infinitesimal* or *geometrically linear theory* of continuum mechanics. In short, $\boldsymbol{P}$ and the actual position vector $\boldsymbol{u} + \boldsymbol{P}$ coincide within an error of $O(\varepsilon)$ as $\varepsilon \to 0$, where ε denotes the supremum of the length of $\boldsymbol{u}$. The Landau order symbol is denoted as O. The Cauchy stress tensor and the first Piola Kirchhoff stress tensor coincide within terms of $O(\delta)$, and the Green strain tensor may be approximated by the symmetric part of the gradient of $\boldsymbol{u}$ within an error of $O(\delta^2)$. Here, δ denotes the magnitude of the gradient of $\boldsymbol{u}$. In the linear theory, terms of the order of $O(\varepsilon)$, $O(\delta)$ and $O(\delta^2)$ are neglected, which will be adopted subsequently. Thus, despite we are formally in the material description, we can refer to the spatial formulations of Lecture 2 of the present book, Irschik (2004), since we need not to distinguish between the different measures for stress and strain in the spatial and the material formulation. We furthermore assume the body to be *linear elastic* and to be subjected to an *actuating eigenstrain*, additionally to the imposed forces and the rigid-body motion. The corresponding mathematical formulation of this constitutive assumption will be given below.

It is the goal of the present Lecture to derive a distribution of actuating eigenstrains, which, when being applied together with the imposed forces and the prescribed rigid-body motion of the boundary ∂B_1, results in zero displacements relative to the floating reference configuration throughout the body and at every time instant,

$$\boldsymbol{u} = \boldsymbol{u}(\boldsymbol{P}, t) = \boldsymbol{0} \;\forall\; \boldsymbol{P}, t. \tag{1}$$

The body then behaves as a rigid body during the whole course of the motion. The problem thus defined is called a *shape-control problem*, or a *displacement compensation problm.*

3 Initial-Boundary-Value Problem with Eigenstrains

As a first step, we state the initial boundary-value problem in the framework of which shape control is to be performed. In the *local form of the equation of balance of linear momentum,* see Eq.(11) of Lecture 2 of the present book, Irschik (2004), we first need to compute the acceleration $\boldsymbol{a}$ of the particle under consideration. This *absolute acceleration* in Eq.(11) is defined as the material time derivative of the *absolute velocity* $\boldsymbol{v}$ as observed in the inertial frame. In order to avoid ambiguities, we call this rate an *inertial time rate* in the following. An inertial time rate will be abbreviated by a superimposed dot, $\boldsymbol{a} = \dot{\boldsymbol{v}}$. In computing $\boldsymbol{v}$, we wish to make use of the above introduced material formulation for the displacement relative to the *floating* reference configuration, $\boldsymbol{u} = \boldsymbol{u}(\boldsymbol{P}, t)$, as well as from the time derivative of $\boldsymbol{u}$ as observed in the co-moving frame, $\partial_P \boldsymbol{u} / \partial t$. In the latter rate, we keep the place $\boldsymbol{P}$ fixed, compare Lecture 3 of the present book, Irschik et al. (2004). We subsequently use the short-handed notation $\partial_P(\,)/\partial t = (\,),_t$, where we talk about a *co-moving time rate*. We are thus concerned with the question of the consequences of a change of frame. This question is connected to the aspects of *objectivity*, *observer-invariance* and the *principle of material frame indifference*. In short, objectivity means a class of transformation properties that applies to certain physical entities, and the principle of material frame indifference states the independence of constitutive formulations from the frame of reference. In the following, for the sake of brevity, we do not go into the details of these important topics. For a recent comprehensive presentation, we refer to Chapter 4 of Haupt (2000), see also Lecture 1 of the present book, Belyaev (2004). The following derivations rest upon *Euler's fundamental formula for relative motions*, which asserts how the inertial time rate of a vector is related to the time rate of change observed in the co-moving frame, see e.g. Sect. 143 of Truesdell and Toupin (1960) and Chapters 1 and 8 of Ziegler (1998). The actual position vector $\boldsymbol{p}$ of the particle in the co-moving frame, drawn from the origin A of the spatial coordinate system used in this frame, reads

$$\boldsymbol{p} = \boldsymbol{P} + \boldsymbol{u}(\boldsymbol{P}, t)\,. \tag{2}$$

Euler's fundamental formula then asserts that

$$\dot{\boldsymbol{p}} = \boldsymbol{p},_t + \boldsymbol{\omega} \times \boldsymbol{p}\,. \tag{3}$$

The vector of *angular velocity* of the co-moving frame with respect to the inertial frame is denoted by $\boldsymbol{\omega}$. From Eq.(2), we have

$$\boldsymbol{p},_t = (\boldsymbol{P} + \boldsymbol{u}(\boldsymbol{P}, t)),_t = \boldsymbol{u},_t\,. \tag{4}$$

Since we are in the linear theory, we may replace the last term on the right hand side of Eq.(3) by $\boldsymbol{\omega} \times \boldsymbol{P}$ within an error of $O(\varepsilon)$, where ε denotes the supremum of the length of $\boldsymbol{u}$. Substituting Eq.(4), it is seen that Eq.(3) becomes

$$\dot{\boldsymbol{p}} = \boldsymbol{u},_t + \boldsymbol{\omega} \times \boldsymbol{P}\,. \tag{5}$$

Eventually, applying Euler's fundamental formula to Eq.(5), we obtain:

$$\ddot{\boldsymbol{p}} = \boldsymbol{u},_{tt} + 2\,\boldsymbol{\omega} \times \boldsymbol{u},_t + \boldsymbol{\alpha} \times \boldsymbol{P} + \boldsymbol{\omega} \times (\boldsymbol{\omega} \times \boldsymbol{P})\,, \tag{6}$$

again within an error of $O(\varepsilon)$. The vector of *angular acceleration* is denoted by $\boldsymbol{\alpha} = \dot{\boldsymbol{\omega}} = \boldsymbol{\omega},_t$.

The absolute velocity $\boldsymbol{v}$ of the particle is obtained from Eq.(3) by adding the absolute velocity $\boldsymbol{v}_A$ of the origin of the spatial coordinate system of the co-moving frame

$$\boldsymbol{v} = \boldsymbol{v}_A + \dot{\boldsymbol{p}}\,, \tag{7}$$

and for the absolute acceleration of a particle one obtains

$$\boldsymbol{a} = \dot{\boldsymbol{v}} = \boldsymbol{a}_A + \boldsymbol{u},_{tt} + 2\,\boldsymbol{\omega} \times \boldsymbol{u},_t + \boldsymbol{\alpha} \times \boldsymbol{P} + \boldsymbol{\omega} \times (\boldsymbol{\omega} \times \boldsymbol{P})\,, \tag{8}$$

where the absolute acceleration of the origin of the co-moving frame is denoted by $\boldsymbol{a}_A = \dot{\boldsymbol{v}}_A$. Note that the vectors $\boldsymbol{\omega}$, $\boldsymbol{\alpha}$ and $\boldsymbol{a}_A$ are given to us as functions of time in advance, since they follow from the prescribed rigid-body motion of the boundary ∂B_1.

The local form of the equation of balance of linear momentum now can be written as

$$\rho\,(\boldsymbol{u},_{tt} + 2\,\boldsymbol{\omega}\times\boldsymbol{u},_t) = \boldsymbol{b}^* + div\,\boldsymbol{T}, \tag{9}$$

where we have used the abbreviation

$$\boldsymbol{b}^* = \boldsymbol{b} - \rho\,(\boldsymbol{a}_A + \boldsymbol{\alpha}\times\boldsymbol{P} + \boldsymbol{\omega}\times(\boldsymbol{\omega}\times\boldsymbol{P})). \tag{10}$$

In Eq.(9), *div* stands for the divergence operator with respect to $\boldsymbol{P}$. The imposed body force is $\boldsymbol{b}$, the stress tensor is denoted by $\boldsymbol{T}$, and the mass density is written as ρ, all of these entities being expressed in the material formulation as functions of the place $\boldsymbol{P}$ in the floating reference configuration. Thus, strictly speaking, ρ is the mass density in the floating reference configuration, while $\boldsymbol{T}$ stands for the first Piola-Kirchhoff stress tensor. Since we deal with the geometrically linear theory, in which we assume the gradients of $\boldsymbol{u}$ to be small, these specifications however are not of essence here, see again the above discussion of the orders of error introduced in the linear theory. We assume imposed couples as well as couple stresses to be absent, such that the equation of balance of angular momentum asserts that the stress tensor of the linear theory is symmetric: $\boldsymbol{T} = \boldsymbol{T}^T$. Note that $\boldsymbol{b}^*$ in Eq.(10) is entirely known. The part $-\rho\,(\boldsymbol{a}_A + \boldsymbol{\alpha}\times\boldsymbol{P} + \boldsymbol{\omega}\times(\boldsymbol{\omega}\times\boldsymbol{P}))$ in Eq.(10) represents the *inertial force due to the rigid-body motion.*

The stress is taken to be linearly related to the strain by a *generalised Hooke's law*:

$$\boldsymbol{T} = \boldsymbol{C}\,[\boldsymbol{E}] + \boldsymbol{T}^a, \tag{11}$$

where the strain $\boldsymbol{E}$ in the geometrically linear theory is approximated by the symmetric part of the gradient of $\boldsymbol{u}$ by

$$\boldsymbol{E} = \frac{1}{2}\,(\,grad\,\boldsymbol{u} + grad\,\boldsymbol{u}^T) = sym\,grad\,\boldsymbol{u}, \tag{12}$$

the symmetric part of a tensor being abbreviated by *sym*, and *grad* denoting the gradient operator with respect to $\boldsymbol{P}$. The fourth order tensor of elastic parameters is written as $\boldsymbol{C}$, and $\boldsymbol{C}\,[\boldsymbol{E}]$ stands for the second order tensor that represents the linear mapping of $\boldsymbol{E}$ by means of $\boldsymbol{C}$. Furthermore, the tensor of second order $\boldsymbol{T}^a$ in Eq.(11) represents the so-called *actuation stress*, which is due to an *eigenstrain* acting at the particle under consideration. E.g., in case of a linear *thermal expansion strain*, there is

$$\boldsymbol{T}^a = -\,\boldsymbol{C}\,[\boldsymbol{\alpha}]\,\theta, \tag{13}$$

where θ is the increase in temperature, and $\boldsymbol{\alpha}$ denotes the second-order tensor of coefficients of linear thermal expansion. In case of a *piezoelastic* material, $\boldsymbol{T}^a$ can be expressed as a linear mapping of the *electric field vector*.

In the present Lecture, we take into account the possibility of an anisotropic material behaviour. The following symmetry relations are known to hold for the tensor of elastic parameters $\boldsymbol{C}$ and for the tensor of thermal expansion coefficients $\boldsymbol{\alpha}$:

$$\boldsymbol{C}\,[\boldsymbol{E}]\cdot\bar{\boldsymbol{E}} = \boldsymbol{C}\,[\bar{\boldsymbol{E}}]\cdot\boldsymbol{E},\quad \boldsymbol{\alpha} = \boldsymbol{\alpha}^T, \tag{14}$$

where $\bar{\boldsymbol{E}} = \bar{\boldsymbol{E}}^T$ denotes a symmetric second-order tensor, and the dot product in Eq.(14.1) indicates the contraction of two second-order tensors to a scalar quantity, the *scalar product* of these tensors. E.g., what was denoted as $tr\,(\boldsymbol{T}^T \cdot grad\,\dot{\boldsymbol{u}}) = tr\,(\boldsymbol{T} \cdot grad\,\dot{\boldsymbol{u}})$ in Lecture 2 of the present book, Irschik (2004), in the following is abbreviated by $\boldsymbol{T} \cdot grad\,\dot{\boldsymbol{u}}$, see Eq.(22) below. This abbreviation is admissible, since the scalar product of two second-order tensors is the only product of second-order tensors that will be needed subsequently. For fundamentals of the constitutive equations of elasticity and thermoelasticity, we refer to Gurtin (1972), Carlson (1972), and to Chapter 6 of Ziegler (1998).

The set of field equations for our problem is formed by Eqs.(9), (11) and (12). In addition to the field equations, we consider the following *boundary conditions*. Since the boundary ∂B_1 is assumed to move together with the floating reference configuration, we have

$$\partial B_1:\ \boldsymbol{u} = 0 \tag{15}$$

at every instant of time. This means that

$$\partial B_1:\ \boldsymbol{u},_t = 0\,. \tag{16}$$

At the remaining part of the boundary, ∂B_2, a traction $\boldsymbol{s}$ per unit area is prescribed:

$$\partial B_2:\ \boldsymbol{T} \cdot \boldsymbol{n} = \boldsymbol{t}_{(\boldsymbol{n})} = \boldsymbol{s}\,. \tag{17}$$

This boundary condition follows from the *Cauchy fundamental theorem on stresses*, see Eq.(10) of Lecture 2 of the present book, Irschik (2004).

We furthermore assume that the *initial displacements* and *initial velocities* vanish everywhere in B:

$$t = 0:\ \boldsymbol{u} = 0,\ \boldsymbol{u},_t = 0\,, \tag{18}$$

see Eq.(5). Using Eq.(12.2), it is seen that these latter relations are associated with vanishing initial values of the strain and its co-moving rate:

$$t = 0:\ \boldsymbol{E} = 0,\ \boldsymbol{E},_t = sym\,grad\,\boldsymbol{u},_t\,, \tag{19}$$

since the partial derivative with respect to time and the gradient with respect to $\boldsymbol{P}$ commute.

4 Dynamic Deformation Compensation

In 1885, Neumann proved the uniqueness of the solution of the above linear elastodynamic problem for the special case of $\boldsymbol{a}_A = 0$, $\boldsymbol{\omega} = 0$, $\boldsymbol{\alpha} = 0$, $\boldsymbol{b} = 0$, $\boldsymbol{s} = 0$ and $\boldsymbol{T}^a = 0$. He showed that the latter homogeneous problem has a trivial (vanishing) solution for the displacement $\boldsymbol{u}$. For a contemporary presentation of the proof of Neumann, see Chandrasekharaiah and Debnath (1994). In the present Lecture, we assume $\boldsymbol{a}_A$, $\boldsymbol{\omega}$, $\boldsymbol{\alpha}$, $\boldsymbol{b}$ and $\boldsymbol{s}$ not to vanish and to be prescribed, but we extend the technique developed by Neumann in order to derive additional distributions of $\boldsymbol{T}^a$, such that the displacements are compensated throughout the body and at every time instant, Eq.(1). In other words, it is required that the displacements $\boldsymbol{u}$ due to $\boldsymbol{a}_A$, $\boldsymbol{\omega}$, $\boldsymbol{\alpha}$, $\boldsymbol{b}$, $\boldsymbol{s}$ and $\boldsymbol{T}^a$ in total do vanish. As was mentioned already above, this shape control problem for $\boldsymbol{T}^a$ in general does not have an unique solution.

By integration over the body in the floating reference configuration, we assume that the observer in the co-moving frame studies the following *generalised time-dependent Neumann function* N:

$$2\,N(t) = \int_B (\mathbb{C}[\boldsymbol{E}] \cdot \boldsymbol{E} + \rho\,\boldsymbol{u},_t \cdot \boldsymbol{u},_t)\,dV\,. \tag{20}$$

Note that this function initially vanishes, $N(t=0)=0$, see Eqs.(18.2) and (19.1).

Seen from the co-moving frame, the time-rate of N reads:

$$N,_t = \int_B (\mathcal{C}[\boldsymbol{E}] \cdot \boldsymbol{E},_t + \rho\, \boldsymbol{u},_t \cdot \boldsymbol{u},_{tt})\, dV\,, \tag{21}$$

where we have interchanged the integration over the volume in the floating reference formulation with the co-moving time rate, and Eq.(14.1) has been taken into account. We note that neither the elementary volume dV nor the mass density ρ of a particle in the floating reference configuration do change with time. From the generalised Hooke law, Eq.(11), from the kinematic relation of Eq.(12), and since $(\boldsymbol{T}-\boldsymbol{T}^a)$ is symmetric, we obtain that

$$\mathcal{C}[\boldsymbol{E}] \cdot \boldsymbol{E},_t = (\boldsymbol{T}-\boldsymbol{T}^a) \cdot \boldsymbol{E},_t = (\boldsymbol{T}-\boldsymbol{T}^a) \cdot sym\, grad\, \boldsymbol{u},_t = (\boldsymbol{T}-\boldsymbol{T}^a) \cdot grad\, \boldsymbol{u},_t\,. \tag{22}$$

The following identity is easily verified:

$$(\boldsymbol{T}-\boldsymbol{T}^a) \cdot grad\, \boldsymbol{u},_t = -\,div\, (\boldsymbol{T}-\boldsymbol{T}^a) \cdot \boldsymbol{u},_t + div\, ((\boldsymbol{T}-\boldsymbol{T}^a) \cdot \boldsymbol{u},_t)\,. \tag{23}$$

Using the divergence theorem, we find that

$$N,_t = \int_B (-\,div\, (\boldsymbol{T}-\boldsymbol{T}^a) \cdot \boldsymbol{u},_t + \rho\, \boldsymbol{u},_t \cdot \boldsymbol{u},_{tt})\, dV + \int_{\partial B_2} ((\boldsymbol{T}-\boldsymbol{T}^a) \cdot \boldsymbol{n}\,) \cdot \boldsymbol{u},_t\, dA\,. \tag{24}$$

The surface integral extends over ∂B_2 only, since $\boldsymbol{u},_t$ vanishes at ∂B_1, see Eq.(16). Substituting Eqs.(9) and (17), we can put Eq.(24) into the form

$$N,_t = \int_B (-\rho\, \boldsymbol{u},_{tt} - 2\, \boldsymbol{\omega} \times \boldsymbol{u},_t + \boldsymbol{b}^* + div\, \boldsymbol{T}^a + \rho\, \boldsymbol{u},_{tt}) \cdot \boldsymbol{u},_t\, dV$$

$$+ \int_{\partial B_2} (\boldsymbol{s} - \boldsymbol{T}^a\, n) \cdot \boldsymbol{u},_t\, dA$$

$$= \int_B (\boldsymbol{b}^* + div\, \boldsymbol{T}^a) \cdot \boldsymbol{u},_t\, dV + \int_{\partial B_2} (\boldsymbol{s} - \boldsymbol{T}^a\, n) \cdot \boldsymbol{u},_t\, dA\,. \tag{25}$$

Hence, when the following relations are satisfied,

$$div\, \boldsymbol{T}^a = -\,\boldsymbol{b}^*\,, \tag{26}$$

$$\partial B_2\!: \ \boldsymbol{T}^a\, \boldsymbol{n} = \boldsymbol{s}\,, \tag{27}$$

then it is found that $N,_t = 0$ in Eq.(25), such that $N(t) = const.$ Due to the initial conditions stated in Eqs.(18) and (19), we thus obtain that N in Eq.(20) vanishes throughout the motion,

$$N(t) = N(t=0) = 0\ \forall\, t\,. \tag{28}$$

Now, as long as $\boldsymbol{E} = sym\, grad\, \boldsymbol{u} \neq \boldsymbol{0}$ and $\boldsymbol{u},_t \neq \boldsymbol{0}$, N in Eq.(20) is a non-negative function, since in the linear theory we can assume that $\boldsymbol{C}\,[\boldsymbol{E}] \cdot \boldsymbol{E} > \boldsymbol{0}$ for $\boldsymbol{E} \neq \boldsymbol{0}$, see Sect. 21 of Gurtin (1972), and since the mass density of the body is a positive quantity, such that $\rho\, \boldsymbol{u},_t \cdot \boldsymbol{u},_t > 0$ for $\boldsymbol{u},_t \neq \boldsymbol{0}$. Hence, it follows from Eqs.(20) and (28) that the displacement $\boldsymbol{u}$ indeed must vanish, if Eqs.(26) and (27) are valid.

We conclude that, given the initial conditions of Eq.(18), any set of actuation stresses satisfying Eqs.(26) and (27) guarantees that the relative displacement $\boldsymbol{u}$, i.e. the vibrations relative to the floating reference configuration, do vanish exactly, such that Eq.(1) is satisfied. What must be required, however, is that the trivial solution thus reached is stable, i.e. that in case of small non-trivial initial conditions the relative displacement remains in the neighbourhood of the state $\boldsymbol{u} = 0$, in some sense to be precisely defined. This problem of dynamic stability however is beyond the present linear formulation.

In order to develop a rational solution strategy for solving Eqs.(26) and (27), we consider the following quasi-static problem:

$$div\,\hat{\boldsymbol{T}} + \boldsymbol{b}^* = 0, \tag{29}$$

$$\partial B_2:\ \hat{\boldsymbol{T}} \cdot \boldsymbol{n} = \boldsymbol{s}. \tag{30}$$

The superimposed hat refers to a quasi-static solution. A solution of Eqs.(29) and (30) is called a *statically admissible solution*, since it represents a solution which satisfies the original equation of balance of momentum, Eq.(9), and the boundary condition of traction, Eq.(17), in the static limit. It is quasi-static, since $\boldsymbol{b}^*$ and $\boldsymbol{s}$ are time-dependent in general. In order to compute a statically admissible stress distribution $\hat{\boldsymbol{T}}$, we assume suitable boundary conditions to hold at the part ∂B_1 of the boundary. E.g., we may use the original homogeneous boundary condition of Eq.(15). For a constitutive equation, we subsequently use a linear relation between stress and strain without actuation stress, with a suitable elastic stiffness.

By comparison of Eqs.(26) and (27) with Eqs.(29) and (30), the actuation stresses for exact compensation of the relative displacements now can be chosen as

$$\boldsymbol{T}^a = \hat{\boldsymbol{T}} + \boldsymbol{T}_0^a, \tag{31}$$

where $\boldsymbol{T}_0^a$ satisfies the homogeneous set of Eqs.(26) and (27):

$$div\,\boldsymbol{T}_0^a = 0, \tag{32}$$

$$\partial B_2:\ \boldsymbol{T}_0^a \cdot \boldsymbol{n} = 0. \tag{33}$$

Note that the part $\hat{\boldsymbol{T}}$ in Eq.(31) follows from the solution of a quasi-static boundary value problem, which is much simpler to handle than the original initial-boundary-value problem stated above. Moreover, the part $\boldsymbol{T}_0^a$ reveals the non-uniqueness of the inverse problem under consideration. In order to obtain a solution for Eqs.(32) and (33), we again may use the original homogeneous boundary condition of Eq.(15). We furthermore may assign a linear relation between stress and strain with a non-vanishing actuation stress, in order to reach a non-trivial solution for $\boldsymbol{T}_0^a$. In the subsequent example, $\boldsymbol{T}_0^a$ is set to zero.

5 Example

For a validation of the solution strategy presented in Sections 3 and 4, the following numerical example problem is considered. A quadratic domain, see Fig. 1, is in a state of plane strain; the non-dimensionalised coordinates of its corners are $P_1:(0/1)$, $P_2:(1/1)$, $P_3:(1/0)$ and $P_4:(0/0)$. The boundary $P_1 - P_4 - P_3$ undergoes a prescribed horizontal rigid-body displacement $r_1(t) = U_1\,f(t)\,H(t)$, with $f(t) = (t/\tau)^2$, where U_1 denotes the amplitude, τ is a characteristic time, and $H(t)$ is the Heaviside function. There are no imposed body forces in this example problem, and the edges $P_1 - P_2 - P_3$ and are free of stress. The elastic vibrations of the rectangular domain due to the prescribed rigid-body motion is intended to be eliminated by means of an additional eigenstrain. Exemplary, we use a thermal expansion strain in the following numerical computations, which are performed by means of the Finite Element code ABAQUS.

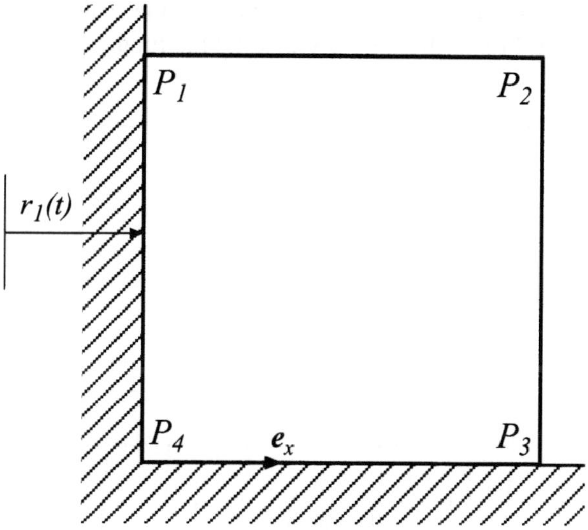

Figure 1. Quadratic domain with horzontal support excitation.

In the first step of the above presented solution strategy, $\boldsymbol{b}^*$ of Eq.(10) is computed. These forces are horizontal, $\boldsymbol{b}^* = b^* \boldsymbol{e}_x$, the corresponding component following to

$$b^* = -\rho\, r_{1,tt} = -\hat{b}_s H(t), \quad \hat{b}_s = 2\rho\, U_1 / \tau^2 . \tag{34}$$

The statically admissible stress $\hat{\boldsymbol{T}}$ due to this fictitious body force can be easily computed, compare Eqs.(29) and (30). It has the form $\hat{\boldsymbol{T}} = -\hat{\boldsymbol{T}}_s H(t)$, where $\hat{\boldsymbol{T}}_s$ is the static stress due to the horizontal body force $\hat{b}_s$. The required distribution of actuation stress $\boldsymbol{T}^a$ follows from Eq.(31) , where we choose $\boldsymbol{T}^a_0 = \boldsymbol{0}$. In case of a thermal actuation stress, $\boldsymbol{T}^a$ has the form

$$\boldsymbol{T}^a = -\boldsymbol{C}[\boldsymbol{\alpha}]\, \hat{\theta}\, H(t) , \tag{35}$$

where $\hat{\theta}$ is a constant reference temperature, and $\boldsymbol{\alpha}$ is the tensor of thermal expansion coefficients, see Eq.(13). The proper spatial distribution of thermal expansion coefficients for the specific problem then becomes, $\hat{\boldsymbol{E}}_s$ is the static strain due to $\hat{b}_s$,

$$\boldsymbol{\alpha} = \boldsymbol{C}^{-1}[\hat{\boldsymbol{T}}_s]\, \hat{\theta}^{-1} = \hat{\boldsymbol{E}}_s \hat{\theta}^{-1} . \tag{36}$$

The finite element validation was done using the commercial code ABAQUS Standard V 6.3 and a self-developed C++ code for pre-processing. The quadratic domain was subdivided into 3600 plane strain elements of type CPE4R for the computations, and the material properties of steel were assigned. At first, the static strain $\hat{\boldsymbol{E}}_s$ due to $\hat{b}_s$ was computed. Using the C++ code, the proper thermal expansion tensors $\boldsymbol{\alpha}$ were calculated according to Eq.(36) and assigned to the corresponding finite elements. In the next steps the finite element code was used to perform dynamic computations with the latter anisotropic thermal expansion tensors, where two problems were considered. In the force problem, elastic vibrations due to the transient body forces Eq.(34) were computed, and in the eigenstrain problem vibrations due to the transient actuation stresses given in Eq.(35) were simulated, where the coupling of thermal and mechanical fields was neglected. Since the negative actuation stress was applied, the displacements in the force and the eigenstrain problem should be equal.

As a characteristic result, the step response functions of the scaled horizontal displacement of point P_2 is presented in Fig. 2, the time is scaled by the fundamental vibration period T_1 .

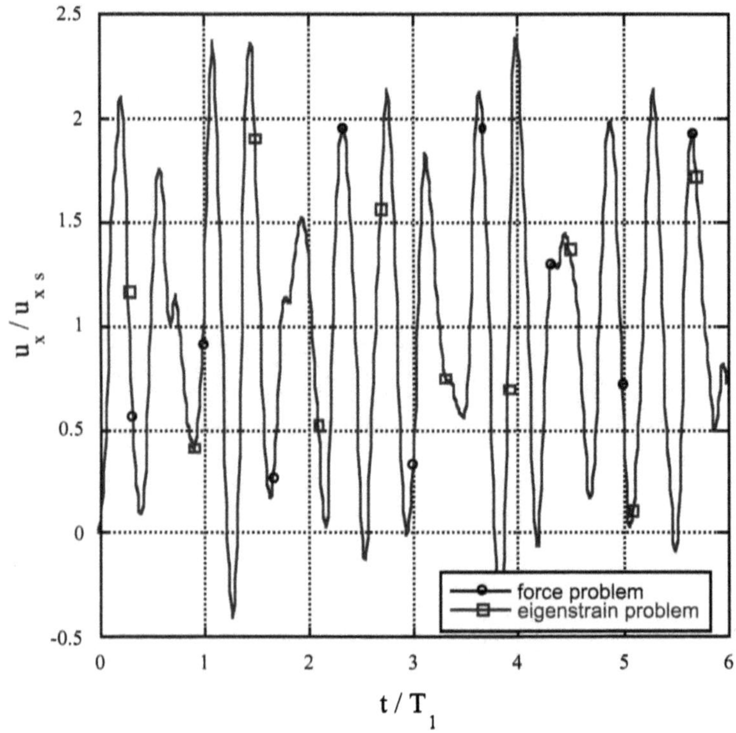

Figure 2. Comparison of the step responses.

The step responses of the force problem and the thermal eigenstrain problem do coincide with a high accuracy, which gives evidence for the validity of the presented method of compensating vibrations in elastic solids and structures in the presence of rigid-body motions.

References

Belyaev, A.K. (2004). Basics of Continuum Mechanics. In: Irschik, H. and Schlacher, K., eds., *Advanced Dynamics and Control of Structures and Machines*. CISM Courses and Lectures No. 444, Wien–New York: Springer-Verlag.

Carlson, D. E. (1972). Linear Thermoelasticity. In Flügge, S., ed., *Handbuch der Physik, Vol. VIa/2*, Berlin: Springer-Verlag.

Chandrasekharaiah, D. S. and Debnath, L. (1994). *Continuum Mechanics*. Boston: Ac. Press.

Gurtin, M. E. (1972). The linear theory of elasticity. In Flügge, S., ed., *Handbuch der Physik, Vol. VIa/2*, Berlin: Springer-Verlag.

Haftka, R.T., and Adelman, H.M. (1985). An analytical investigation of shape control of large space structures by applied temperatures. *AIAA Journal* 23:450-457.

Haupt, P. (2002). *Continuum Mechanics and Theory of Materials*. Berlin: Springer-Verlag.

Irschik, H. (2002). A Review on Static and Dynamic Shape Control of Structures by Piezoelectric Actuation. *Engineering Structures* 24:5-11.

Irschik, H. (2004). A Treatise on the Equations of Balance and on the Jump Relations in Continuum Mechanics. In: Irschik, H. and Schlacher, K., eds., *Advanced Dynamics and Control of Structures and Machines*. CISM Courses and Lectures No. 444, Wien–New York: Springer-Verlag.

Irschik, H., Holl, H. J. and Hammelmüller, F. (2004). The Rayleigh-Ritz Technique and the Lagrange Equations in Continuum Mechanics: Formulations for Material and Non-Material Volumes. In: Irschik, H. and Schlacher, K., eds., *Advanced Dynamics and Control of Structures and Machines*. CISM Courses and Lectures No. 444, Wien–New York: Springer-Verlag.

Irschik, H., and Pichler, U. (2001). Dynamic shape control of solids and structures by thermal expansion strains. *Journal of Thermal Stresses* 24:565-576.

Irschik, H., and Pichler, U. (to appear 2004). An Extension of Neumann's Method for Shape Control of Force-Induced Elastic Vibrations by Eigenstrains. *International Journal of Solids and Structures*.

Irschik, H., Krommer, M., and Pichler, U. (2001). Collocative Control of Beam Vibrations with Piezoelectric Self-Sensing Layers. In Gabbert, U., and Tzou, H.S.., eds., *Proceedings of the IUTAM-Symposium on Smart Structures and Structronic Systems*, Dordrecht 2001. Dordrecht: Kluwer, 315-322.

Irschik, H., Krommer, M., and Pichler, U. (2003). Dynamic shape control of beam-type structures by piezoelectric actuation and sensing. *Int. Journal Applied Electromagnetics and Mechanics* 17(1-3):251-258.

Irschik H., Nader M., Zehetner C. (2003). Exact Cancellation of Vibrations in Elastic Structures Performing Large Rigid Body Motions (Sectional Key Note Lecture). *Proc. of 10th International Congress on Sound and Vibration,* Stockholm, Swede*n*, Vol.7: 3487- 3498.

Irschik, H., and Ziegler, F., (2001). Eigenstrain without Stress and Static Shape Control of Structures. *AIAA-Journal* 39:1985-1999.

Nader, M., Gattringer, H., Krommer, M., and Irschik, H., (2003). Control of Flexural Vibrations of Circular Plates by Shaped Piezoelectric Actuation. *ASME Journal of Vibration and Acoustics* 125:88-94.

Nader M., Irschik H. (2003). Suppression of Force-Induced Circular Plate Vibrations by Thermal Expansion Strains. *Proc. of 5th International Congress on Thermal Stresses and Related Topics,* Blacksburg, Virgina, USA, MA-3-4-1 - MA-3-4-4.

Truesdell, C., Toupin, R. (1960). The Classical Field Theories. In: *Handbuch der Physik*, Vol. III/1, Springer-Verlag.

Ziegler, F., (1998). Mechanics of Solids and Fluids (2nd edition, 2nd corr. print). New York: Springer-Verlag.

Acknowledgement:

The present Lecture Note is a contribution to the K+ Linz Center of Excellence in Mechatronics (LCM), Strategic Project 4.4. Support of LCM by the Austrian K+ Fund and the Government of Upper Austria is gratefully acknowledged.

Computational Dynamics of an Elasto-Plastic Structural Element With Rigid-Body Degrees-of-Freedom.

Johannes Gerstmayr[1)], Hans Irschik[1)] and Markus Dibold[2)]

1 Johannes Kepler University of Linz, FWF-Project No. P15195 and LCM, Linz, Austria
2 Kplus Linz Center of Excellence in Mechatronics (LCM), Linz, Austria

Abstract. In the present Lecture, we study the motion of a single elasto-plastic body that represents a moving element of a structure or machine, where we present equations that are applicable to three-dimensional motions and bodies of arbitrary shape. We assume the displacements and strains of the body to be small with respect to a floating reference configuration, and we present the corresponding small-strain elasto-plastic-constitutive relations. We then point out the necessity of refined computational procedures for obtaining the plastic parts of strain in the case of a reversed loading, a problem often to be encountered in practice. In a Rayleigh-Ritz procedure, the flexible coordinates, which are coupled to the rigid-body degrees of freedom via the equations of motion, must be brought into connection with the plastic parts of strain. Often, the influence of the plastic parts of strain upon the motion of the body can not be neglected. We sketch an advantageous iterative numerical procedure for computing the plastic parts of strain, and we eventually discuss their influence upon the equations of motion in more detail.

1 Introduction

The present Lecture is concerned with the dynamics of elasto-plastic bodies that undergo large rigid-body motions with superimposed small deformations. This problem is of an obvious actuality in various fields, e.g. for multi-body robotic systems designed for maintenance and repair under catastrophic operating conditions. Catastrophic situations are often accompanied by dangerous environmental influences, such as high temperature. These influences may lower the mechanical stiffness of the robotic system in a considerable manner. Plastic strains then must be taken into account for accurately computing the motion of the system. In a cyclic motion of a robotic system or machine, reversed plasticity may take place in some of its elements. The question arises, whether shake-down will occur, or whether low cycle rupture must be expected. The study of elasto-plastic multi-body systems is also of a particular interest for crash-simulations and in earthquake engineering. In the latter field, permanent deformations and rigid-body degrees of freedom have to be encountered.

Only few contributions in the literature so far have included plasticity into the dynamics of flexible systems with rigid-body degrees-of-freedom. The seemingly first contribution taking into account elasto-plastic deformations in flexible multi-body systems is a paper by Ambrósio and Nikravesh (1992). A multi-body dynamics study on a car-crash including comparisons to a real experiment was presented by Ambrósio (2001). Recently, elasto-plastic beam-type bodies with distributed plastic zones and reversed plasticity were successfully studied at the University of Linz. Vibrations of single elasto-plastic beams performing large guided rigid-body rotations, thereby exhibiting ductile damage, were treated by Gerstmayr, Holl and Irschik (2001). The interaction of the plastic zones with the vibrations and the rotation of an elasto-plastic pendulum was taken into account by Gerstmayr and Irschik (2003). Control strategies were incorporated by Gerstmayr and Irschik (2001). A powerful computational algorithm for beam-type elasto-plastic multi-body systems in a plane motion was presented by Gerstmayr (2003).

In the present Lecture, we restrict our considerations to the motion of a single elasto-plastic body that represents a moving element of a structure or machine. The problem of a single body is fundamental for the development of computer codes that can simulate the motion of multi-body dynamic systems. In extension of the above cited references on beam-type systems, we subsequently present equations that are applicable to the three-dimensional motion of single bodies of arbitrary shape. We assume the displacements and strains of the body to be *small* with respect to *a floating reference configuration.* The assumption of small strains not only does simplify the formulation considerably, it also covers a fairly wide range of operations of mechanical systems weakened by environmental influences.

The present Lecture is organized as follows: In Section 2, we give a brief outline of the foundations of the floating frame of reference formulation, and we present the *small-strain elasto-plastic constitutive relations* written as a function of the place in the floating reference configuration. We shortly discuss the conceptual problems that are present in the small-strain approximation with respect to the principle of material frame indifference. We then point out the necessity of refined computational procedures for obtaining the plastic parts of strain in the case of a reversed loading, a problem often to be encountered in robotic systems. In a *Rayleigh-Ritz procedure*, the flexible coordinates, which are coupled to the rigid-body degrees of freedom via the equations of motion, must be brought into connection with the plastic parts of strain. Often, the influence of the plastic parts of strain upon the motion of the body can not be neglected, for examples see Gerstmayr (2003) and Gerstmayr and Irschik (2003). In Section 3 below, we sketch a convenient numerical procedure for computing the plastic parts of strain. Since the latter are coupled to the total strains, an iterative procedure is indispensable. Finally, in Section 4, we discuss the influence of the plastic parts of strain upon the equations of motion in more detail.

2 Elasto-Plastic Constitutive Relations

The motion of the body under consideration is described by means of two frames, one of them being inertial, the other one being in motion relative to the first one. The latter is assumed to be fixed in an *undeformed reference configuration* that moves together with the body, thereby performing large rigid-body motions. In this *co-moving frame*, the undeformed *floating reference configuration* appears to be at rest. For a detailed discussion of the *floating frame of reference formulation* in the dynamics of elastic multi-body systems, we refer to Chapter 5 of the book by Shabana (1998). See also the considerations presented in Lecture 5 of the present book, Irschik, Pichler, Nader and Zehetner (2004). In the present context, we lay emphasis upon the following remarks: Note that, since the deformation of the body is referred to the floating reference configuration, we basically deal with a *material description of continuum mechanics*, in which the mechanical entities such as actual position, displacement, strain and stress are written as a function of time and of the place in the floating reference configuration. For the material and spatial formulations of continuum mechanics, see Lectures 1 and 2 of the present book, Belyaev (2004) and Irschik (2004), respectively.

In the small strain regime, referring to the floating frame, the *constitutive relation* of an *elasto-plastic body* can be written in the classical additive form,

$$\boldsymbol{T} = \boldsymbol{C}[\boldsymbol{E} - \boldsymbol{E}_p] , \tag{1}$$

see e.g. Chapter 2 of Simo and Hughes (1998). Here, $\boldsymbol{T}$ is the stress tensor, $\boldsymbol{C}$ denotes the fourth-order tensor of elastic parameters, $\boldsymbol{E}_p$ is the plastic part of strain, and

$$\boldsymbol{E} = \frac{1}{2}\left(grad\, \boldsymbol{u} + grad\, \boldsymbol{u}^T \right) = sym\, grad\, \boldsymbol{u} \tag{2}$$

is the small strain tensor, the transpose of a tensor being denoted by the superscript T. The symmetric part of a second order tensor is denoted by *sym*, the gradient with respect to the place in floating reference configuration is written as *grad*, and $\boldsymbol{u}$ is the displacement vector from the floating reference configuration. The partial strain field $\boldsymbol{E} - \boldsymbol{E}_p$ in Eq.(1) is sometimes called the elastic part of strain. However, $\boldsymbol{E} - \boldsymbol{E}_p$ in general fails to follow from a displacement field according to Eq.(2). The symbol $\boldsymbol{C}[\boldsymbol{E}]$ stands for the second order tensor representing the linear mapping of $\boldsymbol{E}$ by means of $\boldsymbol{C}$. The total stress in Eq.(1) thus is expressed as a linear mapping of the partial strain field $\boldsymbol{E} - \boldsymbol{E}_p$. This decomposition of a plastic part of strain is admissible, since we assume the total strain $\boldsymbol{E}$ to be small. Proper extensions of Eq.(1) to the large deformation theory still represent an active field of research, see Chapter 7 of Simo and Hughes (1998), and Chapter 11 of Haupt (2000) for some comprehensive expositions. For a critical discussion of various decomposition methods in the case of large deformations, we refer to Chapter 1.12 of Palmov (1998).

The formulation of Eq.(1) suggests the following splitting of the stress into an *elastic part* $\boldsymbol{T}_E$ and a *dissipative stress* $\boldsymbol{T}_D$, such that $\boldsymbol{T} = \boldsymbol{T}_E + \boldsymbol{T}_D$, with

$$\boldsymbol{T}_E = \boldsymbol{C}[\boldsymbol{E}] , \tag{3}$$

$$\boldsymbol{T}_D = -\boldsymbol{C}[\boldsymbol{E}_p] . \tag{4}$$

where the elastic part of stress refers to the total strain $\boldsymbol{E}$.

Note the remarkable analogy between the plastic parts of strain $\boldsymbol{E}_p$ in Eqs.(1)-(4) and the thermal expansion strain that appears in Hooke's law for a thermally loaded linear elastic body, see Carlson (1972) and Parkus (1976) for the linear theory of thermo-elasticity. Plastic parts of strain, likewise to thermal expansion strains, represent sources of the deformation that do appear in the constitutive equations. Such a type of source is called an *eigenstrain*, see Mura (1991), or an initial strain. See also Lecture 5 of the present book, Irschik, Pichler, Nader and Zehetner (2004).

A stress of the type of the dissipative stress $\boldsymbol{T}_D$ that is present even when $\boldsymbol{E}$ vanishes, is called an eigenstress, a self-stress or an initial stress. From a mechanics point of view, dynamic problems of small strain elasto-plasticity thus do belong to the class of *dynamic eigenstrain problems*, see e.g. Irschik, Fotiu and Ziegler (1993) for a fundamental solution of the latter class. Methods developed to compute displacements and stresses in linear elastic bodies due to eigenstrains can be utilised in order to compute the displacements and stresses due to the plastic parts of strain. This analogy has been successfully applied in order to treat elasto-plastic vibrations of structures about an equilibrium position, see Irschik (1986), Irschik and Ziegler (1995), and it has motivated the above cited work on the dynamics of elasto-plastic beam-type systems with rigid-body degrees of freedom.

The following problem should be addressed in connection with the application of Eq.(1): Since the floating frame is in an arbitrary motion relative to the inertial frame, care must be taken about the consistency of the constitutive formulations with respect to the two frames, as this is required by the *principle of material frame indifference*, see Chapters 4 and 7 of Haupt (2000) for a recent comprehensive discussion. This principle states that a constitutive relation should be invariant under a change of the frame. The constitutive formulation of Eq.(1) does not satisfy the principle in a strict manner, since it refers to the small strain tensor $\boldsymbol{E}$ of Eq.(2). Moreover, the flow-rules of plasticity bring into the play the time rate of change of $\boldsymbol{E}$ and $\boldsymbol{E}_p$, which, when seen from the co-moving frame, are performed as *co-moving partial derivatives* with respect to time, holding the place in the floating reference configurations fixed. Small-strain relations involving the co-moving partial derivatives of $\boldsymbol{E}$ and $\boldsymbol{E}_p$ also do not satisfy the principle of material frame indifference. One therefore should use elasto-plastic constitutive equations that are material frame indifferent, thus automatically accounting for the case of large deformations. This of course would be associated with a substantial increase in the numerical effort, which does not appear to be justified for small deformations. In order to motivate the use of the more convenient formulation of Eq.(1), it is recalled that the stress $\boldsymbol{T}$ and the small strain $\boldsymbol{E}$ in Eq.(1) represent proper approximations (with respect to the floating reference configuration) of the second Piola-Kirchhoff stress tensor and the Green strain tensor, respectively. For the order of approximation see Gurtin (1972); compare also Lecture 5 of the present book, Irschik, Pichler, Nader and Zehetner (2004). Constitutive formulations connecting the second Piola-Kirchhoff stress tensor and the Green strain tensor do satisfy the principle of material frame indifference. The error introduced by small strain plasticity into the floating frame of reference formulation can also be nicely studied by comparison to a formulation that uses a reference configuration fixed in the inertial frame, a so-called *absolute coordinate formulation*, see Gerstmayr (2003a). In the latter reference it has been shown that the absolute coordinate formulation, which satisfies the principle of material frame indifference, deviates from the floating frame of reference formulation by quadratic terms of small deformations relative to the floating reference configuratione. Hence, for small displacements and small strains with respect to the floating reference configuration, it appears to be justified to approximate the constitutive equations by the classical formulation of small-strain elasto-plasticity, Eq.(1), using a floating reference configuration that is as close to the actual configuration as possible.

We now shortly specify the small strain elasto-plastic constitutive equations. In case of an *isotropic* tensor of elasticities $\boldsymbol{C}$, we have

$$\boldsymbol{C} = \lambda\, \mathbf{1}\,\mathbf{1} + 2\,\mu\, \boldsymbol{I}, \tag{5}$$

where the second order identity tensor is denoted as $\mathbf{1}$, the fourth order tensor $(\mathbf{1}\,\mathbf{1})$ represents the dyadic product of $\mathbf{1}$ with itself, and the fourth order identity tensor is denoted as $\boldsymbol{I}$, see Chapter 2 of Simo and Hughes (1998). The Lame elasticity parameters are denoted as μ and λ. The parameter μ is equal to the shear modulus G. The Lame parameters are related to Young's modulus Y and Poisson's ratio ν by

$$\mu = G = \frac{Y}{2\,(1+\nu)}, \quad \lambda = \frac{\nu\, Y}{(1+\nu)(1-2\nu)} . \tag{6}$$

Classical plasticity is associated with the deviatoric part $\boldsymbol{T}'$ of the total stress $\boldsymbol{T}$,

$$\boldsymbol{T}' = \boldsymbol{T} - \tfrac{1}{3}(tr\,\boldsymbol{T})\,\mathbf{1} = \boldsymbol{T}_E + \boldsymbol{T}_D - \tfrac{1}{3}(tr\,\boldsymbol{T}_E + tr\,\boldsymbol{T}_D)\,\mathbf{1} . \tag{7}$$

The trace of a second order tensor is indicated by *tr*. The *Huber-v. Mises yield condition* reads

$$f(\boldsymbol{T}') = tr\,(\boldsymbol{T}' \cdot \boldsymbol{T}') - \tfrac{2}{3}\,k^2 \le 0 , \tag{8}$$

where k is the yield stress. The associative *flow rule of Levy and St. Venant* states that the co-moving time rate of $\boldsymbol{E}_p$ is perpendicular to the yield surface,

$$\boldsymbol{E}_{p,t} = \beta \frac{\partial f(\boldsymbol{T}'(\boldsymbol{T}))}{\partial \boldsymbol{T}} = \gamma\, \boldsymbol{T}' , \tag{9}$$

where β and γ are proper plastic multiplieres. The co-moving partial derivative with respect to time, holding the place in the floating reference configurations fixed, is denoted by $(\,)_{,t}$. In both, the elastic and the plastic range, the Kuhn-Tucker conditions must be fulfilled,

$$\gamma \ge 0, \quad f(\boldsymbol{T}') \le 0, \quad \gamma\, f(\boldsymbol{T}') = 0 . \tag{10}$$

In accordance with these relations, it follows that, when

$$f < 0 \Rightarrow \gamma = 0 , \tag{11}$$

such that we are in the elastic range. Now suppose that $f = 0$ in Eq.(8). When

$$f_{,t} < 0 \text{ and } \gamma = 0 , \tag{12}$$

then elastic unloading from the yield surface takes place. Plastic loading occurs for

$$f_{,t} = 0 \text{ and } \gamma > 0 . \tag{13}$$

The special case $f_{,t} = 0$ and $\gamma = 0$ is called a neutral loading. For a comprehensive discussion, see Chapter 2 of Simo and Hughes (1998).

At this point, we emphasize that a careful design of the corresponding numerical algorithm is required in order to take into account the peculiar spatial distribution of plastic strains $\boldsymbol{E}_p$ that may be encountered in the *plastic zones* of the body when a *reversed* motion or force-loading of an elasto-plastic body takes place. A plastic zone is a spatial region of the body with a non-vanishing distribution of plastic strains $\boldsymbol{E}_p$. Both, the spatial extension of a plastic zone in the body, as well as the distribution of $\boldsymbol{E}_p$ within the zone in general do vary during the course of the motion. It is essential to notice that the plastic strains then may be distributed in an extremely complex manner across the plastic zones. In order to visualise this latter fact, Fig.1 shows the distribution of plastic strains $\boldsymbol{E}_p$ in an elasto-plastic element of a slider-crank mechanism after several cycles of reversed motion, computed by the method presented in Gerstmayr and Irschik (2003) and Gerstmayr (2003). A three-dimensional view upon the plastic parts of strain in the mid-plane of the element is given in Fig.1, the in-plane coordinates referring to axial and thickness extension (length versus height). The wavy form of this distribution with several sharp mountain-like peaks and valleys is due to reversed plasticity. In order to consider the influence of this distribution upon the overall behaviour of the body, a fine discretisation is required for $\boldsymbol{E}_p$ in the computational procedure. However, this does not mean that the body must be discretised by an equally fine mesh of Finite Elements, say, in order to accurately compute the displacements and stresses.

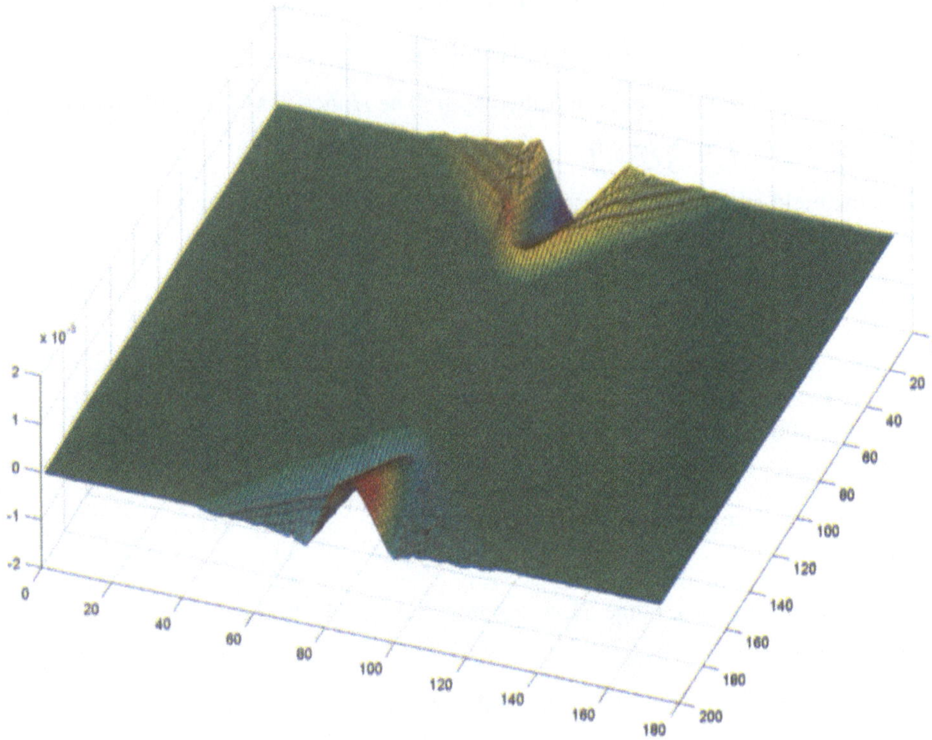

Figure 1: Plastic part of strain over length and height of an element of a slider-crank mechanism after several cycles of reversed plasticity.

3 A Numerical Procedure for Computing the Plastic Part of Strain

In a dynamic problem, the change of plastic strains within every timestep of a numerical computation procedure has to be comparatively small in order to get accurate solutions. Therefore, the use of a fixed-point iteration for the determination of plastic strains is of interest, which means that no elasto-plastic tangent matrix has to be re-computed in every timestep, but that the plastic multipliers are computed by means of a simple return mapping or predictor corrector scheme. The increase of plastic strains within every timestep can be formulated as

$$\int_t^{t+\Delta t} \boldsymbol{E}_{p,\tau}\, d\tau = \boldsymbol{E}_p(t+\Delta t) \pm \boldsymbol{E}_p(t) = \Delta\boldsymbol{E}_p(t+\Delta t) \approx \Delta t\, \boldsymbol{E}_{p,t}(t+\Delta t) \,, \tag{14}$$

where Δt is the time-step. The integral in Eq.(14) is approximated by means of the implicit Euler method, Eq.(14.2). The increment of plastic strain is computed at the end of the timestep, therefore the yield condition can be fulfilled at the end of a timestep (which is the initial value for the next timestep). We introduce the plastic multiplier for the whole timestep

$$\bar{\gamma} = \gamma \Delta t \,. \tag{15}$$

The increment of plastic strain may be approximated by means of the flow rule, Eq. (9), as

$$\Delta \boldsymbol{E}_p(t+\Delta t) \approx \bar{\gamma}\, \boldsymbol{T}'(t+\Delta t)\,. \tag{16}$$

An iterative procedure for the computation of plastic strains is suggested in the following, where the initial values are given at i=0, the last iteration is i and the actual computed iteration is i+1. Initial values for the deviatoric part of stress are given by the solution of the last timestep, the initial value for the increment of plastic strain is zero

$$\boldsymbol{T}'^{(0)}(t+\Delta t) = \boldsymbol{T}'(t)\,, \quad \Delta \boldsymbol{E}_p^{(0)}(t+\Delta t) = 0\,. \tag{17}$$

At the end of the timestep the stress for the iteration (i) can be estimated as

$$\begin{aligned} \boldsymbol{T}_{est}^{(i+1)}(t+\Delta t) &= \mathcal{C}[E^{(i)}(t+\Delta t) - E_p(t) - \Delta \boldsymbol{E}_p^{(i)}(t+\Delta t)] \\ &\approx \mathcal{C}[E^{(i)}(t+\Delta t) - E_p(t) - \bar{\gamma}^{(i+1)}\, \boldsymbol{T}'^{(i)}(t+\Delta t)]\,. \end{aligned} \tag{18}$$

Inserting the estimate given in Eq.(18.2) into the yield condition, Eq. (9), leads to a quadratic equation for the unknown plastic multiplier $\bar{\gamma}^{(i+1)}$. This leads to an exact fulfillment of the yield condition, Eq.(8), compared to the standard procedure using elasto-plastic tangent matrices. The iterative procedure (i=0, 1, ...) takes the following 4 steps:

1) solve the elastic problem ($\boldsymbol{T}^{(i)}$, $\boldsymbol{E}^{(i)}$) using an estimate for plastic strains $\boldsymbol{E}_p^{(i)}$, see Section 4 below for the equations of motion
2) check the Kuhn-Tucker conditions; compute the residual; if residual < tolerance, then go to step 4
3) compute plastic multipliers from the yield condition, Eq. (9), taking the estimate for the stress $\boldsymbol{T}_{est}^{(i+1)}$, Eq. (18); go back to step 1
4) take plastic strains as initial value for next timestep, go to next timestep

In the case of 2-dimensional problems, the estimate for the stress, Eq.(18), can be formulated by means of simple matrix operations using a formulation by Simo and Hughes (1998). The quadratic equation for the yield condition can be written down explicitly, and may be used for an efficient simulation of elasto-plastic multibody systems, for a discussion see Gerstmayr (2003b).

4 Equations of Motion in Case of a Given Plastic Strain

We now turn to the equations of motion of the single body, which form a *set of partial differential equations*, imposed by the local form of the mechanical balance equations in the inertial frame, see Lectures 1 and 2 of the present book, Belyaev (2004) and Irschik (2004). It is a common computational strategy to put these partial differential equations into the approximate form of a *set of ordinary differential equations in time*, which then is integrated numerically. In the floating frame of reference formulation, this approximate form is reached by means of a *Ritz approximation* for the displacement vector $\boldsymbol{u}$ from the floating reference configuration, see also Lecture 3 of the present book, Irschik, Holl and Hammelmüller (2004). In the following, the components of mechanical vectors and second-order tensors, such as $\boldsymbol{u}$ and

$\boldsymbol{T}_D$, are assumed to be expressed in a Cartesian coordinate system that is fixed in the co-moving frame, and that is called the *floating system*. The components of the symmetric mechanical tensors of second order considered above are written as column matrices containing their components in the floating coordinate system. This mapping has become standard in the Finite Element method, and it is listed e.g. in Chapter 4.3 of Simo and Hughes (1998). The symbols for mechanical vectors and tensors written subsequently thus are understood as the (3x1) column matrix of Cartesian components of a mechanical vector, or as the (6x1) column matrix of Cartesian components of a mechanical second order tensor, respectively. Notations and multiplication rules of matrix algebra are used, and quite naturally there emerge matrices that are not (3x1) or (6x3), respectively. In what follows, we try to extend the floating frame of reference formulation given by Shabana (1998) with respect to plastic parts of strain.

In the sense of the Ritz approximation, the displacement vector $\boldsymbol{u}$ is approximated as

$$\boldsymbol{u} = \boldsymbol{u}(\boldsymbol{P}, t) = \boldsymbol{S}\,\boldsymbol{q}_f, \tag{19}$$

where $\boldsymbol{S}$ is a matrix of time-invariant Ritz Ansatz-functions, depending on the position vector $\boldsymbol{P}$ of the particle in the floating reference configuration, and $\boldsymbol{q}_f = \boldsymbol{q}_f(t)$ is a column matrix of time-dependent generalised coordinates, called the *flexible coordinates*. The vector $\boldsymbol{P}$ is the position vector of the place of a material particle in the floating reference configuration, expressed in the floating system. Hence, the vector $\boldsymbol{P}$ does not change with time. In order to describe the motion of the floating reference configuration with respect to the inertial frame, we need to know the position vector of the origin of the floating system, which conveniently is formulated by its Cartesian components in a coordinate system fixed in the inertial frame, the latter being called the *inertial system*. Furthermore, in order to locate the orientation of the floating system with respect to the inertial one, angles of rotation need to be introduced. For this sake, several celebrated formulations are at our disposal, such as the three Euler angles or the four Euler parameters, the latter being connected by an additional constraint equation. For a comprehensive exposition on this reference kinematics, see Chapter 2 of Shabana (1998). Let the column vector of the Cartesian components of the origin of the floating system, expressed in the inertial system, be denoted as $\boldsymbol{r}^* = \boldsymbol{r}^*(t)$, and let $\boldsymbol{\theta} = \boldsymbol{\theta}(t)$ denote the column matrix of the Euler parameters or angles. The *generalized coordinates* describing the motion of the body then are formed by the column matrix

$$\boldsymbol{q} = \boldsymbol{q}(t) = [\boldsymbol{r}^{*T} \;\; \boldsymbol{\theta}^T \;\; \boldsymbol{q}_f^T]^T, \tag{20}$$

the transpose of a matrix being denoted by the superscript T. In the following, the representation of a mechanical vector as a column matrix in the inertial frame is denoted by the superscript (*). Occasionally, we also will use this superscript in matrices needed for the further transformation between the two coordinate systems. The actual position vector of a material particle, when expressed as a column vector in the inertial system, becomes

$$\boldsymbol{p}^* = \boldsymbol{p}^*(\boldsymbol{P}, \boldsymbol{q}(t)) = \boldsymbol{r}^* + \boldsymbol{A}^*(\boldsymbol{P} + \boldsymbol{u}), \tag{21}$$

where $\boldsymbol{A}^* = \boldsymbol{A}^*(\boldsymbol{\theta})$ denotes the rotation matrix needed for transforming vectors written in the floating system into the inertial one. The absolute velocity vector of a particle, when expressed as a column matrix in the inertial frame, follows as

$$\boldsymbol{v}^* = \frac{\partial \boldsymbol{p}^*(\boldsymbol{P}, \boldsymbol{q}(t))}{\partial t} = \frac{\partial \boldsymbol{p}^*}{\partial \boldsymbol{q}}\dot{\boldsymbol{q}} = \boldsymbol{L}\,\dot{\boldsymbol{q}}. \tag{22}$$

The non-square matrix $\boldsymbol{L} = \partial \boldsymbol{p}^* / \partial \boldsymbol{q}$ represents the Jacobian matrix that contains the partial derivatives of the 3 components of $\boldsymbol{p}^*$ with respect to the generalized coordinates, holding the place $\boldsymbol{P}$ in the floating reference configuration fixed. This gives

$$\boldsymbol{L} = \boldsymbol{L}(\boldsymbol{P}, \boldsymbol{q}(t)) = [\mathbf{1} \quad (-\boldsymbol{A}^* \, Skw(\boldsymbol{P} + \boldsymbol{S}\,\boldsymbol{q}_f)\,\boldsymbol{G}^*) \quad (\boldsymbol{A}^* \boldsymbol{S})]\,. \tag{23}$$

In Eq.(23), $\mathbf{1}$ denotes the (3x3) identity matrix, and *Skw* stands for the skew-symmetric (3x3) matrix formed from the 3 Cartesian components of its arguments. The possibly non-square matrix $\boldsymbol{G}^* = \boldsymbol{G}^*(\boldsymbol{\theta})$ connects the column matrix of the time derivatives $\dot{\boldsymbol{\theta}}$ with the vector of angular velocity $\boldsymbol{\omega}^*$ of the floating reference configuration, as expressed in the inertial system,

$$\boldsymbol{\omega}^* = \boldsymbol{\omega}^*(t) = \boldsymbol{G}^* \,\dot{\boldsymbol{\theta}}\,. \tag{24}$$

Likewise to $\boldsymbol{A}^*$, the proper form of $\boldsymbol{G}^*$ can be found in the textbooks of dynamics. The column matrix

$$\dot{\boldsymbol{q}} = \dot{\boldsymbol{q}}(t) = \frac{\partial \boldsymbol{q}(t)}{\partial t} = [\dot{\boldsymbol{r}}^{*T} \quad \dot{\boldsymbol{\theta}}^T \quad \dot{\boldsymbol{q}}_f^T]^T \tag{25}$$

in Eq.(22) contains the generalized velocities, representing the time derivatives of $\boldsymbol{q}$. Introducing the formulation for the absolute velocity, Eq.(22), into the formula for the kinetic energy $\mathcal{K}$ and integrating over the body in the floating reference configuration, we obtain an expression of the form

$$\mathcal{K} = \frac{1}{2}\,\dot{\boldsymbol{q}}^T \boldsymbol{M}\,\dot{\boldsymbol{q}}\,. \tag{26}$$

see Eq.(16) of Lecture 2 of the present book, Irschik (2004), for a definition of $\mathcal{K}$. In Eq.(26), the mass matrix is given by the following integral over the volume $\mathcal{V}$ of the body in the reference configuration:

$$\boldsymbol{M} = \int_{\mathcal{V}} \boldsymbol{L}^T \boldsymbol{L}\,\rho\, dv\,. \tag{27}$$

The mass density is denoted by ρ, and it may be usually treated as a constant in solid structures.

Applying the above Ritz-type formulation in the context of the *Lagrange equations*, see Lecture 3 of the present book, Irschik, Holl and Hammelmüller (2004), the equations of motion of the elasto-plastic body are eventually obtained as

$$\boldsymbol{M}\,\ddot{\boldsymbol{q}} = \boldsymbol{Q}_v + \boldsymbol{Q}_{ext} + \boldsymbol{Q}_{int}\,, \tag{28}$$

forming the desired set of ordinary differential equations in time. For the derivation of Eq.(28), and for details about the form of the mass matrix $\boldsymbol{M}$, the reader is referred to the thorough exposition given in Chapter 5 of Shabana (1998). When the coordinates $\boldsymbol{q}$ are not independent from one another, e.g. in the case of the four Euler parameters, or when there are other constraint equations to be considered, e.g. joints or contacts between adjacient bodies, these constraints must be taken into account in Eq.(28) in the form of *additional constraint forces*. This can be conveniently done using the method of *Lagrange multipliers*. For details of the latter formulation, the reader is referred to Chapter (3.4) of Shabana (1998).

In Eq.(28), the generalized force $\boldsymbol{Q}_v$, the so-called *quadratic velocity vector*, results from the differentiation of the kinetic energy $\mathcal{K}$ of Eq.(26) with respect to time and with respect to the generalized coordinates, and it accounts for the gyroscopic and the Coriolis effects,

$$Q_v = -\dot{M}\dot{q} + [\frac{\partial}{\partial q}\mathcal{K}]^T . \tag{29}$$

The partial derivative of $\mathcal{K}$ with respect to q in Eq.(23) is meant as the row matrix containing the derivatives of $\mathcal{K}$ with respect to the single components of q. The quadratic velocity *vector* Q_v is non-linearly coupled in both, the generalized coordinates and velocities, see again Chapter 5 of Shabana for the detailed form of Q_v. The generalized force Q_{ext} in Eq.(28) refers to the external forces, and it can be computed from a comparison with the *virtual work of the external forces,* which is done when the generalized coordinates are varied according to the rule of forming a variation, see Eq.(5) of Lecture 3 of the present book, Irschik, Holl and Hammelmüller (2004). The variation of the position vector of a particle in its actual configuration follows to

$$\delta p^* = \frac{d}{d\lambda}[p^*(q + \lambda\,\delta q)]\Big|_{\lambda=0} = [\frac{\partial p^*}{\partial(q + \lambda\,\delta q)}\frac{\partial(q + \lambda\,\delta q)}{\partial\lambda}]\Big|_{\lambda=0} = L\,\delta q , \tag{30}$$

with the Jacobian matrix L of Eq.(23). The virtual work of the external forces is obtained from the formula given for the power of the external forces, cf. Eq.(16) of Lecture 2 of the present book, Irschik (2004), by replacing the velocity v by the variation δp^*, such that we may write

$$\delta W_{ext} = \int_V b^{*T}\,\delta p^*\,dv + \oint_S t_{(n)}^{*T}\,\delta p^*\,da = Q_{ext}^T\,\delta q . \tag{31}$$

Inserting Eq.(24), the generalized forces accounting for the external forces then follow as

$$Q_{ext}^T = \int_V b^{*T} L\,dv + \oint_S t_{(n)}^{*T} L\,da . \tag{32}$$

The generalized forces accounting for the internal forces again are obtained from a comparison of virtual work terms. According to Eqs.(3) and (4), the stress is split into an elastic and a dissipative part. We first consider the elastic part, T_E. In the formula for the power of the internal forces stated in Eq.(17) of Lecture 2 of the present book, Irschik (2004), we replace the velocity by the variation of the displacement vector δu from the floating reference configuration. We then obtain the *virtual work of the elastic part of stress*, $\delta W_{int,E}$, where we are allowed to express the components of the matrices in the floating system for convenience. The gradient of δu can be replaced by its symmetric part, since T_E is symmetric. Consistent with Eq. (19), we may write

$$\delta u = S\,\delta q_f, \quad \mathrm{sym}\,grad\,\delta u = \hat{S}\,\delta q_f, \tag{33}$$

where sym $grad\,\delta u$ is mapped as a (6x1) column matrix, and the square matrix $\hat{S}$ emerges from the application of the respective tensor operations to the vector δu. Hence,

$$\delta W_{int,E} = -\int_V T_E^T \hat{S}\,dv\,\delta q_f = Q_{int,E}^T\,\delta q , \tag{34}$$

such that the generalised forces due to the elastic part of stress follow by comparison to

$$Q_{int,E} = [0^T \quad 0^T \quad -\int_V T_E^T \hat{S}\,dv]^T . \tag{35}$$

In analogy, we may write for the *virtual work of the dissipative part of stress* T_D that

$$\delta W_{int,D} = -\int_V T_D^T \hat{S}\, dv\, \delta q_f = Q_{int,D}^T\, \delta q\,. \tag{36}$$

We thus obtain

$$Q_{int,D} = [0^T \;\; 0^T \;\; -\int_V T_D^T \hat{S}\, dv]^T. \tag{37}$$

An important distinction has to be made when computing the generalised forces $Q_{int,D}$ and $Q_{int,E}$ in Eqs.(37) and (35), respectively. When we map the symmetric fourth order tensor of elasticities C as a (6x6) matrix of components in the floating system according to the rules given in Chapter 4.3 of Simo and Huges (1998), the elastic stress can be written as

$$T_E = C\hat{S}q_f. \tag{38}$$

Hence, the elastic part of the generalised internal forces follows in the simple form

$$Q_{int,E} = [0^T \;\; 0^T \;\; -q_f^T K^T]^T, \tag{39}$$

see Eq.(29), where the *elastic stiffness-matrix* K is given by

$$K = \int_V \hat{S}^T C\hat{S}\, dv. \tag{40}$$

No simple expression like Eq.(38), which is linear in the flexible coordinates, is available for the dissipative part of stress in Eq.(37), since the latter represents a mapping of the plastic part of strain E_p according to Eq.(4), $T_D = -C[E_p]$, the time-evolution of E_p in turn following according to the constitutive relations and the numerical algorithm presented in Sections 2 and 3 above. As was exemplary demonstrated in Fig. 1, the distribution of E_p within the plastic zones may be highly complex in the case of a reversed loading. On the other hand side, plastic strains exist only within the plastic zones, such that the volume integration in Eq.(37) does not extend over the whole body. Hence, in order to be integrated properly, the dissipative stress and the plastic strain in Eq.(37) must be evaluated in a fine net of integration points throughout the plastic zones, the latter changing their extent in the course of time.

We end this exposition on the role of plasticity in computational algorithms for bodies with rigid-body degrees-of-freedom by recalling that the Ritz-Ansatz technique introduced in Eq.(19) is quite directionally linked to the *Finite Element* formulation, which is widely used in engineering practice. The above formulas e.g. might be understood as the equations describing a single finite element. That E_p has to be evaluated in a fine mesh of points of course does not mean that the body must be subdivided into a large number of Finite Elements.

References

Ambrósio, J.A.C. (2001). Geometric and material nonlinear deformations in flexible multibody systems. In: Ambrósio, J.A.C. and Kleiber, M., eds., *Computational Aspects of Nonlinear Structural Systems with Large Rigid Body Motion*, Nato Science Series, IOS Press 2001.

Ambrósio, J.A.C., and Nikravesh, P.E. (1992). Elasto-plastic deformations in multibody dynamics. *Nonlinear Dynamics*, Vol. 3.

Belyaev, A.K. (2004). Basics of Continuum Mechanics. In: Irschik, H., and Schlacher, K., eds., *Advanced Dynamics and Control of Structures and Machines*. CISM Courses and Lectures No. 444, Wien-New York: Springer-Verlag.

Carlson, D.E. (1972). Linear Thermoelasticity. In: *Handbuch der Physik*, Vol. VIa/2. Berlin: Springer-Verlag.

Gerstmayr, J. (2003). An adaptive method for the dynamics of elasto-plastic multibody systems. *Journal of Mechanics Based Design of Structures and Machines,* 31: 201-227.

Gerstmayr J. (2003a). Comparison of the absolute nodal coordinate and the floating frame of reference formulation by means of a simplified strain formulation. *Proceedings of DETC'03 ASME Design Engineering Technical Conferences*, Chicago, Illinois, USA, CD-ROM Proceedings, ASME Paper No. VIB-48306.

Gerstmayr J. (2003b). The absolute nodal coordinate formulation with elasto-plastic deformations. *Proceedings of ECCOMAS Thematic Conference on Advances in Multibody Dynamics 2003*, Lissabon, Portugal, Paper No. MB2003-007.

Gerstmayr, J., Holl, H. J. and Irschik, H. (2001). Development of plasticity and damage in vibrating structural elements performing guided rigid-body motions, *Archive of Applied Mechanics*, 71: 135-145.

Gerstmayr, J., and Irschik, H. (2001). Control of an elasto-plastic pendulum. *Proceedings of DETC'01, ASME Design Engineering Technical Conferences*, Pittsburg, PE; USA; CD-Rom Proc., ASME Paper No.VIB-21600.

Gerstmayr,J., and Irschik,H. (2002). Vibrations of the elasto-plastic pendulum, *International Journal of Nonlinear Mechanics* 38: 111-122.

Gurtin, M. E. (1972). The Linear Theory of Elasticity. In: *Handbuch der Physik*, Vol. VIa/2. Berlin: Springer-Verlag.

Haupt, P. (2000). *Continuum Mechanics and Theory of Materials*. Berlin: Springer - Verlag.

Irschik, H. (2004). A Treatise on the Equations of Balance and on the Jump Relations in Continuum Mechanics. In: Irschik, H., and Schlacher, K., eds, *Advanced Dynamics and Control of Structures and Machines*. CISM Courses and Lectures No. 444, Wien-New York: Springer-Verlag.

Irschik, H., Pichler, U., Nader, M., and Zehetner, Ch. (2004). Compensation of Deformations in Elastic Solids and Structures in the Presence of Rigid-Body Motions. In: Irschik, H., and Schlacher, K., eds, *Advanced Dynamics and Control of Structures and Machines*. CISM Courses and Lectures No. 444, Wien-New York: Springer-Verlag.

Irschik, H., Holl, H.J., Hammelmüller, F. (2004). The Rayleigh-Ritz Technique and the Lagrange Equations in Continuum Mechanics: Formulations for Material and Non-Material Volumes. In: Irschik, H., and Schlacher, K., eds., *Advanced Dynamics and Control of Structures and Machines*. CISM Courses and Lectures No. 444, Wien-New York: Springer-Verlag.

Irschik, H., and Ziegler, F. (1995). Dynamic processes in structural thermo-viscoplasticity. *Applied Mechanics Reviews*, 48: 301-315.

Irschik, H., Fotiu, P., and Ziegler, F. (1993). Extension of Maysel's formula to the dynamic eigenstrain problem. *Journal of Mechanical Behavior of Materials*, 5: 59-66.

Irschik, H. (1986). Biaxial Dynamic Bending of Elastoplastic Beams. *Acta Mechanica*, 62: 155-167.

Mura, T. (1991). *Micromechanics of Defects in Solids*, Second Ed., Kluwer Academic Publisher.

Palmov, V. (1998). *Vibrations of Elasto-Plastic Bodies*. Berlin-Heidlberg: Springer-Verlag.

Parkus, H. (1976). *Thermoelasticity*, 2nd Ed. Wien-New York: Springer-Verlag.

Shabana, A.A., (1998). *Dynamics of Multibody Systems*, 2nd Ed. Cambridge: University Press.

Simo, J.C., and Hughes, T.J.R. (1998). *Computational Inelasticity*. New York: Springer-Verlag.

Acknowledgement:

Support of the present work in the framework of FWF-Project No.P15195 and LCM-Project 4.4 is gratefully acknowledged.

High Frequency Dynamics of Engineering Structures

Alexander K. Belyaev

Department of Mechanics and Control, Polytechnic University of St. Petersburg, Russia

Abstract. This paper considers the high frequency dynamics which links vibration theory and thermodynamics. It is shown that the high frequency dynamics is the high frequency limit of vibration theory and the low frequency limit of thermodynamics. Two approaches to the high frequency dynamics of complex engineering structures are proposed. The first one is high frequency structural analysis. Its boundary value is derived from vibration theory under the assumption that the structures which have high modal overlap possess some thermodynamic properties. The second approach is the vibrational conductivity approach to high frequency dynamics which is governed by a boundary value problem of the heat conduction type. The latter has been modified to incorporate a heat sink term so that the spatial resonant absorption of vibration is taken into account.

At high frequencies the structures are shown to behave like mechanical systems with a continuous spectrum of natural frequencies and considerable spatial decay. The effect of the secondary systems' vibration on the vibrational field in the structure becomes evident. The secondary systems act as dynamic absorbers on the primary structure which, in the case of high modal overlap, provides considerable spatial absorption of high frequency vibration. The secondary systems may be damaged since they absorb vibrational energy at their natural frequencies. This provides the main motivation for interest in high frequency dynamics. A comparison of the properties of high frequency vibration with those of low frequency vibration and thermal vibration is provided.

1 Introduction

There exist many different types of dynamics. The types of dynamics related to this study are shown in Figure 1 together with their frequency range of application. Historically, the dynamics of rigid bodies (I. Newton, 1687 and L. Euler, 1736) was the first, and, from the point of view of vibration theory, this dynamics deals with zero eigenfrequencies. The dynamics of solids was developed a century later. Many names might be mentioned, however it was M. Duhamel, 1843, see Timoshenko, 1983, who began the systematic study of the vibration of solids. The frequencies, which this dynamics deals with, can be referred to as the natural frequencies of the solid under consideration. Thermodynamics should also be considered as a relevant dynamics in which the frequencies of thermal motions are essentially higher than the frequencies of mechanical vibrations. Classical thermodynamics was developed by J. Maxwell, 1867 and L. Boltzmann, 1872. They proposed a kinetic-statistical approach to thermal processes known as the "mechanical theory of heat" which linked thermodynamic processes to dynamics, cf. Truesdell, 1984. Quantum thermodynamics as developed by M. Planck, 1900, see e.g. Riedi 1976, should also be considered as a relevant dynamics. The quantum effects are considerable only at very high frequencies, at least at the frequencies essentially higher than those of the thermal motion of the molecules.

The dynamics of rigid body can be derived from vibration theory since the rigid body motions correspond to zero eigenfrequencies within vibration theory. The equations governing the dynamics of solids can be obtained from rigid body dynamics provided that the solid is modelled by a regular spatial array of elemental mass-spring systems. In fact, one of the earliest works on the vibration of distributed, continuous systems performed by J. Lagrange in 1788 was based upon a discrete element model, see Fahy, 1985. This means that the dynamics of rigid bodies and the dynamics of solids can be considered as two neighbouring dynamics. Classical thermodynamics and quantum thermodynamics may also be considered as two neighbouring dynamics. For example, the Planck formula for radiation links the Rayleigh-Jeans formula at low radiation frequencies and the Wien formula at high radiation frequencies when the quantum effects are considerable, cf. Riedi 1976.

To the best of the author's knowledge, all attempts to derive the dynamics of solids directly from thermodynamics and visa versa have been rather awkward. One of a few successful applications of the vibration theory of solids to thermodynamics is the Debye theory of the low-temperature lattice heat capacity, e.g. Riedi 1976. In this theory the discrete lattice was approximated as an elastic continuum in the long wavelength region and an appropriate number of long wavelength modes were taken to derive an equation for the heat capacity. The difficulties encountered when attempting to link the dynamics of solids to thermodynamics indicate that there exists a gap between these theories which "reserves" a place for a new dynamics. The latter is in fact a low frequency limit of thermodynamics and a high frequency limit of the dynamics of solids. The new dynamics that fills this gap will be referred to here as high frequency dynamics, and the corresponding frequencies will be referred to as high frequencies of mechanical vibration or, for short, high frequencies, see Figure 1.

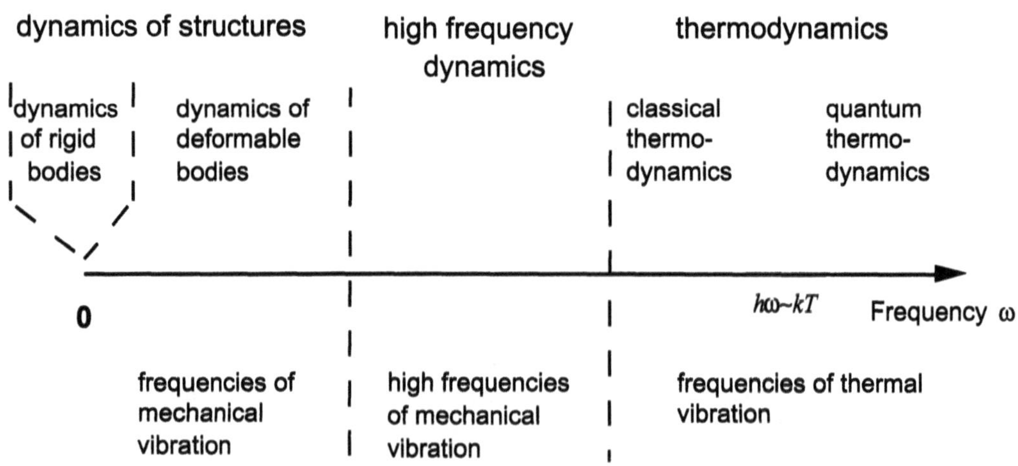

Figure 1. The different types of dynamics including high frequency dynamics

From a historical perspective, the development of dynamics is a kind of evolution of our knowledge about dynamic processes in natural science. Similar to the theory of human evolution, a missing link is seen to exist in the evolution of dynamics. At present this statement is no more than a mere allegation, however this allegation presents us with some conclusions. If high frequency dynamics is the high frequency limit of vibration theory and the low frequency limit of thermodynamics then the conventional methods of vibration theory and thermodynamics can be applied there. In particular, the governing equations of high frequency dynamics can be derived from these two different types of dynamics. Deriving the governing equations and properties of the new dynamics is a challenging problem since this implies the coupling of mechanical and thermal properties.

An attempt will be made here to derive the boundary value problem for high frequency dynamics of complex engineering structures. The main conclusions are of the basic character and they are certainly valid for such traditional structural elements as rods, beams, shells, and solids in general. The dynamics of complex structures is of great practical interest for engineers and provides a very challenging and interesting problem. The analysis however will not be restricted to complex structures, the necessary references and discussions for rods, beams and plates will also be provided.

In order to prove the existence of the new dynamics one has to consider the following: (i) what new qualities does this new dynamics possess compared with vibration theory and thermodynamics? (ii) the governing equations must be derived; (iii) the boundaries of the high frequency dynamics should be indicated. Answering these points is the objective of the present study.

2 Typical for complex structures

Complex mechanical structures like ships, buildings, spacecraft and aircraft are actually assemblies of substructures. Attempts to pedantically describe all details and peculiarities of real structures are doomed to failure for the following reasons. Firstly, at higher frequencies, the existence of many inherently uncontrolled factors plays a principle role. Review papers by Ibrahim 1987 and Fahy 1994 give a very deep insight into the problem of uncertainties in dynamics. In structural dynamics, uncertainties arise from stiffness, mass and damping fluctuations caused by variations in material properties as well as variations resulting from manufacturing and assembly. The latter factor causes vagueness in the boundary conditions for each structural member since the high-frequency dynamic properties of joints between structural members are especially uncertain. Secondly, the essential heterogeneity and presence of complicated interiors, i.e. secondary systems, have to be taken into account. Thirdly, even if it were possible to obtain an "exact" boundary-value problem in which all the complexity of the structure was taken into account and we could solve this problem, the very interpretation of this "exact" result would present great difficulty. The reason for this is that the field of vibration of a complex structure (for instance, under a broad-band excitation) is a very complicated function of time and spatial coordinates since a great many modes are excited in the structure. Summarising, one can say that the blind extension of conventional methods of modal analysis to higher frequencies at the expense of computational cost reaches a deadlock because the results become unreliable and "indigestible".

Another typical feature of complex structures is that all modern structures are weakly coupled. Any engineering structure is in fact an assembly of substructures attached to one another or to the framework only at several points. As a result, the complex structures possess such low global dynamic rigidity that only the first, the second and seldom the third global resonances are observed in such structures. Beyond the region of these few global resonances the vibrations localise within the structural members or groups of structural members.

On the other hand, the review of the literature in structural dynamics suggests that present-day analytical methods are too idealised and do not adequately reflect the inherent complexity of structures. Complex engineering structures are often taken to consist of a few structural members of well-studied shapes (shells, beams etc.) otherwise lumped-mass models are applied. The consideration of secondary systems is commonly avoided in the literature. This avoidance is even more surprising taking into account the fact that the secondary systems actually comprise the major portion of the structural members of a unit.

3 Derivation of high frequency dynamics from vibration theory

In this part we derive the boundary problem of high frequency dynamics by means of modal analysis and the Hamilton variational principle of dynamics, see Palmov 1979, Belyaev and Palmov 1986 and Belyaev 1991.

We consider an actual mechanical structure V which is assembled from a great number of structural members, each having volume V_n and boundary surface S_n $(n=1,2,...N)$. The structural members are attached to one another or to the framework. The sizes of the structural members V_n are much smaller than those of the entire structure V. Nevertheless, the structural members may be complex dynamic system themselves, e.g. various devices.

Consider a representative structural member V_n, for short, a substructure V_n. Because of the localization of vibration within the structural members, the vector of the absolute displacement of the substructure $\boldsymbol{u}_n(\boldsymbol{r},t)$ may be sought in the form of an expansion in terms of the substructure's normal modes $\boldsymbol{u}_{nk}(\boldsymbol{r})$

$$\boldsymbol{r} \in V_n; \quad \boldsymbol{u}_n(\boldsymbol{r},t) = \sum_{k=1}^{\infty} \boldsymbol{u}_{nk}(\boldsymbol{r})\, q_{nk}(t) + \boldsymbol{U}(\boldsymbol{r},t) \tag{3. 1}$$

where $q_{nk}(t)$ is the k-th generalised coordinates of the substructure V_n , $\boldsymbol{r}$ is the reference position and t is time. The function $\boldsymbol{U}(\boldsymbol{r},t)$ is usually introduced in problems of mathematical physics to improve the convergence in the vicinity of a boundary, cf. Mindlin and Goodman, 1950. The reason for introducing $\boldsymbol{U}(\boldsymbol{r},t)$ in equation (3.1) is quite different. The objective of the present study is to reveal the basic laws governing the high frequency vibration in complex structures which is possible to realize only in the framework of an integral approach. To this end, the boundary value problem for $\boldsymbol{U}$ must be as simple as possible, that is, it is reasonable to require maximum smoothness of the function $\boldsymbol{U}(\boldsymbol{r},t)$ with respect to the spatial coordinate $\boldsymbol{r}$ within the whole structure.

Now the question of how to specify the substructures' normal modes arises. The normal modes are known to be sensitive to the boundary conditions, the latter being vague for any substructure, see Part 2. However, fortunately for structural dynamicists any set of normal modes is known to be complete and as shown in the treatise on the substructure synthesis method by Meirovitch 1980, the normal modes in energy space of substructures are not required to satisfy any boundary conditions in the internal boundaries of substructures. Hence, the normal modes may be chosen according to any suitable principle. For our analysis, the convenience of interpretation of the obtained results is of primary priority. For this reason, the normal modes are so specified that they vanish on the substructural boundaries S_n. In this case the function $\boldsymbol{U}(\boldsymbol{r},t)$ coincides with the actual displacement on the substructural boundaries S_n and besides, it is supposed to be a smooth function of $\boldsymbol{r}$ within the entire structure. Hence $\boldsymbol{U}(\boldsymbol{r},t)$ may be referred to as the displacement of the framework or the displacement of the primary structure.

The substructural normal modes $\boldsymbol{u}_{nk}(\boldsymbol{r})$ are chosen as nontrivial solutions of the following boundary-value problem

$$\begin{aligned} &\boldsymbol{r}\in V_n \quad \nabla\cdot\left[\boldsymbol{C}:\left(\nabla\boldsymbol{u}_{nk}\right)\right]+\rho\omega_{nk}^2\boldsymbol{u}_{nk}=0 \\ &\boldsymbol{r}\in S_n \quad \boldsymbol{u}_{nk}=0 \end{aligned} \tag{3.2}$$

Here $\rho(\boldsymbol{r})$ is the density, $\boldsymbol{C}(\boldsymbol{r})$ is the tensor of elastic moduli, ω_{nk} is k-th eigenfrequency of the substructure V_n, the symbols $\cdot$ and $:$ denote the scalar and double scalar product, respectively, and ∇ is the Hamilton operator. Since the normal modes of any substructure are known to be orthogonal, we normalise them to produce an orthonormal set of normal modes within each substructure

$$\boldsymbol{r}\in V_n; \qquad \int_{V_n}\rho\boldsymbol{u}_{nk}\cdot\boldsymbol{u}_{ns}\,dV=\delta_{ks}; \quad \int_{V_n}\left(\nabla\boldsymbol{u}_{nk}\right):\boldsymbol{C}:\left(\nabla\boldsymbol{u}_{ns}\right)dV=\omega_{nk}^2\,\delta_{ks} \tag{3.3}$$

The basic idea of our analysis is as follows. Equation (3.1) ensures that the actual vibrational field is split. From a mechanical point of view, the function $\boldsymbol{U}$ describes the gross effects which are relevant to the overall dynamic behaviour while the modal sum represents the local effects, such as vibration localization, parameter uncertainties, weak coupling and heterogeneities. From a thermodynamical point of view, $\boldsymbol{U}$ is a macroparameter, while the modal sum describes the microstates. High frequency vibration exhibits some thermodynamic properties however the number of degrees of freedom (microstates) is not large enough to ignore the modal vibrations and to specify the "vibrational state" in terms of a few macroscopic parameters. From a mathematical point of view, the method of multiple scales, Nayfeh 1981, is applied where $\boldsymbol{U}$ is a slow variable while the modal sum represents the fast variables.

The kinetic energy of the structure with consideration to the expansion (3.1) and the normal modes' properties (3.3) is as follows

$$T=\frac{1}{2}\sum_{n=1}^{N}\int_{V_n}\rho\dot{\boldsymbol{u}}_n\cdot\dot{\boldsymbol{u}}_n\,dV=\frac{1}{2}\int_V\rho\dot{\boldsymbol{U}}\cdot\dot{\boldsymbol{U}}\,dV+\frac{1}{2}\sum_{n=1}^{N}\sum_{k=1}^{\infty}\left[\dot{q}_{nk}^2+\dot{q}_{nk}\int_{V_n}\rho\boldsymbol{u}_{nk}\cdot\dot{\boldsymbol{U}}\,dV\right] \tag{3.4}$$

In view of the smoothness of $\boldsymbol{U}(\boldsymbol{r},t)$ and the essentially heterogeneous nature of the structure the following approximation appears to be valid

$$\int_V\rho\dot{\boldsymbol{U}}\cdot\dot{\boldsymbol{U}}\,dV=\sum_{n=1}^{N}\left(\dot{\boldsymbol{U}}\right)_n\cdot\left(\dot{\boldsymbol{U}}\right)_n\int_{V_n}\rho dV=\sum_{n=1}^{N}\langle\rho\rangle\left(\dot{\boldsymbol{U}}\right)_n\cdot\left(\dot{\boldsymbol{U}}\right)_n V_n=\int_V\langle\rho\rangle\dot{\boldsymbol{U}}\cdot\dot{\boldsymbol{U}}\,dV \tag{3.5}$$

where $\langle \rho \rangle = \frac{1}{V}\int_V \rho dV$ is an average density of the structure, i.e. an overall parameter. The latter equality in (3.5) corresponds to the standard transition from the Riemann-Stieltjes sum to the corresponding integral, which is admissible for large N. In the last sum appearing in equation (3.4) $\boldsymbol{U(r,t)}$ may be placed beyond the integral since it is a smooth function of $\boldsymbol{r}$, while $\rho(\boldsymbol{r})$ is a rapidly changing function of $\boldsymbol{r}$. If we introduce the average displacement of the centre of mass of the substructure V_n when it vibrates due to the normal mode $\boldsymbol{u}_{nk}(\boldsymbol{r})$, i.e.

$$\langle \boldsymbol{u}_{nk} \rangle = \frac{1}{\langle \rho \rangle_n V_n} \int_{V_n} \rho \boldsymbol{u}_{nk} dV ,$$

expression (3.4) can be rewritten in the following form

$$T = \int_V \langle \rho \rangle \dot{\boldsymbol{U}} \cdot \dot{\boldsymbol{U}} \, dV + \frac{1}{2} \sum_{n=1}^{N} \sum_{k=1}^{\infty} \left[\dot{q}_{nk}^2 + \langle \rho \rangle \langle \boldsymbol{u}_{nk} \rangle \cdot \dot{\boldsymbol{U}} V_n \dot{q}_{nk} \right] \tag{3.6}$$

The equation for the potential energy is obtained similarly, cf. Belyaev and Palmov 1986, and Belyaev 1991,

$$\Pi = \frac{1}{2} \int_V (\nabla \boldsymbol{U}) : \langle \boldsymbol{C} \rangle : (\nabla \boldsymbol{U}) \, dV + \frac{1}{2} \sum_{n=1}^{N} \sum_{k=1}^{\infty} \omega_{nk}^2 q_{nk}^2 \tag{3.7}$$

Here $\langle \mathbf{C} \rangle$ is an average tensor of elastic moduli, i.e. an overall parameter. The normal modes $\boldsymbol{u}_{nk}(\boldsymbol{r})$ form a complete set of functions. Hence, the function $\boldsymbol{U(r,t)}$ in equation (3.1) is "redundant", and one cannot derive a unique boundary problem for $\boldsymbol{U(r,t)}$. In order to obtain a unique boundary problem for $\boldsymbol{U(r,t)}$ one needs an additional condition. This condition has actually been imposed by equation (3.7), where it has been assumed that

$$\sum_{n=1}^{N} \sum_{k=1}^{\infty} q_{nk} \int_{V_n} (\nabla \boldsymbol{U}) : \boldsymbol{C} : (\nabla \boldsymbol{u}_{nk}) \, dV = 0 \tag{3.8}$$

Equation (3.7) implies that the primary structure and the internal degrees of freedom (i.e. secondary systems) are orthogonal in potential energy and exchange energy through the kinetic energy alone. Thus the primary structure and secondary systems excite each other kinematically rather than through the strength factors.

The work of the external loads is as follows

$$W = \sum_{n=1}^{N} \sum_{k=1}^{\infty} p_{nk} q_{nk} + \int_V \boldsymbol{h} \cdot \boldsymbol{U} \, dV + \int_S \boldsymbol{f} \cdot \boldsymbol{U} \, dS \tag{3.9}$$

where $\boldsymbol{h}$ and $\boldsymbol{p}$ are body and surface forces, respectively, and generalised forces p_{nk} are given by

$$p_{nk} = \int_{V_n} \boldsymbol{h} \cdot \boldsymbol{u}_{nk} \, dV \tag{3.10}$$

Assuming δU and δq_{nk} are independent variations and applying the Hamilton variational principle yields the following boundary value problem

$$\boldsymbol{r} \in V_n; \quad \nabla \cdot \left[\langle \boldsymbol{C} \rangle : (\nabla \boldsymbol{U}) \right] - \langle \rho \rangle \left[\ddot{\boldsymbol{U}} + \sum_{k=1}^{\infty} \langle \boldsymbol{u}_{nk} \rangle \ddot{q}_{nk} \right] + \boldsymbol{h} = 0 \tag{3.11}$$

$$\boldsymbol{r} \in V_n; \quad \ddot{q}_{nk} + 2\psi_{nk}\omega_{nk}\dot{q}_{nk} + \omega_{nk}^2 q_{nk} = p_{nk} \pm \langle \rho \rangle \langle \boldsymbol{u}_{nk} \rangle \cdot \ddot{\boldsymbol{U}} V_n; \ k = 1,2,\ldots\infty \tag{3.12}$$

$$\boldsymbol{r} \in S; \quad \boldsymbol{N} \cdot \left[\langle \boldsymbol{C} \rangle : (\nabla \boldsymbol{U}) \right] = \boldsymbol{f} \tag{3.13}$$

where $\boldsymbol{N}$ is the unit vector of the outward normal to the surface S. The modal damping is introduced into equation (3.12) via a dimensionless damping factor ψ_{nk}. Equation (3.11) governs the dynamics of the primary structure while expression (3.12) is in fact an infinite number of equations for the modal coordinates of the structural members. Further, equations (3.11) and (3.12) are coupled through inertial terms.

It is worth mentioning some alternative approaches leading to similar boundary value problems governing high frequency vibrations in engineering structures. In papers Palmov, 1969 and Belyaev, 1992, the primary structure is modelled by an elastic carrier medium while the secondary systems are represented by oscillators attached to the carrier medium. The general three-dimensional medium of complex structure has been reported by Palmov 1969. Modelling of extended complex structures by a Cosserat rod with a microstructure, cf. Belyaev, 1992, permits the consideration of high frequency longitudinal, torsional and bending vibrations. Complex structures modelled by one-dimensional random media are studied in papers by Belyaev 1993 and 1994. This approach has the merit that uncertainties in mechanical parameters and boundary conditions, as well as the heterogeneity of the complex structures are considered in the framework of the same approach. The Dyson integral equation has been applied in this papers which allows one to handle even essentially heterogeneous structures.

4 Time-reduced boundary value problem

Since the main objective of the present study is to study the dynamic properties of structures at high frequencies, it is convenient to consider the analysis in the frequency domain. Let us assume that the external loads can be represented by their spectral representations

$$\boldsymbol{h}(\boldsymbol{r},t) = \int_{\pm\infty}^{+\infty} \hat{\boldsymbol{h}}(\boldsymbol{r},\omega)\, e^{i\,\omega t}\, d\omega; \quad \boldsymbol{f}(\boldsymbol{r},t) = \int_{\pm\infty}^{+\infty} \hat{\boldsymbol{f}}(\boldsymbol{r},\omega)\, e^{i\,\omega t}\, d\omega \tag{4.1}$$

where $\hat{\boldsymbol{h}}(\boldsymbol{r},\omega)$ and $\hat{\boldsymbol{f}}(\boldsymbol{r},\omega)$ are corresponding spectra. Such a representation is known to be suitable for the analysis of deterministic, random, stationary, nonstationary, linear and nonlinear vibration. Thus the solution of the boundary problem (3.11)-(3.13) is sought in the following form

$$\boldsymbol{U}(\boldsymbol{r},t) = \int_{-\infty}^{+\infty} \hat{\boldsymbol{U}}(\boldsymbol{r},\omega)\, e^{i\,\omega t}\, d\omega; \quad q_{nk}(t) = \int_{-\infty}^{+\infty} \hat{q}_{nk}(\omega)\, e^{i\,\omega t}\, d\omega \tag{4.2}$$

Substituting equations (4.1) and (4.2) into equations (3.11)-(3.13) results in the following boundary problem for the spectra

$$r\in V_n;\quad \nabla\cdot\left[\langle C\rangle:\left(\nabla\hat{U}\right)\right]+\langle\rho\rangle\omega^2\left(\hat{U}+\sum_{k=1}^{\infty}\langle u_{nk}\rangle\hat{q}_{nk}\right)+\hat{h}=0 \tag{4.3}$$

$$r\in V_n;\quad \left(-\omega^2+2i\psi_{nk}\omega_{nk}\omega+\omega_{nk}^2\right)\hat{q}_{nk}=\hat{p}_{nk}+\omega^2\langle\rho\rangle\langle u_{nk}\rangle\cdot\hat{U}\,V_n \tag{4.4}$$

$$r\in S;\quad N\cdot\left[\langle C\rangle:\left(\nabla\hat{U}\right)\right]=\hat{f} \tag{4.5}$$

Using equation (4.4) to obtain an expression for $\hat{q}_{nk}$ and substituting it into equation (4.3), the governing differential equation for high-frequency vibration of the primary structure is obtained as:

$$\nabla\cdot\left[\langle C\rangle:\left(\nabla\hat{U}\right)\right]+\omega^2 M(\omega)\cdot\hat{U}+\hat{h}_e=0 \tag{4.6}$$

Here

$$\hat{h}_e=\hat{h}+\langle\rho\rangle\,\omega^2\sum_{k=1}^{\infty}\frac{\hat{p}_{nk}\langle u_{nk}\rangle}{\pm\omega^2+2i\psi_{nk}\omega_{nk}\omega+\omega^2} \tag{4.7}$$

$$M(\omega)=\langle\rho\rangle\left[I+\langle\rho\rangle\sum_{k=1}^{\infty}\frac{\langle u_{nk}\rangle\langle u_{nk}\rangle V_n}{\pm\omega^2+2i\psi_{nk}\omega_{nk}\omega+\omega_{nk}^2}\right] \tag{4.8}$$

where I is the unit tensor and $\langle u_{nk}\rangle\langle u_{nk}\rangle$ denotes a vector dyad. Equation (4.6) differs from the conventional time-reduced vibration equation of an elastic body only in that the tensor $M(\omega)$ replaces the mass density and the spectrum of effective body force $\hat{h}_e$ appears instead of h. The tensor $M(\omega)$ is seen to be dependent upon both the overall (averaged mass density $\langle\rho\rangle$) and the local (ω_{nk}, ψ_{nk}, $\langle u_{nk}\rangle$, V_n) parameters, and is symmetric, cf. (4.8). In addition to this, real engineering structures have such complicated compositions that it is impossible to indicate the axis of anisotropy for the substructures' spectral characteristics. The tensor is thus assumed to be isotropic, see Belyaev and Palmov, 1986, and Belyaev, 1991, i.e.

$$M(\omega)=M(\omega)\,I \tag{4.9}$$

where

$$M(\omega)=\frac{1}{3}M(\omega):I=\langle\rho\rangle\left[1+\frac{1}{3}\langle\rho\rangle\,\omega^2\sum_{k=1}^{\infty}\frac{\langle u_{nk}\rangle\cdot\langle u_{nk}\rangle V_n}{-\omega^2+2i\psi_{nk}\omega_{nk}\omega+\omega_{nk}^2}\right]. \tag{4.10}$$

This parameter is crucial for further analysis. It reflects the inertial and spectral properties of the complex structure and for this reason it may be referred to as the generalised mass of the complex structure.

As follows from equation (4.10) $M(\omega)$ is the sum of an infinite number of modal resonance curves corresponding to the vibration of a single degree of freedom system; this is shown schematically in Figure 2.

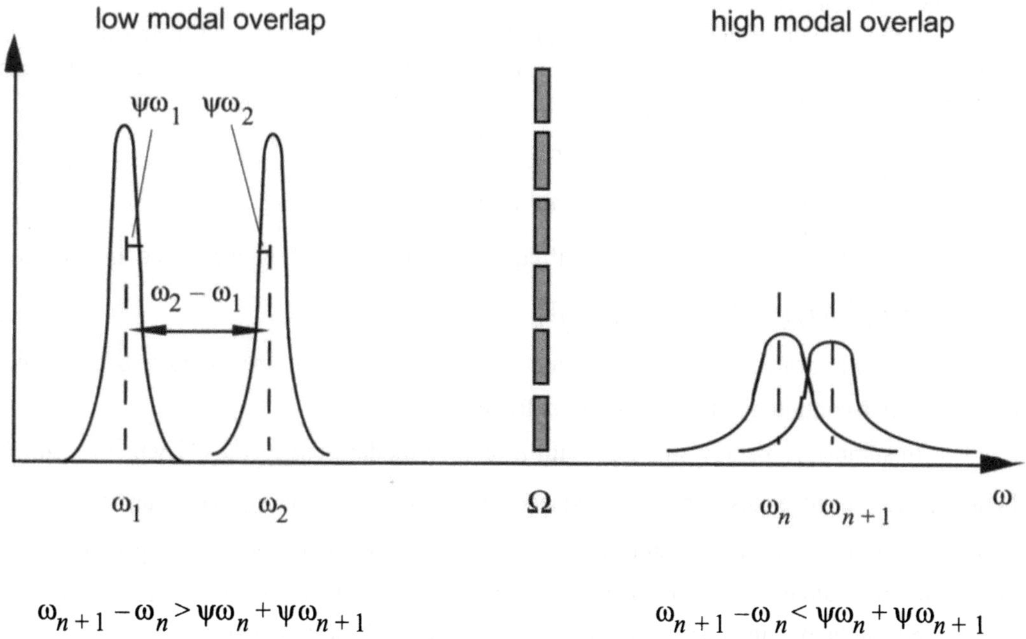

Figure 2. Frequency domains of low and high modal overlap

Two frequency domains with distinct dynamic properties are observed there: (i) a low frequency domain with rather distant modal resonance curves, and (ii) high frequency domain with high modal overlap. The width of each resonance curve is known to be $2\psi_{nk}\omega_{nk}$ at the "half-power" level. Provided that the width of the resonance curve is large compared to the eigenfrequency separation (so-called high modal overlap), i.e.

$$\Delta\omega_{nk} = \omega_{nk+1} \pm \omega_{nk} < \psi_{nk}\omega_{nk} + \psi_{nk+1}\omega_{nk+1} \tag{4.11}$$

the modal resonance curves merge. In this case $M(\omega)$ is a smooth function of frequency. In the domain of high frequency overlap, i.e. $\omega_{nk+1} \approx \omega_{nk}$, inequality (4.11) may be rewritten as

$$\frac{\omega_{nk}}{\Delta\omega_{nk}} > \frac{1}{2\psi_{nk}} \qquad \text{or asymptotically for large } k \qquad \frac{\omega}{\Delta\omega} > \frac{1}{2\psi} \tag{4.12}$$

where $\Delta\omega$ is the eigenfrequency separation.

The left hand side of equation (4.12) $(\omega / \Delta\omega)$ increases as the frequency increases while the right hand side of the equation (4.12) $(1 / 2\psi)$ is frequency-independent. Therefore for each structural member there exists a frequency domain wherein the modal resonance curves are overlapped. This frequency domain of high modal overlap is referred to as the high frequency region, cf. Belyaev, 1991. An experimental study of honeycomb plates by Clarkson and Ranky, 1983, has confirmed the existence of the frequency domain wherein the modal resonance curves are located so densely that one cannot indicate a particular modal resonance curve. The same effect in a large engineering structure such as a nuclear power plant has been reported by Krutzik, 1988.

In the domain of high modal overlap the modal curves merge, and their sum appears to be a locally smooth frequency function. In this case the modal sum in equation (4.10) can be treated as an integral over the high-frequency region rather than a sum, i.e.

$$M(\omega) = \langle\rho\rangle\left[1 + \omega^2\int_0^\infty \frac{\Phi(\alpha)d\alpha}{-\omega^2 + 2i\psi\alpha\omega + \alpha^2}\right] \tag{4.13}$$

Here a locally smooth function of the eigenfrequency distribution Φ is introduced as follows

$$\Phi(\omega_{nk})\,\Delta\omega_{nk} = \frac{1}{3}\langle\rho\rangle\langle u_{nk}\rangle\cdot\langle u_{nk}\rangle V_n \tag{4.14}$$

Equation (4.13) indicates that in the high-frequency domain ($\omega > \Omega$) the structure behaves like a system with a continuous spectrum of eigenfrequencies.

In summary, for any structure with a relative modal density $d_{nk} = \left(\Delta\omega_{nk}\right)^{\pm 1}$ and a damping ψ_{nk} there exists a critical frequency Ω. For frequencies greater than the critical frequency, i.e. $\omega > \Omega$, the structure exhibits the behaviour of a mechanical system with a continuous spectrum of natural frequencies. The value for the critical frequency Ω is not well defined and can be obtained only asymptotically from the following equation

$$\Omega = \left(2\psi_{nk}d_{nk}\right)^{\pm 1} \tag{4.15}$$

cf. eq.. (4.14). This parameter is specific for each structure and is very much dependent upon the size of the structure, namely, the larger the structure, the lower its critical frequency Ω.

The spectral properties of the complex structures are assumed to be gross parameters since the spectral characteristics of the structure can be obtained only as a result of certain experiments involving excitation over narrow-bands or single frequency excitation of the whole structure. Equation (4.13) may be rewritten as follows

$$M(\omega) = \langle\rho\rangle\left[1 - i\,\kappa(\omega)\right]^2 \tag{4.16}$$

where $\kappa(\omega)$ are nondimensional frequency-dependent parameters. Comparison of equations (4.13) and (4.16) yields

$$\kappa(\omega) = \omega^3\int_\Omega^\infty \frac{\psi\alpha\Phi(\alpha)d\alpha}{\left(\alpha^2 - \omega^2\right)^2 + 4\psi^2\alpha^2\omega^2} \tag{4.17}$$

When the damping ψ in the latter equation is finite, then $\kappa(\omega)$ is finite as well. The parameter $\kappa(\omega)$ however remains finite even for vanishingly small material damping. In order to prove this, the integral in equation (4.13) is estimated using methods of random vibration theory, see Bolotin 1984, for small damping ($\psi << 1$) and with a locally smooth function $\Phi(\alpha)$

$$\kappa(\omega) = \frac{1}{2}\pi\omega\,\Phi(\omega) \tag{4.18}$$

From this last formula one can see that $\kappa(\omega)$ and consequently the absorption of high frequency vibration does not depend on the damping ψ. Actually, this coefficient is absent in equation (4.18). For vanishingly small damping the value of absorption is determined by the distribution function $\Phi(\alpha)$, that is, the internal degrees of freedom act as dynamic absorbers on the carrier structure. In the frequency domain of high modal overlap the resonance curves of the internal degrees of freedom merge yielding considerable spatial absorption of vibration over the whole high-frequency region.

In summary, the boundary value problem that governs three-dimensional high frequency vibration in complex structures takes the following form

$$\begin{aligned} &\textit{in volume } V && \nabla\cdot\left[\langle C\rangle : \left(\nabla\hat{U}\right)\right] + \omega^2\langle\rho\rangle\left[1 - i\,\kappa\right]^2\hat{U} + \hat{h}_e = 0 \\ &\textit{on boundary } S && N\cdot\left[\langle C\rangle : \left(\nabla\hat{U}\right)\right] = \hat{f} \end{aligned} \tag{4.19}$$

We may formulate now the basic distinctions between low and high frequency vibration. These are, in fact, the distinctions between vibration theory and high frequency dynamics since it is these theories which govern the vibration in the above frequency regions. In the low frequency region the global natural frequencies of the structure form a discrete spectrum which is strongly influenced by the boundary conditions. From the viewpoint of high frequency vibration, a complex mechanical structure appears to be an infinite body with considerable spatial absorption of vibration. The continuous spectrum of eigenfrequencies and insensitivity to the boundary conditions on the remote boundary surface confirm this conclusion. In other words: low-frequency vibration incorporates the entire complex structure, while high-frequency vibration propagates from the source of excitation.

5 Rationale for the description of high frequency dynamics by the methods of thermodynamics

Up to now only the relation of high frequency dynamics to vibration theory has been discussed and the boundary value problem of the high frequency structural analysis was derived by means of vibration theory methods. The aim of the forthcoming parts is to derive a boundary value problem for high frequency vibration by the methods of thermodynamics.

The idea of such a description is based on the obvious analogy between the thermal motion of molecules and high frequency vibration. On the one hand the temperature is a thermodynamic macroparameter, while on the other it is a measure of the vibrational energy of molecules. However in thermodynamics, the concept of temperature can only be introduced once the assumption was made that the number of microstates (i.e. degrees of freedom) in the system is large. The basic idea of thermodynamics is to ignore the microscopic nature and to observe that the thermodynamic state of a system may be completely specified in terms of small numbers of macroscopic quantities. This is the approach of thermodynamics, one of the objectives of which is to find the relationships between the measurable quantities. It is well known from the equilibrium statistical thermodynamics that the relative fluctuation, i.e. the ratio of the root mean square fluctuation of the energy to the mean energy is of order $1/\sqrt{N}$ where N is the number of degrees of freedom. Since a cubic centimetre of a solid contains some 10^{20} atoms, the relative fluctuation is about 10^{-10}. For this reason, temperature alone is sufficient to describe the field of thermal vibration. It is very instructive to cite the thermal physicist's point of view: "the concept of temperature is not required in mechanics and is only meaningful for systems containing many particles", see Riedi, 1976. This remark is absolutely correct. It can however be improved in such a way that the concept of temperature may be extended to mechanical systems with a number of degrees of freedom which is too large to be analysed by the conventional methods of vibration theory, but too small to apply the concept of energy equipartition. This is exactly the case of high frequency dynamics.

The ideology of Statistical Energy Analysis (SEA) may be considered as a rationale for the use of the methods of thermodynamics to high frequency dynamics. Numerous alternative approaches and modifications have been suggested within the framework of SEA, see to this aim two comprehensive reviews by Fahy, 1974, 1994. In the modal approach to SEA the structure is viewed in terms of substructures. According to the central result of SEA the power flow P_{12} between two coupled substructures is proportional to the difference of the vibrational energy of each substructure, in the notation of Scharton and Lyon, 1968, and Fahy, 1994,

$$P_{12}=M_{12}\left[E_1 n_1^{\pm 1} \pm E_2 n_2^{\pm 1}\right] \tag{5.1}$$

Here M_{12} is a modal-average power transfer coefficient, E_i and n_i are the total energy and the modal density of the substructure i ($i=1,2$), respectively. Provided that the modal energy of the substructure e_i is introduced as follows

$$e_1=E_1 n_1^{\pm 1} \; ; \quad e_2=E_2 n_2^{\pm 1} \tag{5.2}$$

equation (6.1) may be rewritten as

$$P_{12}=M_{12}\left[e_1 \pm e_2\right] \tag{5.3}$$

As the eigenfrequencies, natural modes and damping are described in a statistical sense in the modal approach, so the distribution of amplitudes, phases, directions and decay are supposed to be of a random nature in the wave approach to SEA, see Maidanik and Dickey, 1989, and Langley, 1992. In the wave approach the vibrational field is modelled as a superposition of travelling waves and the wave intensity is directly related to modal energy (6.2) which again results in expression (6.3). A derivation of eq. (5.3) in a general case of weakly-coupled structures with internal friction can be found for example in Belyaev, 2002.

As seen from equation (6.3), the fundamental SEA relation between power flow and modal energy difference is fully analogous to the Fourier law in thermodynamics which states that heat flow is proportional to temperature difference. This analogy has allowed Eichler 1965 to reduce the vibration propagation in complex structures to a discrete transfer scheme, no differential equation of the heat conduction type, however, has been suggested.

A boundary value problem of the heat conduction type was first applied by Palmov, 1974 to the analysis of random vibration in complex structures. Differential equations of the parabolic type were proposed later by Belov and Rybak 1975, Belov, Rybak and Tartakovskii,

1977, Bulitskaya, Vyalyshev and Tartakovsky, 1983 for one-dimensional layered continua. A parabolic differential equation has been reported by Belyaev 1985 to govern the propagation of a wave packet in a one-dimensional medium having a microstructure. The vibrational conductivity approach to high frequency dynamics of complex structures was proposed in the eighties, see Belyaev and Palmov, 1980 and 1986, see also Belyaev 1994. Nefske and Sung 1989 have reported a power flow finite element analysis of dynamic systems. In their paper the boundary value problem of the heat conduction type was derived and applied to the analysis of beams. Recently various approaches resulting in a differential equation of the parabolic type have been suggested for various structural components, namely, rods, single and coupled beams, plates and circular membranes, see Burrel et al, 1990, Bouthier and Bernhard 1992, Guyader 1993, Carcaterra and Sestieri 1994, Le bot and Jezequel 1993, Langley 1991 and 1995. In the author's opinion, the derivation reported by Langley 1995 should be considered as a general derivation for two-dimensional structural members. Kishimoto and Bernstein 1995 have reported that the energy flow models based upon thermodynamic energy rather than stored energy can be used to predict the energy flow in low-dimensional systems. The fundamental thermodynamic energy of a system (i.e. "temperature") has been shown not to be its stored energy content, but rather its ability to shed "heat". Their approach permits one to overcome some shortcomings of the modal approach to SEA. An interesting application of the vibrational conductivity approach to granular media has been reported by Blekhman 1994 when discussing vibrational rheology of granular media. In order to derive the governing equations of the granular media subjected to high frequency vibration, the granular medium has been supposed to be modelled as a compressible Newtonian fluid with parameters depending upon the local vibration while the imposed high frequency vibration has been computed using the vibrational conductivity approach. This model allows one to reveal and explain some "slow convective" flows of granular media in a vibrating vessel.

The aforementioned observations and analogies can be viewed as a rationale for applying the methods of thermodynamics to the modelling of structures at high frequencies.

6 Vibrational conductivity approach to high frequency dynamics

Use is made here of a direct method which demands a given boundary value problem and deals only with the parameters' identification. Since high frequency vibrations in complex structures are believed to be described by some thermodynamic approach, the heat conduction equation is used. Further, since complex structures behave like heavily damped mechanical systems, cf. Part 4 of the present paper, a "heat sink" term $\pm \alpha S$ is included in the traditional heat conduction equation, to give the following vibrational conductivity equation

$$\textit{in volume } V; \quad \nabla \cdot \left(K \nabla S \right) \pm \alpha S = 0 \tag{6.1}$$

Here analogous to the heat conduction equation S is a "vibrational temperature", K is vibrational conductivity and α is vibrational conductance. The boundary condition of the second kind is as follows

$$\textit{on boundary } B; \quad N \cdot \left(K \nabla S \right) = \Gamma \tag{6.2}$$

where Γ is the vibrational flux prescribed at the surface.

Identification of the parameters of the vibrational conductivity approach implies a comparison of the solution of the boundary value problem, equations (6.1)-(6.2) with a solution from another theory. High frequency dynamics is a high frequency limit of vibration theory, and high frequency vibrations are governed by the boundary value problem (4.19). On the other hand, high frequency dynamics is a low frequency limit of thermodynamics, i.e. the same vibrations are described by the equations governing the vibrational conductivity approach (6.1)-(6.2). Thus, if the same vibrational field is modelled by means of these two approaches, this permits us to identify the parameters of the vibrational conductivity approach, provided that closed form solutions are obtained. The stationary vibrational field in a one-dimensional extended complex structure $0<x<L$ driven at $x=0$ by the force $F\exp(i\omega t)$ is obtained by solving one-dimensional boundary value problem (4.19). The mean value of the square of the amplitude of the vibrational field in one dimension is given by

$$E\left\langle U_x U_x^*\right\rangle = \frac{a^2}{c^2\omega^2\left(1+\kappa^2\right)} \frac{\cosh\dfrac{2\omega\kappa(L\pm x)}{a}}{\sinh\dfrac{2\omega\kappa L}{a}} E\left\langle F_x F_x^*\right\rangle \tag{6.3}$$

The steady state boundary problem of the vibrational conductivity approach in one dimension for the same object is as follows

$$0<x<L\ ; \quad K\frac{d^2 S_x}{dx^2} \pm \alpha S_x = 0 \tag{6.4}$$

$$x=0\ , \quad \pm K\frac{dS_x}{dx} = \Gamma_x\ ; \quad x=L\ , \quad \frac{dS_x}{dx}=0 \tag{6.5}$$

Its solution is given by

$$S_x = \frac{1}{\sqrt{\alpha K}} \frac{\cosh\sqrt{\dfrac{\alpha}{K}}\,(L\pm x)}{\sinh\sqrt{\dfrac{\alpha}{K}}\,L}\Gamma_x \tag{6.6}$$

Equations (6.3) and (6.6) are qualitatively similar and govern the same mechanical system described by means of the two different approaches. Hence

$$\sqrt{\frac{\alpha}{K}} = \frac{2\omega\kappa}{a}; \quad \sqrt{\alpha K}\,\frac{S_x}{\Gamma_x} = \frac{a^2}{c^2\omega^2\left(1+\kappa^2\right)} \frac{E\left\langle U_x U_x^*\right\rangle}{E\left\langle F_x F_x^*\right\rangle} \tag{6.7}$$

These two equations are not sufficient to identify uniquely the parameters of the approach. For unique identification one must analyse a one-dimensional structure with weakly heterogeneous overall mechanical properties. In this case, the boundary value problem is given by

$$0<x<L\ ; \quad \frac{d}{dx}\left[K(x)\frac{dS_x}{dx}\right] \pm \alpha(x)\, S_x = 0 \tag{6.8}$$

$$x=0\ , \quad \pm K(0)\frac{dS_x}{dx} = \Gamma_x\ ; \quad x=L\ , \quad \frac{dS_x}{dx}=0 \tag{6.9}$$

An asymptotic solution is obtained by means of the WKB-method, see Heading 1962. To this end, a new independent variable ξ and a new dependent variable $\Sigma(\xi)$ are introduced as follows

$$\xi=\int_0^x \sqrt{\alpha/K}\,dx\ ;\qquad \Sigma(\xi)=\left(\alpha K\right)^{1/4} S_x \tag{6.10}$$

Substituting equation (6.10) into equation (6.8) yields

$$\frac{d^2\Sigma}{d\xi^2}\pm\left[1+\left(\alpha K\right)^{\pm 1/4}\frac{d^2}{d\xi^2}\left(\alpha K\right)^{1/4}\right]\Sigma=0 \tag{6.11}$$

The parameters of a weakly heterogeneous system are slowly varying functions of ξ. For this reason, the second term in the square brackets may be asymptotically neglected. Thus, a solution that asymptotically satisfies the boundary condition (6.10) is given by

$$S_x=\frac{\Gamma_x}{\left[\alpha(x)\,K(x)\,\alpha(0)\,K(0)\right]^{1/4}}\,\frac{\cosh\int_x^L\sqrt{\alpha/K}\,d\xi}{\sinh\int_0^L\sqrt{\alpha/K}\,d\xi} \tag{6.12}$$

In the case of a complex structure with slowly varying overall parameters, equation (6.3) for an averaged vibrational field takes the following form

$$E\left\langle U_x U_x^*\right\rangle=\frac{E\left\langle F_x F_x^*\right\rangle}{\rho(x)\,a(x)\,\rho(0)\,a(0)\,\omega^2}\,\frac{\cosh 2\omega\kappa\int_x^L\frac{d\xi}{a}}{\sinh 2\omega\kappa\int_0^L\frac{d\xi}{a}} \tag{6.13}$$

A comparison of equations (6.12) and (6.13) yields

$$\left[\alpha(x)\,K(x)\right]^{1/4}=a(x)\,\rho(x)\ ;\qquad \frac{\alpha(x)}{K(x)}=\left(\frac{2\omega\kappa}{a(x)}\right)^2\ ;$$

$$\frac{S_x}{\Gamma_x}\left[\alpha(0)\,K(0)\right]^{1/4}=\frac{E\left\langle U_x U_x^*\right\rangle}{E\left\langle F_x F_x^*\right\rangle}\,\rho(0)\,a(0)\,\omega^2 \tag{6.14}$$

which contains more information than equation (6.7) because of the additional relation at $x=0$. The first two equations yield

$$K=\frac{a^3\rho^2}{2\omega\kappa}\ ;\quad \alpha=2\omega\kappa\,a\,\rho^2 \tag{6.15}$$

This allows one to rewrite the third equation in (6.14) as follows

$$\frac{S_x}{\Gamma_x} = \frac{E\left\langle U_x U_x^* \right\rangle}{E\left\langle F_x F_x^* \right\rangle}\omega^2 \tag{6.16}$$

Parameter Γ implies an external vibrational flow which depends only upon the external loading parameters, namely $E\left\langle F_x F_x^* \right\rangle$ and frequency ω . This allows us to take e.g.

$$\Gamma_x = \omega^{\pm 2} E\left\langle F_x F_x^* \right\rangle \tag{6.17}$$

which yields the following expression

$$S_x = E\left\langle U_x U_x^* \right\rangle \tag{6.18}$$

For the random loading $E\left\langle F_x F_x^* \right\rangle$ is the spectral density of the loading while for deterministic loading it may be viewed as the square of the loading amplitude.

The equations developed for K and α are also valid for the three-dimensional case since they do not depend upon the dimension of the problem. The expressions for S and Γ are expected to be some linear combination of the corresponding one-dimensional ensemble averages since the original problem is linear. The only frame-indifferent linear combination of the one-dimensional solutions is known to be the first invariant of the tensor of ensemble average of displacement

$$S = E\left\langle U_x U_x^* \right\rangle + E\left\langle U_y U_y^* \right\rangle + E\left\langle U_z U_z^* \right\rangle = E\left\langle \boldsymbol{U} \cdot \boldsymbol{U}^* \right\rangle \tag{6.19}$$

and the first invariant of the tensor of ensemble average of stress

$$\Gamma = \omega^{\pm 2} E\left\langle \tau_N \tau_N^* \right\rangle \tag{6.20}$$

where τ_N is the normal stress caused by the external loading. As seen from equation (6.19), the vibrational temperature S may be viewed as the spectral density of random high frequency vibration for the random loading; in the case of deterministic loading S is the ensemble-averaged square of the vibration amplitude. Equations (6.15), (6.19), (6.20) finally determine all the parameters of the approach.

The parameters of the vibrational conductivity approach are seen to be frequency-dependent, therefore, the vibrational temperature S turns out to be a frequency-dependent parameter. Thus, for each frequency ω there exists a field of vibrational temperature $S(\boldsymbol{r},\omega)$. This is in contrast to thermodynamics wherein according to the law of equipartition of energy each degree of freedom has the same amount of energy, i.e. the thermal energy is uniformly distributed through the frequency band of thermal vibration. This difference determines a border between high frequency dynamics and thermodynamics only qualitatively. The problem of how to derive a closed form expression for the frequency separating high frequency dynamics and thermodynamics remains to be tackled. Thus, similar to the border between vibration theory and high frequency dynamics, equation (4.15), the border between high frequency dynamics and thermodynamics is not well defined.

The accuracy of the vibrational conductivity approach has been discussed in papers by Nefske and Sung 1989, Carcaterra and Sestieri 1994, Langley 1991 and 1995. Without going into detail, we mention that the accuracy of the approach increases as the frequency increases, which is quite natural. The main attraction of the vibrational conductivity approach is the scalar governing differential equation of the heat conduction type. In particular, the finite element formulation can be easily achieved by applying commercial finite element codes.

7 Conclusions

We have made an attempt to determine a "realm", the inherent properties and the boundary value problem for a new dynamics referred to as high frequency dynamics. The starting point for the present study was an observation that traditional methods of vibration theory fail at progressively higher frequencies. Statistical Energy Analysis (SEA) has been shown by Dowell and Kubota 1985 to be the limiting form of modal analysis, although the central result of SEA is of apparent thermodynamic character, cf. equation (5.3). This means that there exists a new dynamics which is a high frequency limit of vibration theory and a low frequency limit of thermodynamics. The frequency boundaries separating high frequency dynamics from the "neighbouring" dynamics, namely vibration theory and thermodynamics, have been indicated. We have succeeded in developing a closed form asymptotic expression for the left boundary, see equation (4.15), while the right boundary has been described only qualitatively. The both boundaries are "fuzzy" since the frequency domain of high frequency dynamics overlaps the frequency domains of both vibration theory and thermodynamics.

Two boundary value problems governing high frequency dynamics have been derived. The first one, high frequency structural dynamics, Parts 3 and 4, has been derived by means of the methods of vibration theory under the assumption that the structure possesses some thermodynamic properties, e.g. the primary structure displacement $\boldsymbol{U}$ is what is called a macroparameter in thermodynamics. The second one, vibrational conductivity approach, Parts 5 and 6, is the boundary problem of the heat conduction type modified in that regard that a heat sink term is introduced. The latter is a demand of vibration theory because of the spatial resonant absorption of vibration by secondary structures.

In the high frequency domain structures have been shown to possess both mechanical and thermodynamical properties. At high frequencies the complex structures behave like mechanical systems with a continuous spectrum of natural frequencies and considerable spatial decay. For these reasons, for high frequency vibration the structure appears to be a heavily damped infinite body. In contrast to this, at low frequencies the structure has a discrete spectrum of natural frequencies. The normal modes are not affected by the modal damping but are strongly influenced by the boundary conditions, in particular, on the remote boundary surfaces. The field of high frequency vibration depends upon the frequency, which is in contrast to the thermodynamic law of equipartition of thermal vibration energy.

The secondary systems act as dynamic absorbers on the primary structure which in the case of high modal overlap provides considerable spatial absorption of high frequency vibration. The secondary systems, which are predominantly sensitive electrical and mechanical equipment, may be affected or even damaged, since they absorb vibrational energy at their natural frequencies. This provides the main motivation for considering the high-frequency dynamics of complex structures.

Acknowledgement

The financial support by LCM (Linz Center of Competence in Mechatronics) is sincerely acknowledged.

References

Belov, V.D. and Rybak, S.A. (1975) Applicability of the transport equation in one-dimensional wave-propagation problem. *Journal of Soviet Physics - Acoustics* **21**, 110-114.

Belov, V.D., Rybak, S.A. and Tartakovskii, B.D. (1977) Propagation of vibrational energy in absorbing structures. *Journal of Soviet Physics - Acoustics* **23**, 115-119.

Belyaev, A.K. (1985) Description of a one-dimensional vibrational state with a parabolic equation. *Soviet Applied Mechanics* **21**, 297-301.

Belyaev, A.K. (1991) Vibrational state of complex mechanical structures under broad-band excitation. *International Journal of Solids and Structures* **27**, 811-823.

Belyaev, A.K. (1992) Dynamic simulation of high-frequency vibration of extended complex structures. *Mechanics of Structures and Machines* **20**, 155-168.

Belyaev, A.K. (1993) High-frequency vibration of extended complex structures. *Probabilistic Engineering Mechanics* **8**, 15-24.

Belyaev, A.K. (1994) Nonlinear high-frequency vibration of complex engineering structures. In Thompson, J.M.T. and Bishop S.R., eds., *Nonlinearity and Chaos in Engineering Dynamics,* Chichester: Wiley. 222-229.

Belyaev, A.K. (1994) Vibrational conductivity approach to high-frequency dynamics. *Nuclear Engineering and Design* **150**, 281-286.

Belyaev, A.K. (2003) Combining continuous and discrete energy approaches to high frequency dynamics of structures. In: A.K.Belyaev, A.K. and Guran, A., eds., *Selectic topics on mechatronic and structronic systems.* World Scientific Publishers. 244-291.

Belyaev, A.K. and Palmov, V.A. (1980) *Recent Advances in Structural Dynamics* (M. Petyt editor), 157-168. University of Southampton. Theory of vibroconductivity.

Belyaev, A.K. and Palmov, V.A. (1986) Integral theories of random vibration of complex structures. In Elishakoff, I. and Lyon, R.H., eds., *Random Vibration - Status and Recent Developments* Amsterdam: Elsevier. 19-38.

Blekhman, I.I. (1994) *Vibrational Mechanics* (in Russian) Moscow:Nauka.

Bolotin, V.V. (1984) *Random Vibrations of Elastic Systems*. The Hague: Nijhoff.

Bouthier, O.M. and Bernhard, R.J. (1992) Models of space-averaged energetics of plates. *American Institute of Aeronautics and Astronautics Journal* **30**, 613-623.

Bulitskaya, I.A., Vyalyshev A.I. and Tartakovskii, B.D. (1983) Propagation of vibrational and acoustic energy along a structure with losses. *Journal of Soviet Physics - Acoustics* **29**, 333-334.

Burrell, S.C., Warner, J.L. and Chernuka, M.W. (1990) Power flow finite element analysis applied to a thin circular plate. In *The American Society of Mechanical Engineers,* Paper 90-WA/NCA-10.

Carcaterra, A. and Sestieri, A. (1994) Energy trends in high frequency structural problems. In Fergusson N.S., Wolfe, H.W. and Mei, C., eds., *Structural dynamics: Recent advances*. The Institute of Sound and Vibrations Research, University of Southampton. 482-493.

Clarkson, B.L. and Ranky, M.F. (1983) Modal density of honeycomb plates. *Journal of Sound and Vibration* **91**, 103-118.

Dowell, E.H. and Kubota, Y. (1985) Asymptotic modal analysis and Statistical Energy Analysis of dynamical systems. *Transaction of the American Society of Mechanical Engineers, Journal of Applied Mechanics* **52**, 949-957.

Eichler, E. (1965) Thermal circuit approach to vibrations in coupled systems and noise reduction of a rectangular box. *Journal of the Acoustical Society of America* **37**, 995-1007.

Euler, L. (1736) *Mechanica sive motus scientia analytice exposita.* St. Petersburg.

Fahy, F.J. (1974) Statistical energy analysis - a critical review. *Shock and Vibration Digest* **6**, 14-33.

Fahy, F.J. (1985) *Sound and Structural Vibration.* London:Academic Press.

Fahy, F.J. (1994) Statistical energy analysis: a critical overview. *Philosophical Transaction of the Royal Society London A* **346**, 431-447.

Guyader, J.L. (1993) Methodes de prevision vibroacoustique aux moyennes frequences. *Proceedings of the Institute of Acoustics* **15** 131-142.

Heading, J. (1962) *An Introduction to Phase-Integral Method.* New York: Wiley.

Ibrahim, R.A. (1987) Structural dynamics with parameter uncertainties. *Applied Mechanics Review* **40**, 309-328.

Kishimoto, Y. and Bernstein, D.S. (1995) Thermodynamic modeling of interconnected systems, Part I: conservative coupling, Part II: Dissipative coupling. *Journal of Sound and Vibration* **182**, 23-76.

Krutzik, N.J. (1988) Reduction of the dynamic response by aircraft crash on building structures. *Nuclear Engineering and Design* **110**, 191-200.

Langley, R.S. (1991) Analysis of beam and plate vibrations by using the wave equation. *Journal of Sound and Vibration* **150**, 47-65.

Langley, R.S. (1992) A wave intensity technique for the analysis of high frequency vibrations. *Journal of Sound and Vibration* **159**, 483-502.

Langley, R.S. (1995) On the vibrational conductivity approach to high frequency dynamics for two-dimensional structural components. *Journal of Sound and Vibration* **182**, 637-657.

Le Bot, A. and Jezequel, L. (1993) Energy formulation for one-dimensional problems. *Proceedings of the Institute of Acoustics* **15**, 561-568.

Lyon, R.H. (1975) *Statistical Energy Analysis of Dynamical Systems: Theory and Applications.* Cambridge: MIT Press.

Maidanik, G. and Dickey, J. (1989) Wave derivation of the energetics of driven coupled one-dimensional dynamic systems. *Journal of Sound and Vibration* **139**, 31-42.

Meirovitch, L. (1980) *Computational Methods in Structural Dynamics.* Sijthoff-Noordhoff International Publishers.

Mindlin, R.D. and Goodman, L.E. (1950) Beam vibration with time-dependent boundary conditions. *Transaction of the American Society of Mechanical Engineers, Journal of Applied Mechanics* **17**, 377-380.

Nayfeh, A.H. (1981) *Introduction to Perturbation Techniques.* New York:Wiley.

Nefske, D.J. and Sung, S.H. (1989) Power flow finite element analysis of dynamic systems: basic theory and application to beam. *Journal of Vibration, Acoustics, Stress, and Reliability in*

Design **111**, 94-100.

Newton, I. (1687) *Philosophiae naturalis principia mathematica.* London.

Palmov, V.A. (1969) On a model of medium of complex structure (in Russian). *Prikladnaya Matematika i Mekhanika (Journal of Applied Mathematics and Mechanics)* **33**, 768-773.

Palmov, V.A. (1974) Description of high-frequency vibration of complex dynamic objects by methods of the theory of thermal conduction (in Russian). In Chelomey V.N., ed., *Selected Methods of Applied Mechanics,* Moscow:VINITI. 241-251.

Palmov, V.A. (1979) Integral methods for the analysis of vibration of dynamic structures (in Russian). *Advances in Mechanics* **2**, 3-24.

Palmov, V.A. (1998) *Vibration of Elastoplastic Bodies.* Berlin, Springer.

Riedi, P.C. (1976) *Thermal Physics.* London:The Macmillan Press.

Scharton, T.D. and Lyon, R.H. (1968) Power flow and energy sharing in random vibration. *Journal of the Acoustical Society of America* **43**, 1332-1343.

Timoshenko, S.P. (1983) *History of strength of materials.* New York: Dover.

Truesdell, C.A. (1984) Historical introit. The origin of Rational Thermodynamics. In Truesdell C.A., ed., *Rational Thermodynamics,* New York: Springer-Verlag. 1-48.

Basic Differential Geometry for Mechanics and Control

Kurt Schlacher and Kurt Zehetleitner

Institute of Automatic Control and Control Systems Technology,
Johannes Kepler University of Linz, Austria

Abstract Differential geometry is an old mathematical discipline which contributed and contributes a lot to mathematical physics. Also its use in mechanics, electrodynamics or thermodynamics has a long history. Nevertheless, the introduction of the geometric methods to nonlinear, model based control occurred only about 25 years ago. The driving force for this development was the same as for physics, that is, geometric methods allow us to deal with dynamic systems in a coordinate free manner. Therefore, this contribution presents an overview on this discipline which starts with smooth manifolds, bundles, vector fields and finishes with jet bundles and jet coordinates.

1 Introduction

When engineers investigate systems which are described by a set of ordinary differential equations, they have to face the problem of finding a suitable space which contains the state and the input of the system. Unfortunately, the choice of an Euclidean space is not possible in general, and one has to pass to a more abstract concept. For an example, one cannot describe all the rotations in the 3-dimensional space globally by 3 coordinates, although a local description is always possible. Abstract manifolds overcome this problem, since they behave locally like a piece of an Euclidean space, but they take into account the global nonlinear behavior. This approach implies that one has to extend the calculus from the Euclidean space to manifolds. Choosing special coordinates for a problem allows us to perform all the investigations in the usual way, but one has to face the problem that a canonical choice of coordinates does not exist. Roughly speaking, all the calculations must be meaningful for any coordinate system. This idea leads to a coordinate free representation of dynamic systems. It will turn out that the systems under consideration can be represented with the help of special mathematical objects, called tangent vector fields on abstract manifolds. This implies that differential geometry becomes the main tool for the description and the design of nonlinear control systems.

2 Basics of Manifolds and Bundles

Curves and surfaces in the Euclidean space were studied since the earliest days of geometry. However, the discoveries of Gauss profoundly altered the course of differential geometry and pointed the way to the concept of an abstract manifold. Therefore, we give the basic definitions and present some essential results concerning m-dimensional abstract manifolds. One can construct more complex manifolds from simpler ones. Fibered manifolds are an import case, and it will turn out that the generalizations of the tangent plane of a surface, the so-called tangent bundle, is a special fibered manifold. Therefore, also a short introduction to fibered manifolds and bundles is given in the following subsection. The subsection after the next presents the concept of tangent bundles, tangent vectors for abstract manifolds and introduces important maps on manifolds, called flows.

2.1 Manifolds

A manifold is, roughly speaking, the generalization of an m-dimensional smooth surface in the space $\mathbb{R}^n$ with $n \geq m$. Although one can show that every smooth m-dimensional manifold can be embedded in $\mathbb{R}^{2m+1}$, which was proven by Hassler Whitney (Spivak, 1979), we will give a more abstract definition of a smooth manifold that avoids the reference to any embedding in the space $\mathbb{R}^k$ for some k.

Definition 2.1. A smooth *m-dimensional manifold* is a set $\mathcal{M}$, together with a countable collection of subsets $\mathcal{U}_\alpha \subset \mathcal{M}$, the *coordinate charts*, and one-to-one maps $\phi_\alpha : \mathcal{U}_\alpha \to V_\alpha$ onto connected open subsets $V_\alpha \subset \mathbb{R}^m$, the local *coordinate maps*, such that the following properties are satisfied:

1. The *coordinate charts* cover $\mathcal{M}$, or $\mathcal{M} = \cup_\alpha \mathcal{U}_\alpha$ is met.
2. Let $\mathcal{U}$ denote the overlap $\mathcal{U}_\alpha \cap \mathcal{U}_\beta = \mathcal{U} \neq \{\,\}$. The composite map $\phi_\beta \circ \phi_\alpha^{-1}$: $\phi_\alpha(\mathcal{U}) \to \phi_\beta(\mathcal{U})$ is a smooth (infinitely differentiable) function.
3. For any pair of different points $p \in \mathcal{U}_\alpha$, $q \in \mathcal{U}_\beta$, $p \neq q$ there exist open subsets W_α, W_β such that $\phi_\alpha(p) \in W_\alpha \subset V_\alpha$, $\phi_\beta(q) \in W_\beta \subset V_\beta$ and $\phi_\alpha^{-1}(W_\alpha) \cap \phi_\beta^{-1}(W_\beta) = \{\}$ are met.

Some facts are worth mentioning. The coordinate charts allow us to define a topology for $\mathcal{M}$ by declaring the sets $\phi_\alpha^{-1}(W)$ to be open for any open $W \subset \phi_\alpha(\mathcal{U}_\alpha) \subset \mathbb{R}^m$. In terms of this topology, the third requirement in the definition above says that $\mathcal{M}$ has the Hausdorff separation property. The degree of differentiability of the transition functions $\phi_\beta \circ \phi_\alpha^{-1}$ determines the degree of smoothness of the manifold $\mathcal{M}$. For the sake of simplicity, we will consider only *smooth manifolds* here. Besides the basic coordinate charts $\phi_\alpha : \mathcal{U}_\alpha \to V_\alpha$, one can add additional coordinate charts $\phi : \mathcal{U} \to V \subset \mathbb{R}^m$ subject to the requirement that the transition functions $\phi \circ \phi_\alpha^{-1}$, $\phi_\alpha \circ \phi^{-1}$ are smooth for any α on the overlap $\phi_\alpha(\mathcal{U} \cap \mathcal{U}_\alpha)$. In this case the chart $\phi : \mathcal{U} \to V$ is said to be *compatible* with the basic charts. Furthermore, the *maximal collection* of all *compatible charts* is called an *atlas* of $\mathcal{M}$.

The simplest case of an m-dimensional manifold is the space $\mathbb{R}^m$. Let $x = (x^1, \ldots, x^m)$ be a coordinate system of $\mathbb{R}^m$ and let φ^i be the *coordinate functions* of $\mathbb{R}^m$ which map a point $q \in \mathbb{R}^m$ into $\mathbb{R}$ by $x^i = \varphi^i(q)$. Given an m-dimensional manifold $\mathcal{M}$ with coordinate charts $\mathcal{U}_\alpha$ and coordinate maps ϕ_α, then the composite mapping $\varphi^i \circ \phi_\alpha$ maps $\mathcal{M}$ into $\mathbb{R}$ by $x^i = \varphi^i \circ \phi_\alpha(p)$. We call the functions $x^i(p) = \varphi^i \circ \phi_\alpha(p)$ *coordinate functions* of $\mathcal{M}$. Although we denote different objects, here the coordinates and the coordinate functions, by the same symbol, there should be no confusion. Furthermore, this allows us to dispense with the explicit reference to a local coordinate chart. We will say $x = \left(x^1, \ldots, x^m\right)$ is a local coordinate system of $\mathcal{M}$, which is a shortcut for the property, that there is a local coordinate map $\phi_\alpha : \mathcal{U}_\alpha \to V_\alpha$ with an open subset $V_\alpha \subset \mathbb{R}^m$ and a coordinate chart $\mathcal{U}_\alpha$ such that each $p \in \mathcal{U}_\alpha$ has the local representation $x^i = \phi^i_\alpha(p)$. Of course, we know the representation of p in any other compatible chart by the conditions of Definition 2.1.

Let us consider the m-dimensional unit sphere $\mathbb{S}^m$,

$$\sum_{i=1}^{m+1} \left(x^i\right)^2 = 1$$

then it is straightforward to see that $\mathbb{S}^m$ is a smooth manifold embedded in $\mathbb{R}^{m+1}$, since the sets

$$\mathcal{U}_i = \mathbb{S}^m \cap \left\{x \in \mathbb{R}^{m+1} \mid x^i > 0\right\}, \; \mathcal{U}_{1+m+i} = \mathbb{S}^m \cap \left\{x \in \mathbb{R}^{m+1} \mid x^i < 0\right\}$$

together with the coordinate maps ϕ_i, ϕ_{1+m+i}, which are the projections onto the planes $\left\{x, x^i = 0\right\}$, are *coordinate charts.* Whereas the 2-dimensional cone

$$\left(x^1\right)^2 - \left(x^2\right)^2 - \left(x^3\right)^2 = 0$$

embedded in $\mathbb{R}^3$ is no manifold, because the origin does not have a neighborhood $\mathcal{U}_\alpha$ which can be mapped one-to-one onto an open subset of $\mathbb{R}^2$. A less trivial example of a smooth m-dimensional manifold is the set of all lines through the origin 0 in $\mathbb{R}^{m+1}$ which is called the real projective space $\mathbb{RP}^m$.

Having the concept of manifolds at our's disposal, we may consider smooth functions $f : \mathcal{M} \to \mathbb{R}$ on the m-dimensional manifold $\mathcal{M}$. Let $\mathcal{U}_\alpha$ be a coordinate chart with coordinate map ϕ_α, then $\bar{f}(x) = f\left(\phi_\alpha^{-1}(x)\right)$ is a map $\bar{f} : \phi_\alpha(\mathcal{U}_\alpha) \to \mathbb{R}$. We call f differentiable (smooth) at $p \in \mathcal{U}_\alpha$, iff $\bar{f}$ is differentiable (smooth) at $\phi_\alpha(p)$. It is easy to see that this property does not depend on the choice of $\mathcal{U}_\alpha$. The important set of smooth functions on $\mathcal{M}$ is denoted by $C^\infty(\mathcal{M})$. Obviously, we have $\phi^i_\alpha \in C^\infty(\mathcal{M})$ for a smooth manifold. The reader is asked to reflect twice on this construction. Roughly speaking, we can do calculus like in $\mathbb{R}^m$, if we confine our calculations to suitable charts and make sure that the calculations are meaningful in other charts as well. Now, we are ready to extend this idea to maps between manifolds.

Let $\mathcal{M}, \mathcal{N}$ be two smooth m- and n-dimensional manifolds. A map $f : \mathcal{M} \to \mathcal{N}$ is said to be smooth, iff its local representation is smooth for every coordinate chart. In other words, let $\mathcal{U}_\alpha$, $\mathcal{V}_\beta$ be charts of $\mathcal{M}$ and $\mathcal{N}$ with coordinate maps ϕ_α, φ_β such that $f(\mathcal{U}_\alpha) \subset \mathcal{V}_\beta$ is met, then f is smooth if the composite map $\bar{f} = \varphi_\beta \circ f \circ \phi_\alpha^{-1} : \phi_\alpha(\mathcal{U}_\alpha) \to \varphi_\beta(\mathcal{V}_\beta)$ is smooth. The smooth map f can be used to transfer a smooth function $g \in C^\infty(\mathcal{N})$ from $\mathcal{N}$ back to $\mathcal{M}$ by

$$f^*(g) = g \circ f \in C^\infty(\mathcal{M}). \tag{2.1}$$

The function $f^*(g) \in C^\infty(\mathcal{M})$ is also called the *pullback* of g by f. A mapping $f : \mathcal{M} \to \mathcal{N}$ is called a local *diffeomorphism*, if f has a smooth inverse f^{-1} at least locally. In this case we can transfer functions from $\mathcal{M}$ to $\mathcal{N}$.

Let $x = (x^1, \ldots, x^m)$ and $y = (y^1, \ldots, y^n)$ be coordinate systems for the two smooth m- and n-dimensional manifolds $\mathcal{M}$, $\mathcal{N}$, then we may rewrite the map $f : \mathcal{M} \to \mathcal{N}$ as $y^i = \bar{f}^i(x)$ using these coordinates. Following the ideas above, we are able to define the *rank* of f as the rank of the Jacobian $\left[\frac{\partial}{\partial x^j}\left(\bar{f}^i(x)\right)\right] = \left[\partial_{x^j}\bar{f}^i\right]$. Again, the reader may convince oneself that this definition of the rank is independent of the choice of the charts. Additionally, the following theorem tells us that the maps f between manifolds with constant rank, also called *regular maps*, admit a very simple form.

Theorem 2.2. *Let $\mathcal{M}$, $\mathcal{N}$ be two smooth m- and n-dimensional manifolds. Let k be the rank of the regular map $f : \mathcal{M} \to \mathcal{N}$ at $p \in \mathcal{M}$, then there exist coordinates $x = (x^1, \ldots, x^m)$ near p and coordinates $y = (y^1, \ldots, y^n)$ near $f(p)$ such that f takes the canonical form*

$$y = \left(x^1, \ldots, x^k, \underbrace{0, \ldots, 0}_{n-k}\right)$$

in x, y.

This theorem is an easy consequence of the implicit function theorem, see (Boothby, 1986), (Choquet-Bruhat and DeWitt-Morette, 1982), and will not be proven here.

One-to-one maps between manifolds can be used to parametrize submanifolds, like we parametrize curves and surfaces in $\mathbb{R}^3$. Let $\mathcal{M}$ be a smooth manifold, then a submanifold of $\mathcal{M}$ is a subset $\mathcal{N} \subset \mathcal{M}$ together with a manifold $\bar{\mathcal{N}}$ and a smooth one-to-one map $f : \bar{\mathcal{N}} \to \mathcal{M}$ with maximal rank. In particular, the dimensions of $\mathcal{N}$, $\bar{\mathcal{N}}$ coincide and do not exceed the dimension of $\mathcal{M}$. Unfortunately, this definition of a submanifold admits some irregularities that we wish to avoid. Therefore, we define now the more restrictive regular submanifold.

Definition 2.3. Let $\mathcal{M}$ be a smooth m-dimensional manifold and $\mathcal{N}$ be a submanifold of $\mathcal{M}$. We call $\mathcal{N}$ *a regular* n-dimensional *submanifold*, iff for each $p \in \mathcal{N}$ there exists a chart $\mathcal{U}$, $p \in \mathcal{U}$ with the coordinate map ϕ such that in local coordinates $x = (x^1, \ldots, x^m)$ the conditions

$$x^{n+i} = \phi^{n+i}(q) = 0 \,, \qquad i = 1, \ldots, m-n$$

are met for all $q \in \mathcal{U} \cap \mathcal{N}$.

The coordinates of Definition 2.3 are also called flat coordinates. Furthermore, the parametrization of the submanifold is replaced by the natural inclusion.

Fibered Manifolds and Bundles Let us consider the two smooth manifolds $\mathcal{M}$, $\mathcal{N}$ and a smooth map $f : \mathcal{M} \to \mathcal{N}$. Frequently, one considers the graph of f instead of f itself. The *graph* of f is the new function $\mathrm{gr}_f : \mathcal{M} \to \mathcal{M} \times \mathcal{N}$ defined by $\mathrm{gr}_f(p) = (p, f(p))$, $p \in \mathcal{M}$. The product $\mathcal{M} \times \mathcal{N}$ is called the *total space* because it contains the dependent as well as the independent coordinates and $\mathcal{M}$, the domain, is called the *base space*. This idea can be extended to a new structure.

Definition 2.4. A *fibered manifold* is a triple $(\mathcal{E},\pi,\mathcal{B})$ with the manifolds $\mathcal{E}$, $\dim(\mathcal{E}) = m+n$, $\mathcal{B}$, $\dim(\mathcal{B}) = m$ and a map $\pi : \mathcal{E} \to \mathcal{B}$ that is onto with rank m. The manifold $\mathcal{E}$ is called the *total space*, the map π the *projection* and the manifold $\mathcal{B}$ the *base space*. The subset $\mathcal{F}_p = \pi^{-1}(p)$ of $\mathcal{E}$ is called the *fiber* over $p \in \mathcal{B}$.

In many cases the idea of a fibered manifold without any additional restriction is slightly too general. E.g., different fibers may have totally different topological structures. This problem may be resolved by requiring that the fibered manifolds look rather like a product of manifolds, and the resulting object is called a *bundle*.

Definition 2.5. A *fibered manifold* $(\mathcal{E},\pi,\mathcal{B})$ is a *bundle*, iff there exists a manifold $\mathcal{F}$, called the typical fiber, and a map $\psi : \pi^{-1}(\mathcal{U}_p) \to \mathcal{U}_p \times \mathcal{F}_p$ defined on a neighborhood $\mathcal{U}_p$ of $p \in \mathcal{B}$ such that $\mathrm{pr}_1 \circ \psi = \pi$ (with the projection $\mathrm{pr}_1(p,q) = p$ for $q \in \mathcal{F}_p$) and $\mathcal{F}$, $\mathcal{F}_p$ are diffeomorphic for all p.

In the following we will write $\mathcal{E}$ for the bundle $(\mathcal{E},\pi,\mathcal{B})$ whenever the base space $\mathcal{B}$ and the projection π are clear. On a bundle we can introduce *adapted coordinates* (x,u) at least locally, where x^i, $i = 1,\ldots,m$ are coordinates of the base $\mathcal{B}$ and u^α, $\alpha = 1,\ldots,n$ are coordinates for the typical fiber. We get an even simpler picture, if we look at x as the independent and u as the dependent coordinates. Obviously, we have $\pi(x,u) = x$. A fibered manifold $(\mathcal{E},\pi,\mathcal{B})$ which is diffeomorphic to the bundle $(\mathcal{B}\times\mathcal{F},\mathrm{pr}_1,\mathcal{B})$ is called *trivial*. The well known Möbius band can be represented by a fibered manifold which is, of course, nontrivial. If the typical fiber of a bundle is a linear space, then the bundle is called a vector bundle. An important example of a vector bundle is, e.g., the tangent bundle $\mathcal{T}(\mathcal{M})$ of the manifold $\mathcal{M}$ which will be introduced in the following subsection.

This subsection started with the graph of a map between manifolds. Now, we are ready to give the adequate definition for fibered manifolds.

Definition 2.6. Let $(\mathcal{E},\pi,\mathcal{B})$ be a fibered manifold. A map $\sigma : \mathcal{B} \to \mathcal{E}$ is called a *section* of $(\mathcal{E},\pi,\mathcal{B})$, if it satisfies $\pi \circ \sigma = \mathrm{id}_\mathcal{B}$ on its domain with $\mathrm{id}_\mathcal{B}$ as the identity map on $\mathcal{B}$. The set of all sections of $(\mathcal{E},\pi,\mathcal{B})$ will be denoted by $\Gamma(\mathcal{E})$.

It is worth mentioning that we do not require that a section is defined globally. E.g., there are manifolds that do not admit smooth global constant sections. Let us look at the unit sphere $\mathbb{S}^2 \subset \mathbb{R}^3$, then one can imagine that it is impossible to assign a tangent vector to any point of $\mathbb{S}^2$ in a smooth way, a fact expressed by the Hairy Ball theorem that states, simply spoken, you cannot comb a hairy ball in a smooth way. Nevertheless, we can define a tangent bundle with base $\mathbb{S}^2$ as defined in the next section.

Like we considered maps between manifolds, we can introduce maps for bundles that preserve the bundle structure.

Definition 2.7. Let $(\mathcal{E},\pi,\mathcal{B})$, $(\bar{\mathcal{E}},\bar{\pi},\bar{\mathcal{B}})$ be two bundles. A *bundle map* is a pair $f = (f_\mathcal{B}, f_\mathcal{E})$ of maps $f_\mathcal{B} : \mathcal{B} \to \bar{\mathcal{B}}$, $f_\mathcal{E} : \mathcal{E} \to \bar{\mathcal{E}}$ such that $\bar{\pi} \circ f_\mathcal{E} = f_\mathcal{B} \circ \pi$ is met on the domain of f.

Let (x,u) and $(\bar{x},\bar{u})$ be adapted coordinates of the bundles $(\mathcal{E},\pi,\mathcal{B})$, $(\bar{\mathcal{E}},\bar{\pi},\bar{\mathcal{B}})$, then a bundle map f has the local representation

$$\bar{x} = f_\mathcal{B}(x) \ , \qquad (\bar{x},\bar{u}) = f_\mathcal{E}(x,u) \ ,$$

where $\bar{f}_{\mathcal{B}}$, $\bar{f}_{\mathcal{E}}$ denote the representation of $(f_{\mathcal{B}}, f_{\mathcal{E}})$ in the adapted coordinates. From this representation it is easy to see that a bundle map $f = (f_{\mathcal{B}}, f_{\mathcal{E}})$ allows us to construct the following map

$$\tilde{\sigma} = f_{\mathcal{E}} \circ \sigma : \mathcal{B} \rightarrow \bar{\mathcal{E}} \ ,$$

for a section $\sigma \in \Gamma(\mathcal{E})$. Iff $f_{\mathcal{B}}$ is a diffeomorphism then one can transfer a section $\sigma \in \Gamma(\mathcal{E})$ to a section $\bar{\sigma} \in \Gamma(\bar{\mathcal{E}})$ by

$$\bar{\sigma} = f_* (\sigma) = \tilde{\sigma} \circ f_{\mathcal{B}}^{-1} = f_{\mathcal{E}} \circ \sigma \circ f_{\mathcal{B}}^{-1} . \tag{2.2}$$

We call the map f_* the pushforward and $\tilde{\sigma}$ also the *pushforward* of σ by f.

2.2 Flow, Tangent Vectors and the Tangent Bundle

Before we are able to introduce the idea of tangent vectors on abstract manifolds, we consider special maps on a smooth m-dimensional manifold $\mathcal{M}$, called flows.

Definition 2.8. Let $\mathcal{I} \subset \mathbb{R}$ be an open real interval that contains 0 and let $\mathcal{M}$ be an m-dimensional smooth manifold. A smooth map $\varphi : \mathcal{I} \times \mathcal{M} \rightarrow \mathcal{M}$ is called a local smooth flow on $\mathcal{M}$, iff φ is a smooth map that satisfies

$$\begin{aligned} \varphi_0 (p) &= p \\ \varphi_{\varepsilon^1 + \varepsilon^2} (p) &= \varphi_{\varepsilon^1} \circ \varphi_{\varepsilon^2} (p) = \varphi_{\varepsilon^2} \circ \varphi_{\varepsilon^1} (p) \\ \varphi_{\varepsilon}^{-1} (p) &= \varphi_{-\varepsilon} (p) \end{aligned}$$

for all $\varepsilon, \varepsilon^1, \varepsilon^2 \in \bar{\mathcal{I}} \subset \mathcal{I}$ with a suitable subinterval $\bar{\mathcal{I}}$ and all $p \in \mathcal{U}_q$ with an open neighborhood $\mathcal{U}_q$ of $q \in \mathcal{M}$.

Roughly speaking, a local flow allows two interpretations. Let us fix $\varepsilon \in I$, then $\varphi_\varepsilon : \mathcal{M} \rightarrow \mathcal{M}$ is a diffeomorphism on $\mathcal{M}$. For fixed $p \in \mathcal{M}$, we get a curve $\varphi_\varepsilon (p) : I \rightarrow \mathcal{M}$ in a neighborhood of p. Now the following theorem tells us that the flow allows a very simple representation in special coordinates.

Theorem 2.9. *Let $\varphi_\varepsilon : \mathcal{I} \times \mathcal{M} \rightarrow \mathcal{M}$ be a local smooth flow on an m-dimensional smooth manifold $\mathcal{M}$, and the neighborhood $\mathcal{U}_p$ of $p \in \mathcal{M}$ belongs to the domain of φ, then there exist coordinates $x = (x^1, \ldots, x^m)$ near p such that the representation $\bar{\varphi}_\varepsilon$ of φ_ε in these coordinates is given by*

$$\bar{\varphi}_\varepsilon = \left(x^1 + \varepsilon, x^2, \ldots, x^m\right) \ .$$

This theorem is also known as the *Flow Box* theorem. Since the proof can be found, e.g., in (Boothby, 1986), it is omitted here.

Let φ_ε be a flow on $\mathcal{M}$ like above and $\bar{\varphi}_\varepsilon$ the representation of φ_ε in coordinates x. Using the coordinates we may differentiate φ_ε with respect to ε and get

$$\left. \frac{\mathrm{d}}{\mathrm{d}\tau} \bar{\varphi}_\tau^i (x) \right|_{\tau=\varepsilon} = \bar{v}^i \circ \bar{\varphi}_\varepsilon (x) \ , \qquad \bar{v}^i = \left. \frac{\mathrm{d}}{\mathrm{d}\varepsilon} \bar{\varphi}_\varepsilon^i \right|_{\varepsilon=0} \tag{2.3}$$

with the smooth functions $\bar{v}^i \in C^\infty(\mathbb{R}^m)$. Furthermore, we derive the following system of differential equations from φ_ε by

$$\frac{\mathrm{d}}{\mathrm{d}\varepsilon} x^i = \bar{v}^i(x) \ . \tag{2.4}$$

Therefore, a flow φ_ε generates a system of m ordinary first order differential equations. Inversely, let us assume that a system of m ordinary first order differential equations like (2.4) has a uniquely determined solution at least locally, then this system generates a local flow. This fact follows from the theorem concerning the existence and uniqueness of the initial value problem for explicit ordinary differential equations, see (Choquet-Bruhat and DeWitt-Morette, 1982). The m-tuple $(\bar{v}^1, \ldots, \bar{v}^m)$ is called the *infinitesimal generator* of $\bar{\varphi}_\varepsilon$ in the coordinates x.

To keep the formulas short and readable, we use the Einstein convention for sums of the type

$$a_i b^i = \sum_{i=1}^{m} a_i b^i$$

from now on, whenever the range of the index i is clear and we use the shortcuts

$$\partial_i = \frac{\partial}{\partial x^i} \ , \qquad d_\tau = \frac{\mathrm{d}}{\mathrm{d}\tau}$$

for differential operators.

Until now, the infinitesimal generator, see (2.3), of a flow φ is at our's disposal in local coordinates only. To derive a coordinate independent definition we consider functions $g \in C^\infty(\mathcal{M})$ together with φ_ε. Let $\bar{g}$, $\bar{\varphi}_\varepsilon$ denote the representation of g and φ_ε in local coordinates $x = (x^i)$, $i = 1, \ldots, m$ with coordinate functions ϕ^i, then we can differentiate $\bar{g}(\bar{\varphi}_\varepsilon)$ with respect to ε at a point $p \in \mathcal{U} \subset \mathcal{M}$ and get

$$d_\varepsilon \bar{g}(\bar{\varphi}_\varepsilon(x))\big|_{\varepsilon=0} = \bar{v}^i(x)\,\partial_i \bar{g}(x) = \bar{v}^i(\phi(p))\,\partial_i \bar{g}(\phi(p)) \tag{2.5}$$

with the function $\bar{v}^i$ from (2.3). It is easy to see, that the value of this derivation at p depends on the values of $\bar{v}^i$ at $\phi(p)$ only. Therefore, the infinitesimal generator of a flow corresponds in the coordinates x to a differential operator

$$\bar{v}_x = \bar{v}^i(x)\,\partial_i \tag{2.6}$$

which meets

$$\begin{aligned} \bar{v}_x(\alpha_1 g_1 + \alpha_2 g_2) &= \alpha_1 \bar{v}_x(g_1) + \alpha_2 \bar{v}_x(g_2) \ , && \text{linearity over the real numbers} \\ \bar{v}_x(g_1 g_2) &= \bar{v}_x(g_1)\, g_2 + g_1 \bar{v}_x(g_2) \ , && \text{Leibniz rule} \end{aligned}$$

for all $g_1, g_2 \in C^\infty(\mathcal{M})$ and real numbers α_1, α_2. This property allows us to define the tangent space of a manifold at a point p.

Definition 2.10. Let $\mathcal{M}$ be a smooth m-dimensional manifold, then the tangent space $\mathcal{T}_p(\mathcal{M})$ at $p \in \mathcal{M}$ is the set of all linear mappings v_p, $w_p : C^\infty(\mathcal{M}) \to \mathbb{R}$ which are linear over the real numbers and satisfy the Leibniz rule. The vector space operations are given by

$$(v_p + w_p)(f) = v_p(f) + w_p(f) \ , \qquad (\alpha v_p)(f) = v_p(\alpha f)$$

for all $f \in C^\infty(\mathcal{M})$, $\alpha \in \mathbb{R}$.

A linear map which satisfies the Leibniz rule is called a *derivation.* Now, we are ready to derive the tangent bundle $\mathcal{T}(\mathcal{M})$ of a manifold $\mathcal{M}$ by gluing together the tangent spaces $\mathcal{T}_p(\mathcal{M})$.

Theorem 2.11. *Let $\mathcal{M}$ be a smooth m-dimensional manifold, then the set*

$$\mathcal{T}(\mathcal{M}) = \cup_{p\in\mathcal{M}}\mathcal{T}_p(\mathcal{M})$$

is a bundle $(\mathcal{T}(\mathcal{M}), \tau_\mathcal{M}, \mathcal{M})$, called the tangent bundle *of $\mathcal{M}$, where $\tau_\mathcal{M}$ denotes the projection. The typical fiber $\mathcal{F}$ is an m-dimensional linear vector space.*

The proof of this theorem is based on the extension of the coordinate charts and maps of Definition 2.1 to suitable charts and functions for $\mathcal{T}(\mathcal{M})$. We do not present the proof here, it can be found, e.g., in (Boothby, 1986), (Choquet-Bruhat and DeWitt-Morette, 1982), (Spivak, 1979). Let $x = (x^1, \ldots, x^m)$ be coordinates of $\mathcal{M}$ then the adapted coordinates of $\mathcal{T}(\mathcal{M})$ are denoted by $(x, \dot{x}) = (x^1, \ldots, x^m, \dot{x}^1, \ldots, \dot{x}^m)$ rather than (x, u).

As usual, we call sections of $(\mathcal{T}(\mathcal{M}), \tau_\mathcal{M}, \mathcal{M})$ *tangent vector fields.* The set of all smooth vector fields is denoted by $\Gamma(\mathcal{T}(\mathcal{M}))$. Using local coordinates x, we write for elements v of $\Gamma(\mathcal{T}(\mathcal{M}))$ like in (2.6). Again we want to emphasize, that we do not require the sections to be defined globally.

Let $f : \mathcal{M} \to \mathcal{N}$ be a smooth map between two smooth manifolds $\mathcal{M}$, $\mathcal{N}$. We extend f to a bundle map $(f_\mathcal{B}, f_\mathcal{E})$ by setting $f_\mathcal{B} = f$ and by deriving $f_\mathcal{E} : \mathcal{T}(\mathcal{M}) \to \mathcal{T}(\mathcal{N})$ from the requirement that

$$f_\mathcal{E}(v)(g) = v(f_\mathcal{B}^*(g))$$

holds for all $g \in C^\infty(\mathcal{N})$, $v \in \Gamma(\mathcal{T}(\mathcal{M}))$. It is worth deriving a coordinate expression for $f_\mathcal{E}$. Let $(x^1, \ldots, x^m, \dot{x}^1, \ldots, \dot{x}^m)$, $(y^1, \ldots, y^m, \dot{y}^1, \ldots, \dot{y}^m)$ be adapted coordinates for $\mathcal{T}(\mathcal{M})$, $\mathcal{T}(\mathcal{N})$ then we get for the representation $\bar{f}_\mathcal{E}$ in these coordinates the relations

$$y^i = \bar{f}^i(x) \ , \qquad \dot{y}^i = \partial_j \bar{f}^i(x)\, \dot{x}^j \ ,$$

where $\bar{f}$ denotes the representation of f. If f is a diffeomorphism then we can determine the pushforward for a field $v \in \Gamma(\mathcal{T}(\mathcal{M}))$, see (2.2). Again we write

$$\tilde{v} = f_*(v) = f_\mathcal{E} \circ v \circ f^{-1} \tag{2.7}$$

for the *pushforward* of v by $(f_\mathcal{B}, f_\mathcal{E})$.

Let us take a look back at (2.4). If we set $\varepsilon = t$ in (2.4), with the time t, then it is easy to see that the vector field of (2.6) represents an autonomous dynamic system in a coordinate free manner. Let us consider the bundle $(\mathcal{E}, \pi, \mathcal{B})$ with local adapted coordinates (t) for $\mathcal{B}$ and $\left(t, x^1, \ldots, x^m\right)$ for $\mathcal{E}$. Again, it is straightforward to see that the special vector field $\bar{v} \in \Gamma(\mathcal{T}(\mathcal{E}))$,

$$\bar{v} = \partial_t + v^i(t, x)\,\partial_i$$

represents the dynamic system

$$\begin{aligned} \dot{x}^i &= v^i(t, x)\ , \qquad i = 1, \ldots, m \\ \dot{t} &= 1 \end{aligned} \quad .$$

The smooth control system

$$\dot{x}^i = f^i(x, u)\ , \qquad i = 1, \ldots, m$$

with the state x and the input u can be seen as a local representation of a system on a manifold $\mathcal{M}$ with local coordinates $\left(x^1, \ldots, x^m\right)$. From

$$\bar{v} = f^i(x, u)\,\partial_i$$

we see that $\bar{v}$ describes a submanifold of $\mathcal{T}(\mathcal{M})$ parametrized by the m_u-dimensional input u. Please note that $\bar{v}$ is not a map $C^\infty(\mathcal{M}) \to C^\infty(\mathcal{M})$. There exists a more precise description of this control system, which requires the following fact.

Theorem 2.12. *Let $(\mathcal{E}, \pi, \mathcal{B})$ be a bundle and $f : \bar{\mathcal{B}} \to \mathcal{B}$ be a map between the manifolds $\bar{\mathcal{B}}$, $\mathcal{B}$ with coordinates $(\bar{x})$ for $\bar{\mathcal{B}}$ and adapted coordinates (x, u) for $\mathcal{E}$. Then the set $f^*(\mathcal{E})$,*

$$f^*(\mathcal{E}) = \left\{(\bar{x}, (x, u)) \in \bar{\mathcal{B}} \times \mathcal{E},\ \pi((x, u)) = x = f(\bar{x})\right\}$$

is a bundle, called the pull back bundle $f^*(\mathcal{E})$ of $\mathcal{E}$ by f with the projection $\mathrm{pr}_1 : f^*(\mathcal{E}) \to \bar{\mathcal{B}}$, $(\bar{x}, u) \mapsto (\bar{x})$.

With the bundle $(\mathcal{Z}, \rho, \mathcal{M})$ and local coordinates (x), (x, u) for $\mathcal{M}$, $\mathcal{Z}$ and the projection $\rho(x, u) = (x)$, it is easy to see that $\bar{v} \in \Gamma(\rho^*(\mathcal{T}(\mathcal{M})))$ is met if one sets $\bar{\mathcal{B}} = \mathcal{Z}$ in the theorem above.

These examples show how we can find a clear geometric picture for several types of nonlinear dynamic control systems. Although it is clear from the construction, the reader should convince himself that these geometric pictures are independent of the special choice of coordinates. It is also worth mentioning that there exist also geometric pictures for more complex systems.

2.3 The Cotangent Bundle

Let $(\mathcal{T}(\mathcal{M}), \tau_{\mathcal{M}}, \mathcal{M})$ denote the tangent bundle of a smooth m-dimensional manifold, let $p \in \mathcal{M}$ be a point of $\mathcal{M}$, then the fiber $\mathcal{F}_p = \mathcal{T}_p(\mathcal{M})$ is a linear vector space over p. Like in linear algebra, we may study the linear space of all linear maps $\omega(p) : \mathcal{T}_p(\mathcal{M}) \to \mathbb{R}$. Let us choose the coordinates $\left(x^1, \ldots, x^m\right)$ for $\mathcal{M}$ and $\left(x^1, \ldots, x^m, \dot{x}^1, \ldots, \dot{x}^m\right)$ for $\mathcal{T}(\mathcal{M})$, then a smooth section v of $\mathcal{T}_p(\mathcal{M})$ has the form

$$\bar{v} = \bar{v}^i \partial_i \tag{2.8}$$

with smooth functions $\bar{v}^i$ defined on $\mathcal{M}$, i.e., $\bar{v}^i \in C^\infty(\mathcal{M})$. Let $\bar{\omega}(x)$ denote the map $\omega(p)$ in these coordinates, then $\omega(p)(v)$ is given by

$$\omega(p)(v) = \langle \omega(p), v(p) \rangle = \bar{\omega}_i(x)\,\bar{v}^i(x)$$

for functions $\bar{\omega}_i(x) \in C^\infty(\mathcal{M})$. It is a well known result from linear algebra that the space of all linear functions $\mathcal{T}_p(\mathcal{M}) \to \mathbb{R}$ is an m-dimensional linear vector space, called the dual space, which we denote by $\mathcal{T}_p^*(\mathcal{M})$. Now, we proceed analogously to the tangent bundle und glue the spaces $\mathcal{T}_p^*(\mathcal{M})$ together in a smooth way. This procedure leads to the following theorem.

Theorem 2.13. *Let $\mathcal{M}$ be a smooth m-dimensional manifold, then the set*

$$\mathcal{T}^*(\mathcal{M}) = \cup_{p\in\mathcal{M}} \mathcal{T}_p^*(\mathcal{M})$$

is a bundle $(\mathcal{T}^*(\mathcal{M}), \tau_{\mathcal{M}}^*, \mathcal{M})$, called the cotangent bundle *of $\mathcal{M}$, where $\tau_{\mathcal{M}}^*$ denotes the projection. The typical fiber $\mathcal{F}$ is an m-dimensional linear vector space.*

Again, the reader is referred to the books (Boothby, 1986), (Choquet-Bruhat and DeWitt-Morette, 1982), (Frankel, 1998), (Spivak, 1979) for the proof of this theorem. Like above, let $(x^1, \ldots, x^m)$ be coordinates for $\mathcal{M}$ and $(x^1, \ldots, x^m, \dot{x}^1, \ldots, \dot{x}^m)$ for $\mathcal{T}(\mathcal{M})$, then we denote the coordinates of $\mathcal{T}^*(\mathcal{M})$ by $(x^1, \ldots, x^m, \dot{x}_1, \ldots, \dot{x}_m)$. In these coordinates we write

$$\omega = \bar{\omega}_i \mathrm{d}x^i \tag{2.9}$$

for a section ω, $\dot{x}_i = \bar{\omega}_i$ of $\mathcal{T}^*(\mathcal{M})$. The bilinear map $\langle\cdot,\cdot\rangle_p : \mathcal{T}_p^*(\mathcal{M}) \times \mathcal{T}_p(\mathcal{M}) \to \mathbb{R}$ can be extended to a map $\langle\cdot,\cdot\rangle : \mathcal{T}^*(\mathcal{M}) \times \mathcal{T}(\mathcal{M}) \to C^\infty(\mathcal{M})$ in a straightforward manner. The coordinate expression of this map is $\dot{x}_i \dot{x}^i$, which implies

$$\langle \omega, v \rangle = \bar{\omega}_i \bar{v}^i$$

for the sections (2.8) and (2.9). Furthermore, we have

$$\langle \mathrm{d}x^i, \partial_j \rangle = \delta_j^i$$

with the Kronecker symbol δ_j^i, with $\mathrm{d}x^i$ denoting the sections dual to ∂_j.

Again, let $f : \mathcal{M} \to \mathcal{N}$ be a smooth map between two smooth manifolds $\mathcal{M}$ and $\mathcal{N}$. Then we can pull back a section ω of $\mathcal{T}^*(\mathcal{N})$ to a section $f^*(\omega)$. We also write $f^* : \mathcal{T}^*(\mathcal{N}) \to \mathcal{T}^*(\mathcal{M})$ for this map, which follows from the rule

$$\langle f^*(\omega), v \rangle = \langle f^*(\omega_i)\,\mathrm{d}y^i, f_{\mathcal{E}}(v) \rangle$$

for any $v \in \Gamma(\mathcal{T}(\mathcal{M}))$. The reader should be aware of the fact that in this formula the symbol f^* is used for the pullback of functions $f^*(\omega_i)$ and the pullback of a cotangent vector ω. To derive a coordinate expression for f^*, we take the coordinates $(x^1, \ldots, x^n, \dot{x}_1, \ldots, \dot{x}_n)$, $(y^1, \ldots, y^n, \dot{y}_1, \ldots, \dot{y}_n)$ for $\mathcal{T}^*(\mathcal{M})$, $\mathcal{T}^*(\mathcal{N})$ and get for the representation $\bar{f}_{\mathcal{E}}$ in these coordinates the relations

$$\dot{x}_i = \partial_i \bar{f}^j \dot{y}_j \circ \bar{f} \,.$$

Iff f is a diffeomorphism then we can transfer a section $\mathcal{T}^*(\mathcal{M})$ to a section of $\mathcal{T}^*(\mathcal{N})$. Additionally, we have discovered a difference between sections of vector fields and covector fields. Given the map f, it is always possible to transfer sections of $\mathcal{T}^*(\mathcal{N})$ to sections of $\mathcal{T}^*(\mathcal{M})$, whereas the transfer of sections of $\mathcal{T}(\mathcal{M})$ to sections of $\mathcal{T}(\mathcal{N})$ is possible only if f is a diffeomorphism.

Let us consider a manifold $\mathcal{N}$ with local adapted coordinates $(t, x^1, \ldots, x^n)$ and its cotangent bundle together with the cotangent vectors

$$\bar{\omega}^i = \mathrm{d}x^i - \bar{f}^i(t,x)\,\mathrm{d}t\,, \qquad i = 1, \ldots, n\,.$$

Let us choose a 1-dimensional manifold $\mathcal{M}$ with local coordinate τ and a map $g : \mathcal{M} \to \mathcal{N}$, then the pullback of the covectors above is given by

$$\bar{g}^*(\bar{\omega}^i) = \underbrace{\left(\partial_\tau \bar{g}^i - \partial_\tau \bar{g}^0 \bar{f}^i \circ \bar{g}\right)}\mathrm{d}\tau \in \mathcal{T}^*(\mathcal{M})\,.$$

This expression vanishes, iff the underbraced term vanishes. If we choose the identity map for $t = g^0(\tau) = \tau$, then we derive the following set of differential equations

$$\partial_\tau \bar{g}^i = \bar{f}^i \circ \bar{g}\,, \qquad i = 1, \ldots, n$$

of first order, iff the covectors $g^*(\omega^i)$ vanish. This example shows how we can find another geometric picture for a nonlinear dynamic system. Obviously, this picture is independent of the special choice of coordinates.

2.4 Tensor Bundles

Let $\mathcal{M}$ be a smooth m-dimensional manifold, and let $\mathcal{T}(\mathcal{M})$ and $\mathcal{T}^*(\mathcal{M})$ denote its tangent and cotangent bundle. For a given point $p \in \mathcal{M}$, we define the maps

$$\langle \omega(p), \cdot \rangle : \mathcal{T}_p(\mathcal{M}) \to \mathbb{R}\,, \qquad \langle \cdot, v(p) \rangle : \mathcal{T}_p^*(\mathcal{M}) \to \mathbb{R} \tag{2.10}$$

for $v \in \mathcal{T}(\mathcal{M})$, $\omega \in \mathcal{T}^*(\mathcal{M})$, and consider the map

$$\langle \omega(p), \cdot \rangle \langle \cdot, v(p) \rangle : \mathcal{T}_p(\mathcal{M}) \times \mathcal{T}_p^*(\mathcal{M}) \to \mathbb{R}\,,$$

which is obviously bilinear. We introduce the new symbol $\otimes$, the tensor product, in the following manner

$$(\omega(p) \otimes v(p))(\cdot, \cdot) = \langle \omega(p), \cdot \rangle \langle \cdot, v(p) \rangle$$

for the construction of the bilinear function above. This construction can be extended in a straightforward manner to multilinear functions $(\mathcal{T}_p(\mathcal{M}))^k \times (\mathcal{T}_p^*(\mathcal{M}))^l \to \mathbb{R}$. The reader should convince himself that the space of these multilinear functions is an n^{k+l}-dimensional vector space, denoted by $\mathcal{T}_{l,p}^k(\mathcal{M})$, over the reals with the natural definitions of addition and multiplication. Again, we glue the spaces $\mathcal{T}_{l,p}^k(\mathcal{M})$ together in a smooth way to derive the tensor bundle.

Theorem 2.14. *Let $\mathcal{M}$ be a smooth m-dimensional manifold, then the set*

$$\mathcal{T}_l^k(\mathcal{M}) = \cup_{p \in \mathcal{M}} \mathcal{T}_{l,p}^k(\mathcal{M})$$

is a bundle $\left(\mathcal{T}_l^k(\mathcal{M}), \tau_{l,\mathcal{M}}^k, \mathcal{M}\right)$, *called the* tensor bundle *of* $\mathcal{M}$, *where* $\tau_{l,\mathcal{M}}^k$ *denotes the projection. The typical fiber* $\mathcal{F}$ *is an* m^{k+l}*-dimensional linear vector space.*

For the proof, the reader is asked to consult one of the books (Boothby, 1986), (Choquet-Bruhat and DeWitt-Morette, 1982), (Frankel, 1998), (Spivak, 1979). A section of the bundle $\mathcal{T}_0^k(\mathcal{M})$ is called a *covariant tensor field*, whereas a section of the $\mathcal{T}_l^0(\mathcal{M})$ is called a *contravariant tensor field*. Furthermore, we may indentify $\mathcal{T}_0^1(\mathcal{M})$ with $\mathcal{T}^*(\mathcal{M})$ and $\mathcal{T}_1^0(\mathcal{M})$ with $\mathcal{T}(\mathcal{M})$ in a natural way using (2.10).

Let Φ be a section of $\mathcal{T}_0^k(\mathcal{M})$ given by

$$\Phi = \bigotimes_{i=1}^{k} \omega^i$$

then we may symmetrize the map Φ according to the following rule

$$S(\Phi) = \frac{1}{k!}\sum_{\sigma}\bigotimes_{i=1}^{k}\omega^{\sigma(i)} ,$$

where σ denotes a permutation of $(1,\ldots,k)$ with $(1,\ldots,k) \to (\sigma(1),\ldots,\sigma(k))$. Let $\operatorname{sgn}\sigma$ denote the sign of the permutation σ, then the alternating tensor induced by Φ is given by

$$A(\Phi) = \frac{1}{k!}\sum_{\sigma}\operatorname{sgn}\sigma\bigotimes_{i=1}^{k}\omega^{\sigma(i)} .$$

Obviously, the multilinear map induced by $A(\Phi)$ is a skew symmetric one in any two of its arguments. We would get similar results for sections of $\mathcal{T}_k^0(\mathcal{M})$.

Let $\varphi \in \Gamma\left(\mathcal{T}_0^k(\mathcal{M})\right)$, $\psi \in \Gamma\left(\mathcal{T}_0^l(\mathcal{M})\right)$ be two sections of their tensor bundles, then we may use the tensor product $\otimes$ to construct the new tensor $\varphi \otimes \psi \in \Gamma\left(\mathcal{T}_0^{k+l}(\mathcal{M})\right)$ by

$$\varphi \otimes \psi(v_1,\ldots,v_k,v_{k+1},\ldots,v_{k+l}) = \varphi(v_1,\ldots,v_k)\,\psi(v_{k+1},\ldots,v_{k+l})$$

with $v_i \in \Gamma(\mathcal{T}(\mathcal{M}))$, $i = 1,\ldots,k+l$. Additionally, one can show that the map $\otimes$: $\mathcal{T}_0^k(\mathcal{M}) \times \mathcal{T}_0^l(\mathcal{M}) \to \mathcal{T}_0^{k+l}(\mathcal{M})$ is bilinear and associative. Again, similar results hold for sections of $\mathcal{T}_k^0(\mathcal{M})$ and $\mathcal{T}_l^0(\mathcal{M})$.

Let us confine our considerations to the subspace of alternating covariant tensors fields, denoted by $\wedge^k(\mathcal{T}^*(\mathcal{M}))$ of $\mathcal{T}_0^k(\mathcal{M})$, nevertheless, some similar results would hold for alternating contravariant tensor fields, denoted by $\wedge^k(\mathcal{T}(\mathcal{M}))$.

Definition 2.15. The mapping

$$\wedge : \wedge^k(\mathcal{T}^*(\mathcal{M})) \times \wedge^l(\mathcal{T}^*(\mathcal{M})) \to \wedge^{k+l}(\mathcal{T}^*(\mathcal{M})) ,$$

defined by

$$\varphi \wedge \psi = \frac{(k+l)!}{k!l!}A(\varphi \otimes \psi)$$

for $\varphi \in \Gamma\left(\wedge^k(\mathcal{T}^*(\mathcal{M}))\right)$, $\psi \in \Gamma\left(\wedge^l(\mathcal{T}^*(\mathcal{M}))\right)$, is called the *exterior product* (or *wedge product*) of φ and ψ. This mapping is bilinear and associative.

To simplify the following we identify $\wedge^0(\mathcal{T}^*(\mathcal{M}))$ with $C^\infty(\mathcal{M})$. Furthermore, the reader should convince oneself that the bundles $\wedge^k(\mathcal{T}^*(\mathcal{M}))$ with $k > n$ admit only the zero section because of $\dim\left(\wedge_p^k(\mathcal{T}^*(\mathcal{M}))\right) = 0$ for any $p \in \mathcal{M}$. From now on, we call a section of $\wedge^k(\mathcal{T}^*(\mathcal{M}))$ a k-form. From the natural identification of $\wedge^1(\mathcal{T}^*(\mathcal{M}))$ with $\mathcal{T}^*(\mathcal{M})$, we see that a covector field is a 1-form.

The space of all p-forms with $p = 1, \ldots, m$ on an m-dimensional manifold will be denoted by $\Gamma(\wedge(\mathcal{T}^*(\mathcal{M})))$. The exterior product has a nice behavior for the pullpack of forms. Let $f : \mathcal{M} \to \mathcal{N}$ be a smooth map between the two smooth manifolds $\mathcal{M}$, $\mathcal{N}$, then we have

$$f^*(\alpha \wedge \beta) = f^*(\alpha) \wedge f^*(\beta)$$

for any α, $\beta \in \Gamma(\wedge(\mathcal{T}^*(\mathcal{N})))$, or f^* is a homomorphism of the exterior algebra. The reader should be aware of the fact that the same symbol $\wedge$ is used for the exterior product in $\wedge(\mathcal{T}^*(\mathcal{M}))$ and $\wedge(\mathcal{T}^*(\mathcal{N}))$.

3 Lie Derivatives and the Lie Bracket

In the previous section it has been shown that tangent vector fields are derivations that act on smooth functions, see Definition 2.10 and Theorem 2.11. Let $\mathcal{M}$ be a smooth m-dimensional manifold with local coordinates x. Let v be an element of $\Gamma(\mathcal{T}(\mathcal{M}))$, then the *Lie derivative of a smooth function* $g \in C^\infty(\mathcal{M})$ along the field v at $p \in \mathcal{M}$ is given by

$$L_v(g)(p) = v(g)(p) \tag{3.1}$$

with the map $L_v : C^\infty(\mathcal{M}) \to C^\infty(\mathcal{M})$. Since v is the infinitesimal generator of a (local) flow φ_ε, we may restate (3.1) as

$$L_v(g)(p) = d_\varepsilon\left(g \circ \varphi_\varepsilon(p)\right)\big|_{\varepsilon=0}, \tag{3.2}$$

see again (2.5). The geometric interpretation of (3.2) is quite simple. The expression $L_v(g)(p)$ measures the change of the function g at p for a point that moves along the curve $\varphi_\varepsilon(p)$ with velocity $v \circ \varphi_\varepsilon(p)$. This geometric picture is the key to define the *Lie derivative of a smooth vector field* w along a field v. We set

$$L_v(w)(p) = v(w)(p) = d_\varepsilon\left(w \circ \varphi_\varepsilon(p) - \underbrace{\varphi_{\varepsilon,*}(p)(w(p))}\right)\bigg|_{\varepsilon=0}. \tag{3.3}$$

The only strange thing is the braced term in (3.3). To understand why such a term has to occur here, let us recall that tangent vector fields are sections of the tangent bundle of a manifold. Although their fibers $\mathcal{F}_p$ are diffeomorphic to the typical fiber $\mathcal{F}$, see Definition 2.5 and Theorem 2.11, we must not compare elements of different tangent spaces like we do in $\mathbb{R}^m$. Therefore, we have to take the pushforward of w by $\varphi_\varepsilon(p)$ to compare the vectors in the same tangent space. This explains the braced term. Additionally, this approach shows, how one has to define in principle Lie derivatives of geometric objects like functions, tangent vectors or covectors, p-forms or other tensors. The interested reader is asked to consult the cited literature, e.g., (Frankel, 1998), and the references therein.

Let $f : \mathcal{M} \to \mathcal{N}$ be a diffeomorphism between the two smooth manifolds $\mathcal{M}, \mathcal{N}$, then we have for any $g \in C^\infty(\mathcal{N})$ and $v, w \in \Gamma(\mathcal{T}(\mathcal{M}))$

$$\begin{array}{rclcrcl} v\left(f^*(g)\right) & = & f^*\left(\tilde{v}(g)\right) & \text{or} & L_v\left(f^*(g)\right) & = & f^*\left(L_{\tilde{v}}(g)\right) \\ \widetilde{v(w)} & = & \tilde{v}(\tilde{w}) & \text{or} & \widetilde{L_v(w)} & = & L_{\tilde{v}}(\tilde{w}) \end{array} \tag{3.4}$$

with the pullback $f^*(g) = g \circ f$ of (2.1) and the pushforward $\tilde{v} = f_*(v)$ of (2.7). These relations show that the Lie derivatives have a coordinate free meaning.

Let us introduce the *commutator* or the *Lie bracket* of two vector fields v, w by

$$[v, w](g) = L_v\left(L_w(g)\right) - L_w\left(L_v(g)\right) \tag{3.5}$$

for $g \in C^\infty(\mathcal{M})$. Since the flow φ_ε with the infinitesimal generator v is locally a diffeomorphism, we may write

$$\begin{aligned} L_v\left(L_w(g)\right) - L_w\left(L_v(g)\right) &= d_\varepsilon\left(\left(w(g)\right) \circ \varphi_\varepsilon - \left(\varphi_{\varepsilon,*}w\right)\left(g \circ \varphi_\varepsilon\right)\right)\big|_{\varepsilon=0} \\ &= d_\varepsilon\left(\left(w \circ \varphi_\varepsilon\right) - \left(\varphi_{\varepsilon,*}w\right)\right)\big|_{\varepsilon=0}(g) \\ &= \left(v(w)\right)(g) \end{aligned}$$

because of (3.2), (3.3), (3.4) and therefore, the relation

$$[v, w] = L_v(w) \tag{3.6}$$

because of (3.5). The equation (3.6) shows that the commutator (3.5) is again a vector field on $\mathcal{M}$. From (3.5) and (3.6) it follows that the Lie bracket is a map $\Gamma(\mathcal{T}(\mathcal{M})) \times \Gamma(\mathcal{T}(\mathcal{M})) \to \Gamma(\mathcal{T}(\mathcal{M}))$ that satisfies

$$\begin{array}{ll} \text{a)} & [\lambda v, \mu w] = \lambda\mu[v, w] + \lambda v(\mu) w - \mu w(\lambda) v \\ \text{b)} & [v, w] = -[w, v] \\ \text{c)} & [u, [v, w]] + [v, [w, u]] + [w, [u, v]] = 0 \end{array} \tag{3.7}$$

for all $u, v, w \in \Gamma(\mathcal{T}(\mathcal{M}))$, $\lambda, \mu \in C^\infty(\mathcal{M})$. The relation c) is also called Jacobi identity. The relation a) is a consequence of the fact that Lie derivatives obey the product rule, whereas the relations b), c) follow directly from (3.5). Obviously, (3.5) offers a new interpretation of the Lie derivative (3.3). Additionally, we find a representation of $[v, w]$ in local coordinates x of $\mathcal{M}$ in a straightforward manner. Let $\bar{v}$, $\bar{w}$,

$$\bar{v} = \bar{v}^i(x)\,\partial_i\,, \qquad \bar{w} = \bar{w}^i(x)\,\partial_i \tag{3.8}$$

be the representations of v, w in the coordinates x, then the equation (3.5) implies the relation

$$[\bar{v}, \bar{w}] = \left(\bar{v}\left(\bar{w}^i\right) - \bar{w}\left(\bar{v}^i\right)\right)\partial_i\,. \tag{3.9}$$

As mentioned above, tangent vector fields are derivations and they correspond to linear partial differential equations in local coordinates. For instance, let us consider the fields $\bar{v}$, $\bar{w}$ of (3.8) and the two equations

$$\bar{v}(\bar{g}) = \bar{w}(\bar{g}) = 0 \tag{3.10}$$

in the unknown function $\bar{g} = \bar{g}(x)$. Obviously, these equations imply

$$L_{\bar{w}}(L_{\bar{v}}(\bar{g})) = L_{\bar{v}}(L_{\bar{w}}(\bar{g})) = 0$$

and therefore

$$[\bar{v}, \bar{w}](\bar{g}) = 0 \ . \tag{3.11}$$

Here, we have to distinguish two cases. If

$$[\bar{v}, \bar{w}] = \bar{\lambda}\bar{v} + \bar{\mu}\bar{w} \tag{3.12}$$

is met for functions $\bar{\lambda}, \bar{\mu} \in C^{\infty}(\mathcal{M})$, then (3.10) implies (3.11). Otherwise, the function $\bar{g}$ must fulfill the equation (3.11) in addition, and we say that (3.11) is an *integrability condition* for (3.10). Of course, if (3.12) is not met then we have to look for other integrability conditions by forming all the Lie brackets of the fields $\bar{v}$, $\bar{w}$, $[\bar{v}, \bar{w}]$ and so on. A simple dimensional argument shows that integrability conditions can only appear, if the dimension of $\mathcal{M}$ is greater than 2. This fact explains why the 2-dimensional case is something special.

At present, we have the three interpretations of the Lie bracket at our's disposal, as a Lie derivative of a vector field along another, see (3.3), as the commutator of two derivations, see (3.5), and as integrability condition for two partial differential equations, see (3.11). To see the connection of the Lie bracket with the flow, let us take the two flows φ_{ε^1}, ψ_{ε^2} with infinitesimal generators v, w on a manifold $\mathcal{M}$, then we may construct the map $\varphi_{\varepsilon^1} \circ \psi_{\varepsilon^2}(p)$ and pose the question whether

$$\varphi_{\varepsilon^1} \circ \psi_{\varepsilon^2} = \psi_{\varepsilon^2} \circ \varphi_{\varepsilon^1} \tag{3.13}$$

holds for all $(\varepsilon^1, \varepsilon^2) \in E \subset \mathbb{R}^2$, $0 \in E$. If we take the derivatives with respect to ε^1, ε^2, then we derive immediately the infinitesimal condition

$$[v, w] = 0 \ . \tag{3.14}$$

Therefore, the two flows φ_{ε^1}, ψ_{ε^2} commute, iff the Lie bracket of their infinitesimal generators v, w vanishes.

Now, based on our knowledge of the Lie derivative of a function and a tangent vector we are interested in finding a coordinate expression for the Lie derivative of a covector. This follows immediately from the requirement that the product rule

$$L_v(\langle \omega, w \rangle) = \langle L_v(\omega), w \rangle + \langle \omega, L_v(w) \rangle \tag{3.15}$$

is met for any field $v, w \in \Gamma(\mathcal{T}(\mathcal{M}))$ and $\omega \in \Gamma(\mathcal{T}^*(\mathcal{M}))$. To derive a coordinate expression, we set $\bar{v} = \bar{v}^i\partial_i$, $\bar{w} = \bar{w}^i\partial_i$ and $\bar{\omega} = \bar{\omega}_i \mathrm{d}x^i$. Now, by evaluating seperately the left and right side of (3.15) it follows

$$\bar{v}(\bar{\omega}_i)\bar{w}^i + \bar{\omega}_i\bar{v}(\bar{w}^i) = \langle L_{\bar{v}}(\bar{\omega}_i \mathrm{d}x^i), \bar{w} \rangle + \bar{\omega}_i(\bar{v}(\bar{w}^i) - \bar{w}(\bar{v}^i))$$

and finally by comparing the coefficients we get

$$L_{\bar{v}}(\bar{\omega}_i \mathrm{d}x^i) = \bar{v}(\bar{\omega}_i)\mathrm{d}x^i + \bar{\omega}_j\partial_i\bar{v}^j \mathrm{d}x^i \ . \tag{3.16}$$

Another more geometric picture offers the relation

$$L_v(\omega)(p) = v(\omega)(p) = \mathrm{d}_\varepsilon \left(\varphi_\varepsilon^* (\varphi_\varepsilon(p)) (\omega \circ \varphi_\varepsilon(p)) - \omega(p)\right)\big|_{\varepsilon=0} ,$$

which is the analogon to (3.3) for covectors. Of course, both approaches lead to the same result.

The extension of the Lie derivative to higher order tensors is straightforward and follows from the requirement that the product rule with the tensor product is met. For instance, let us choose $\omega \in \Gamma\left(\mathcal{T}_0^1(\mathcal{M})\right)$, $v \in \Gamma\left(\mathcal{T}_1^0(\mathcal{M})\right)$, then the Lie derivative of $\omega \otimes v$ along $w \in \Gamma(\mathcal{T}(\mathcal{M}))$ follows from the relations

$$L_w(\omega \otimes v) = L_w(\omega) \otimes v + \omega \otimes L_w(v) \ . \tag{3.17}$$

This approach can be extended to higher order tensor fields in a straightforward manner and will therefore not be presented here.

4 Distributions and the Theorem of Frobenius

In the previous sections we have seen that a vector field v on an m-dimensional manifold $\mathcal{M}$ generates, at least locally, a flow φ_t that corresponds to the differential equation

$$d_t \varphi_t(p) = v(\varphi_t(p))$$

or

$$\dot{x}^i = \bar{v}^i(x)$$

in the coordinates $x = \left(x^1, \ldots, x^m\right)$. Obviously, a curve $\gamma_t(p)$ forms a 1-dimensional submanifold $\mathcal{N}$ of $\mathcal{M}$ with $p \in \mathcal{N}$. If we are interested only in $\mathcal{N}$ but not in the special parametrization of the curve by t we restate the problem to find a 1-dimensional submanifold $\mathcal{N}$ of $\mathcal{M}$ such that

$$\mathcal{T}_p(\mathcal{N}) = \mathrm{span}\{v(p)\} \tag{4.1}$$

is met for all $p \in \mathcal{N}$. The symbol span $\{v(p)\}$ denotes the linear subspace of $\mathcal{T}_p(\mathcal{M})$ which is spanned by v. If the image of the curve γ_t equals $\mathcal{N}$, then we say γ_t is an *integral curve* of span $\{v(p)\}$. All integral curves of (4.1) describe the same submanifold $\mathcal{N}$ locally but they differ in their parametrization. It is worth mentioning that the problem of finding an integral curve for a smooth tangent vector field is always locally solvable.

Now we are ready to generalize the problem of finding integral curves to higher dimensional submanifolds. A *distribution* is the assignment of a linear subspace V_p of the tangent space $\mathcal{T}_p(\mathcal{M})$ of an m-dimensional manifold $\mathcal{M}$ at all points $p \in \mathcal{M}$ of its domain. If these subspaces are spanned by d vector fields, we have $V_p = \mathrm{span}\{v_1(p), \ldots, v_d(p)\}$, then we write

$$\Delta_p = \mathrm{span}\{v_1(p), \ldots, v_d(p)\} \ . \tag{4.2}$$

We call the distribution Δ_p *regular* at p, iff the vectors $v_i(p)$ are linearly independent. We say that a regular submanifold $\mathcal{N}$ of $\mathcal{M}$ is an *integral manifold* of the distribution Δ_p, iff

$$\mathcal{T}_p(\mathcal{N}) = \Delta_p \tag{4.3}$$

is met for all $p \in \mathcal{N}$. Although (4.3) is only the extension of (4.1) from 1 to d vectors, surprisingly enough, (4.3) has no solution in general. Let us take a closer look to investigate, why this can happen. From Definition 2.3 of a d-dimensional regular submanifold $\mathcal{N}$ it follows that there exist $n-d$ functionally independent functions f^{d+i} with $i = 1, \ldots, m-d$ such that $f^{d+i}(p) = 0$ is met. Let $v_1, \ldots, v_d$ be the tangent vectors of $\mathcal{N}$, then we have

$$v_j\left(f^{d+i}\right) = 0\ , \quad i = 1, \ldots, m-d, \quad j = 1, \ldots, d\ .$$

These relations imply that

$$[v_j, v_k]\left(f^{d+i}\right) = 0\ , \quad i = 1, \ldots, m-d, \quad j, k = 1, \ldots, d$$

is met. But this can hold, see (3.12), iff

$$[v_j, v_k] = \sum_{i=1}^{d} \lambda^i_{jk} v_i \tag{4.4}$$

is met. The distribution Δ of (4.2) is called *involutive*, if it satisfies (4.4) for functions $\lambda^k_{ij} \in C^\infty(\mathcal{M})$. Now, the following theorem, also called the Theorem of Frobenius, shows that regular and involutive distributions locally have integral manifolds.

Theorem 4.1. *Let $\mathcal{M}$ be a smooth m-dimensional manifold and $\Delta = \operatorname{span}\{v_1, \ldots, v_d\}$ be an involutive distribution which is regular at $p \in \mathcal{M}$, then there exists a d-dimensional integral manifold $\mathcal{N}$ such that*

$$\mathcal{T}_p(\mathcal{N}) = \Delta_p$$

is satisfied. Furthermore, we can find $n-d$ functions f^{d+i} such that

$$f^{d+i}(p) = 0\ , \qquad i = 1, \ldots, m-d$$

is fulfilled.

To show this theorem, at first we consider a regular distribution $\hat{\Delta} = \operatorname{span}\{\hat{v}_1, \ldots, \hat{v}_d\}$ which meets the requirement

$$[\hat{v}_i, \hat{v}_j] = 0\ , \qquad i, j = 1, \ldots, d\ . \tag{4.5}$$

Then all the flows $\varphi^i_{\varepsilon^i}$ commute, see (3.13), (3.14), and it is easy to see that

$$q = \varphi^1_{\varepsilon^1} \circ \cdots \circ \varphi^d_{\varepsilon^d}(p) \tag{4.6}$$

describes the integral manifold in parametrized form. Therefore, it remains to show that a regular involutive distribution Δ has a basis that meets (4.5). Let us choose a representation $\bar{\Delta} = \operatorname{span}\{\bar{v}_1, \ldots, \bar{v}_d\}$ of the distribution $\Delta = \operatorname{span}\{v_1, \ldots, v_d\}$ in coordinates $x = \left(x^1, \ldots, x^m\right)$ such that the matrix $\left[\bar{v}^j_i\right]$ is regular, then we always find a basis

$\{\bar{w}_1, \ldots, \bar{w}_d\}$ of $\bar{\Delta}$ such that $\bar{w}_i^j = \delta_i^j$, $i, j = 1, \ldots, d$ is met, with the Kronecker symbol δ_i^j. From

$$\bar{w}_{ij} = [\bar{w}_i, \bar{w}_j] = \sum_{k=1}^{d} \lambda_{ij}^k \bar{w}_k \qquad \text{with} \qquad \bar{w}_{ij}^k = 0 \,, \qquad k = 1, \ldots, d \,,$$

which stems from the fact that Δ is involutive, it follows that the functions λ_{ij}^k vanish. In these coordinates the integral manifold through the point y is given by $x = \bar{\varphi}_{\varepsilon_1}^1 \circ \cdots \circ \bar{\varphi}_{\varepsilon_d}^d (y)$. These considerations prove Theorem 4.1.

The following theorem, it is a generalization of the Flow Box Theorem 2.9, tells us that we can find coordinates that allow an even simpler representation of an integral manifold.

Theorem 4.2. *Let $\Delta = \operatorname{span}\{v_1, \ldots, v_d\}$ be a regular involutive distribution on a smooth m-dimensional manifold $\mathcal{M}$, and the neighborhood $\mathcal{U}_p$ of $p \in \mathcal{M}$ belongs to the domain of Δ, then there exist coordinates $x = (x^1, \ldots x^m)$ near p such that the integral manifold through $q \in \mathcal{U}_p$ is given by*

$$x^i = y^i + \varepsilon^i \,, \qquad i = 1, \ldots, d \,, \qquad x^{j+d} = y^{j+d} \quad , \qquad j = 1, \ldots, m - d \,,$$

where y denotes the image of q in the coordinates x.

Also the proof of this theorem is omitted here, the interested reader may find it, e.g., in (Boothby, 1986), (Choquet-Bruhat and DeWitt-Morette, 1982), (Spivak, 1979).

Next we show that, given a non-involutive distribution $\Delta = \operatorname{span}\{v_1, \ldots, v_d\}$, one can always construct an involutive distribution containing Δ, by adding suitable vector fields. We define the *involutive closure* $\bar{\Delta}$ of Δ as the smallest (with respect to the dimension) involutive distribution that meets $\Delta \subset \bar{\Delta}$. The determination of $\bar{\Delta}$ is now straightforward. In a first step we add all possible Lie brackets which can be formed by the elements of Δ and are not contained in Δ. If this extended distribution is involutive, then we are done, otherwise we have to repeat this procedure. This procedure stops after a finite number of steps, since an m-dimensional distribution on an m-dimensional manifold is involutive in a trivial manner. Also trivial because of (3.7) is the fact that a 1-dimensional distribution is always involutive. Therefore, the integrability conditions for explicit ordinary differential equations are always fulfilled.

5 The Exterior Derivative

Let $\mathcal{M}$ be a smooth m-dimensional manifold, and let $\mathcal{T}(\mathcal{M})$ and $\mathcal{T}^*(\mathcal{M})$ denote its tangent and cotangent bundle. Let us fix a vector field $v \in \Gamma(\mathcal{T}(\mathcal{M}))$, we have already discussed the map $\langle \cdot, v \rangle : \wedge^1 (\mathcal{T}^*(\mathcal{M})) \to \wedge^0 (\mathcal{T}^*(\mathcal{M})) = C^\infty(\mathcal{M})$ and the Lie derivative of a p-form, which is a map $L_v : \wedge^p (\mathcal{T}^*(\mathcal{M})) \to \wedge^p (\mathcal{T}^*(\mathcal{M}))$. Let us choose adapted coordinates $(x, \dot{x})$ such that we may write v and $f \in C^\infty(\mathcal{M})$ as $\bar{v} = \bar{v}^i \partial_i$ and $\bar{f}$. Obviously, the Lie derivative $v(f)$ can be rewritten as

$$\bar{v}(\bar{f}) = \partial_i \bar{f} \bar{v}^i = \langle \partial_i \bar{f} \mathrm{d}x^i, \bar{v}^i \partial_i \rangle \tag{5.1}$$

with the 1-form $\bar{\omega} = \partial_i \bar{f} \mathrm{d}x^i$. Although neither the functions $\partial_i \bar{f}$ nor the 1-forms $\mathrm{d}x^i$ are independent of the coordinate system, the 1-form ω, it is called the *differential* of f, is independent of the choice of coordinates. Therefore, the differential is a well defined map $\mathrm{d} : \wedge^0 (\mathcal{T}^* (\mathcal{M})) \to \wedge^1 (\mathcal{T}^* (\mathcal{M}))$. Now, we look for an extension of the map d from 0-forms to p-forms, $p > 1$. Let us introduce the special multi-index $I = i_1, \ldots, i_p$ with $i_1 < i_2 < \cdots < i_p$, then we can express any p-form ω in our coordinates as

$$\bar{\omega} = \bar{\omega}_I \mathrm{d}x^I \ , \text{with} \qquad \mathrm{d}x^I = \mathrm{d}x^{i_1} \wedge \mathrm{d}x^{i_2} \wedge \cdots \wedge \mathrm{d}x^{i_p}$$

with functions $\bar{\omega}_I \in C^\infty (\mathcal{M})$. Here, the sum has to be taken over all possible indices I. Now, we extend the map d in our special coordinate system by the construction

$$\mathrm{d}\bar{\omega} = \mathrm{d}\bar{\omega}_I \wedge \mathrm{d}x^I .$$

Astonishingly enough, this construction really leads to the extension of d to higher order p-forms such that the map d is independent of the choice of coordinates. This is the contents of the following theorem.

Theorem 5.1. *Let $\mathcal{M}$ be a smooth m-dimensional manifold. There exists a unique mapping, called the* exterior derivative,

$$\mathrm{d} : \wedge^p (\mathcal{T}^* (\mathcal{M})) \to \wedge^{p+1} (\mathcal{T}^* (\mathcal{M})) \ , \quad p = 0, 1, \ldots \quad ,$$

such that for all $\alpha \in \Gamma (\wedge^p (\mathcal{T}^* (\mathcal{M})))$, $\beta \in \Gamma (\wedge^q (\mathcal{T}^* (\mathcal{M})))$ the following rules are met.

1. $\mathrm{d}(\alpha + \beta) = \mathrm{d}\alpha + \mathrm{d}\beta$.
2. $\mathrm{d}(f)$, $f \in \Gamma (\wedge^0 (\mathcal{T}^* (\mathcal{M}))) = C^\infty (\mathcal{M})$ *is the usual* differential.
3. $\mathrm{d}(\alpha \wedge \beta) = \mathrm{d}\alpha \wedge \beta + (-1)^p \alpha \wedge \mathrm{d}\beta$.
4. $\mathrm{d}^2 (\alpha) = \mathrm{d}(\mathrm{d}\alpha) = \mathrm{d} \circ \mathrm{d} (\alpha) = 0$.

Again, for the proof of this theorem the reader is referred to the books (Boothby, 1986), (Choquet-Bruhat and DeWitt-Morette, 1982), (Frankel, 1998), (Spivak, 1979). The exterior derivative has a nice behavior with respect to the pullback. Let $f : \mathcal{M} \to \mathcal{N}$ be a smooth map between the two smooth manifolds $\mathcal{M}$, $\mathcal{N}$, then we have

$$\mathrm{d}f^* (\omega) = f^* (\mathrm{d}\omega)$$

for any $\omega \in \Gamma (\wedge (\mathcal{T}^* (\mathcal{M})))$. The reader should be aware of the fact that the same symbol d is used for the exterior derivative for $\wedge (\mathcal{T}^* (\mathcal{M}))$ and for $\wedge (\mathcal{T}^* (\mathcal{N}))$.

The extension of the map $\langle \cdot, v \rangle$ to higher order p-forms is given by the interior product.

Definition 5.2. Let $\mathcal{M}$ be a smooth m-dimensional manifold and v be a section of $\mathcal{T} (\mathcal{M})$. The map i_v,

$$i_v : \wedge^p (\mathcal{T}^* (\mathcal{M})) \to \wedge^{p-1} (\mathcal{T}^* (\mathcal{M})) \ , \quad p = 1, 2, \ldots \ ,$$

called the *interior product*, is defined by the following rules:

1. $i_{v+w} = i_v + i_w$, $w \in \Gamma (\mathcal{T} (\mathcal{M}))$,
2. $i_{fv} = f i_v$, $f \in C^\infty (\mathcal{M})$,

3. $i_v(f) = 0$ for $f \in C^\infty(\mathcal{M})$,
4. $i_v(\omega) = \langle \omega, v \rangle$ for $\omega \in \Gamma(\wedge^1(\mathcal{T}^*(\mathcal{M})))$,
5. $i_v(\omega) = \omega(v, \cdot, \ldots, \cdot)$ for $\omega \in \Gamma(\wedge^p(\mathcal{T}^*(\mathcal{M})))$, $p > 1$.

Another notation for the interior product is given by

$$v \rfloor \omega = i_v(\omega)$$

for $v \in \Gamma(\mathcal{T}(\mathcal{M}))$, $\omega \in \Gamma(\wedge^p(\mathcal{T}^*(\mathcal{M})))$, which can easily be extended to other tensors. E.g., we get

$$v \rfloor \omega \otimes w = (v \rfloor \omega) \otimes w$$

for $w \in \Gamma(\mathcal{T}(\mathcal{M}))$. The reader should convince oneself that the interior product fulfills the rule of the following theorem.

Theorem 5.3. *Let $\mathcal{M}$ be a smooth m-dimensional manifold and v be a section of $\mathcal{T}(\mathcal{M})$. The* interior product $i_v : \wedge^p(\mathcal{T}^*(\mathcal{M})) \to \wedge^{p-1}(\mathcal{T}^*(\mathcal{M}))$ *meets the relation*

$$i_v(\alpha \wedge \beta) = i_v(\alpha) \wedge \beta + (-1)^p \alpha \wedge i_v(\mathrm{d}\beta)$$

for all $\alpha \in \Gamma(\wedge^p(\mathcal{T}^*(\mathcal{M})))$, $\beta \in \Gamma(\wedge^q(\mathcal{T}^*(\mathcal{M})))$.

Now we are ready to present the extension of (5.1) to higher order p-forms, which is given by

$$v(\omega) = \mathrm{d}(i_v(\omega)) + i_v(\mathrm{d}\omega) \ . \tag{5.2}$$

The relation (5.2) is also called *Cartan's magic formula.*

The Theorem of Frobenius 4.1 tells us whether a d-dimensional regular distribution $\Delta = \mathrm{span}\{v_1, \ldots, v_d\}$ of a smooth m-dimensional manifold admits at least locally a d-dimensional integral manifold $\mathcal{N}$. Let us derive a second version of this theorem based on 1-forms. Analogously to a distribution, we call the assignment of a linear subspace

$$\Delta_p^* = \mathrm{span}\left\{\omega^1(p), \ldots, \omega^r(p)\right\} \tag{5.3}$$

of the cotangent space $\mathcal{T}_p^*(\mathcal{M})$ at all points $p \in \mathcal{M}$ of its domain a *codistribution.* Again, we say that the codistribution Δ_p^* *is regular* at p, iff the covectors $\omega^i(p)$ are linearly independent. Let us consider the distribution Δ of (4.2). The special codistribution $\Delta^\perp = \mathrm{span}\left\{\omega^{d+1}, \ldots, \omega^m\right\}$ given by

$$\left\langle \omega^i, v_j \right\rangle = 0 \ , \quad i = d+1, \ldots, m \ , \quad j = 1, \ldots, d \tag{5.4}$$

is called the *annihilator* $\Delta^\perp$ of Δ. Now we are ready to present the dual version of the Theorem of Frobenius.

Theorem 5.4. *Let $\mathcal{M}$ be a smooth m-dimensional manifold and Δ be a regular codistribution with $\Delta^\perp = \mathrm{span}\left\{\omega^{d+1}, \ldots, \omega^m\right\}$ near a point p which meets*

$$\mathrm{d}\omega^i = \alpha_j^i \wedge \omega^j \ , \quad i, j = d+1, \ldots, m \ , \tag{5.5}$$

for suitable 1-forms α_j^i. Then one can find $m - d$ functions f^i near p such that

$$\Delta^\perp = \mathrm{span}\left\{\mathrm{d}f^{d+1}, \ldots, \mathrm{d}f^m\right\}$$

is met.

This theorem follows from Theorem 4.1, the relations (5.4) and the formula

$$\mathrm{d}\omega\left(v, w\right) = v\left(\langle\omega, w\rangle\right) - w\left(\langle\omega, v\rangle\right) - \langle\omega, [v, w]\rangle \tag{5.6}$$

with $v, w \in \Gamma\left(\mathcal{T}\left(\mathcal{M}\right)\right)$, $\omega \in \Gamma\left(\wedge^1\left(\mathcal{T}\left(\mathcal{M}\right)\right)\right)$. The reader should convince oneself that the condition

$$\bigwedge_{j=d+1}^{m} \omega^j \wedge \mathrm{d}\omega^i = 0 \tag{5.7}$$

is equivalent to (5.5), but (5.7) is sometimes easier to check.

It has already be shown that alternating tensors or p-forms have a very rich structure. One of their most important properties is that they describe geometric objects which we can integrate. Again, the reader is asked to consult one of the books (Boothby, 1986), (Choquet-Bruhat and DeWitt-Morette, 1982), (Frankel, 1998), (Spivak, 1979), since we can present here the most important results for our purpose only.

Although the theory of integration can be extended to more complex cases, we assume from now on that the manifolds under consideration can be oriented in the following way.

Definition 5.5. Let $\mathcal{M}$ be a smooth m-dimensional manifold. $\mathcal{M}$ is said to be *orientable*, if there exists a non-vanishing smooth m-form Ω. In this case, we say $\mathcal{M}$ is oriented by Ω.

Given $\Delta = \operatorname{span}\{v_1, \ldots, v_m\}$, $v_i \in \Gamma\left(\mathcal{T}\left(\mathcal{M}\right)\right)$ such that $\Delta_p = \mathcal{T}_p\left(\mathcal{M}\right)$ then we can get $\Omega\left(v_1, \ldots, v_n\right) > 0$ or $\Omega\left(v_1, \ldots, v_n\right) < 0$. Therefore, we can distinguish two orientations on $\mathcal{M}$ with respect to Ω. Obviously, the required volume form for the special manifold $\mathbb{R}^m$ is given by $\Omega = \mathrm{d}x^1 \wedge \ldots \wedge \mathrm{d}x^m$ and $(\partial_1, \ldots, \partial_m)$ is a positively oriented basis of $\mathbb{R}^m$.

Of further interest are manifolds with boundaries. Let us consider the manifold $\mathbb{R}^m$ together with the half space $H^m = \left\{x = \left(x^1, \ldots, x^m\right) \in \mathbb{R}^m,\ x^m \geq 0\right\}$, the simplest manifold with boundary. The subspace $\partial H^m = \{x \in H^m,\ x^m = 0\}$ is called the boundary of H. Since ∂H^m is a submanifold of H^m, as well as of $\mathbb{R}^m$, ∂H^m inherits the submanifold topology as following. A set U of ∂H^m is said to be open, iff $U = V \cap \partial H^m$ is met for an open set V of H^m or $\mathbb{R}^m$. The map $\left(x^1, \ldots, x^{m-1}\right) \to \left(x^1, \ldots, x^{m-1}, 0\right)$ allows us to identify ∂H^m with $\mathbb{R}^{m-1}$. Roughly speaking, we may use Definition 2.1 for a manifold with boundary, iff we replace there $\mathbb{R}^m$ by H^m. This fact allows us to define the boundary of an abstract manifold by the boundary of the coordinate charts and we get the following result.

Theorem 5.6. *Let $\mathcal{M}$ be a smooth m-dimensional manifold with boundary, then the inclusion map $i : \partial\mathcal{M} \to \mathcal{M}$ is an imbedding (i is one-to-one,* $\operatorname{rank}(i) = m - 1$, *and the topology of $i\left(\partial\mathcal{M}\right)$ equals the submanifold topology.*

The reader is asked to consult the cited literature for the case of a non-smooth boundary. It is straightforward to see that the manifold H^m is orientable, since we can always choose $\Omega = \mathrm{d}x^1 \wedge \ldots \wedge \mathrm{d}x^m$ (see Definition 5.5) for $\mathbb{R}^m$. Given a point $p \in \partial H^m$ and a tangent vector $v \in \Gamma\left(\mathcal{T}_p\left(\mathbb{R}^m\right)\right)$, then v is called an inward (outward) pointing vector, iff $v^m > 0$ ($v^m < 0$) is met for $v = v^i\partial_i$. Obviously, $v^m = 0$ implies $v \in \Gamma\left(\mathcal{T}\left(\partial H^m\right)\right)$. Let us choose the outward pointing vector $-\partial_m$. A suitable choice for the volume form on ∂H^m follows from the relation $\mathrm{i}_{-\partial_m}(\Omega) = (-1)^m \mathrm{d}x^1 \wedge \ldots \wedge \mathrm{d}x^{m-1}$. This approach can be extended to abstract manifolds.

Theorem 5.7. *Let $\mathcal{M}$ be a smooth m-dimensional orientable manifold with boundary $\partial\mathcal{M} \neq \{\}$, then the orientation of $\mathcal{M}$ induces an orientation on $\partial\mathcal{M}$.*

From now on, we will use the orientation for $\partial\mathcal{M}$ as described above. Based on this convention, we are ready to present Stoke's Theorem.

Theorem 5.8. *Let $\mathcal{M}$ be a smooth, compact, m-dimensional, oriented manifold with boundary and let ω be a smooth $(m-1)$-form defined on $\mathcal{M}$, then*

$$\int_{\mathcal{M}} \mathrm{d}\omega = \int_{\partial\mathcal{M}} \omega$$

is fulfilled.

Again, for the proof we refer the reader to the cited literature. Now, let us state some examples for Theorem 5.8. When $m = 1$ and $\mathcal{M}$ is the interval $[a, b]$, then Stoke's Theorem is the fundamental theorem of calculus

$$\int_a^b \mathrm{d}f = f(a) - f(b) \ .$$

Now, let us choose a compact, oriented 2-dimensional submanifold $\mathcal{M} \subset \mathbb{R}^2$ together with the smooth 1-form $\omega = a_i(x^1, x^2)\,\mathrm{d}x^i$. Stoke's Theorem says in this case

$$\int_{\mathcal{M}} (\partial_1 a_2 - \partial_2 a_1)\,\mathrm{d}x^1 \wedge \mathrm{d}x^2 = \int_{\partial\mathcal{M}} (a_1 \mathrm{d}x^1 + a_2 \mathrm{d}x^2) \ .$$

The reader is invited to compare this result with the well known Green's Theorem of advanced calculus. Let us repeat the same procedure for a compact, oriented, 3-dimensional submanifold $\mathcal{M} \subset \mathbb{R}^2$ together with the smooth 2-form $\omega = a^i(x^1, x^2, x^3)\,\mathrm{i}_{\partial_i}(\Omega)$, $\Omega = \mathrm{d}x^1 \wedge \mathrm{d}x^2 \wedge \mathrm{d}x^3$, then Stoke's theorem asserts

$$\int_{\mathcal{M}} \partial_i a^i \Omega = \int_{\partial\mathcal{M}} \omega \ ,$$

which is nothing else than the *divergence theorem* of advanced calculus also known as Gauss' Theorem. Finally, we take a compact, oriented 2-dimensional manifold $\mathcal{M}$. We consider $\mathcal{M}$ as a piece of a surface embedded in $\mathbb{R}^3$ and bounded by a simple curve. Given the smooth 1-form $\omega = a_i(x^1, x^2)\,\mathrm{d}x^i$, we may consider its restriction to $\mathcal{M}$ or $\partial\mathcal{M}$. Here, Stoke's theorem asserts

$$\int_{\mathcal{M}} (\partial_i a^j - \partial_j a^i)\,\mathrm{d}x^i \wedge \mathrm{d}x^j = \int_{\partial\mathcal{M}} \omega \ ,$$

where the sum is taken for $i < j$. Obviously, this case is nothing else than the *Stoke's theorem* of advanced calculus.

A special case of Theorem 5.4 is $\mathrm{d}\omega^i = 0$, since the relation (5.5) is met in a trivial manner. Now, let ω denote an arbitrary p-form. We say that ω is *closed*, if $\mathrm{d}\omega = 0$ is met. The p-form ω is said to be *exact*, if there exist a $(p-1)$-form α such that $\mathrm{d}\alpha = \omega$ is fulfilled. Obviously, every exact form is closed, but the converse is not true in general. Let us consider the unit circle $\mathbb{S}^1 \subset \mathbb{R}^2$ and the 1-form

$$\omega = \frac{-x^2 \mathrm{d}x^1 + x^1 \mathrm{d}x^2}{\left(x^1\right)^2 + \left(x^2\right)^2} = \mathrm{d}\left(\arctan \frac{x^2}{x^1}\right)$$

on $\mathbb{R}^2 \backslash \{0\}$. The pull back of ω to $\mathbb{S}^1$ by $f^1(\varphi) = x^1 = \cos\varphi$, $f^2(\varphi)\, x^2 = \sin\varphi$ is $f^*(\omega) = \mathrm{d}\varphi$. But φ is not a C^1 function on $\mathbb{S}^1$. The unit circle has the obvious property that it is not contractible to a point. The following theorem, also called Poincaré's lemma, tells us that for contractible sets, closed forms are also exact.

Theorem 5.9. *Let $\mathcal{M}$ be a smooth, m-dimensional manifold and $\mathcal{U}$ an open subset diffeomorphic to $\mathbb{R}^m$, then all closed p-forms are exact.*

Again, the reader is asked to consult one of the books (Boothby, 1986), (Choquet-Bruhat and DeWitt-Morette, 1982), (Frankel, 1998), (Spivak, 1979) for the proof of this theorem.

6 Riemannian Manifolds

Let us consider again covariant tensor fields in $\mathcal{T}_0^k(\mathcal{M})$, where we denote the subspace of symmetric covariant tensor fields by $\Gamma\left(S^k\left(\mathcal{T}^*(\mathcal{M})\right)\right)$. A Riemannian manifold is, roughly speaking, an m-dimensional smooth manifold together with a symmetric, positive definite section of $S^2\left(\mathcal{T}^*(\mathcal{M})\right)$.

Definition 6.1. A smooth m-dimensional manifold $\mathcal{M}$ on which a field g of smooth symmetric, positive definite, bilinear forms is defined, is called a *Riemannian manifold* with *Riemannian metric* g.

The simplest case of a Riemannian manifold is $\mathbb{R}^m$ together with $g = \delta_{ij} \mathrm{d}x^i \otimes \mathrm{d}x^j$, which describes the usual inner product

$$(v, w) = g(v, w) = v^i \delta_{ij} w^j$$

for $v, w \in \Gamma\left(\mathcal{T}\left(\mathbb{R}^m\right)\right)$.

On a Riemannian manifold $\mathcal{M}$, we may define an inner product $(\cdot,\cdot) : \mathcal{T}(\mathcal{M})^2 \to \mathbb{R}$ with the usual properties simply by $(v, w) = g(v, w)$ with $v, w \in \Gamma\left(\mathcal{T}(\mathcal{M})\right)$. Obviously, the metric allows us to measure the length of tangent vectors or the length of a curve $\gamma(\tau) : \mathbb{R} \to \mathcal{M}$ by

$$L = \int_a^b \left(\left(\partial_1 \gamma^i\right) \partial_i, \left(\partial_1 \gamma^i\right) \partial_i\right)^{1/2} \mathrm{d}\tau \,.$$

Additionally, one can show that any connected differentiable manifold can be made into a metric space, i.e., a space on which a metric is defined.

To simplify the following, let us choose coordinates $x = \left(x^1, \ldots, x^m\right)$ for $\mathcal{M}$. The metric tensor in these coordinates is given by $\bar{g} = \bar{g}_{ij} \mathrm{d}x^i \otimes \mathrm{d}x^j$, $\bar{g}_{ij} = \bar{g}_{ji}$ for suitable functions $\bar{g}_{ij}$ defined on $\mathcal{M}$. Now, it is straightforward to see that the relation

$$\bar{p}_i = \bar{g}_{ij} \bar{v}^j \tag{6.1}$$

defines a map $\mathcal{T}(\mathcal{M}) \to \mathcal{T}^*(\mathcal{M})$ such that

$$\langle \bar{p}, \bar{w} \rangle = \bar{p}_i \bar{w}^i = \bar{v}^i \bar{g}_{ij} w^j = (\bar{v}, \bar{w})$$

is met for all $v, w \in \Gamma(\mathcal{T}(\mathcal{M}))$, $p \in \Gamma(\mathcal{T}^*(\mathcal{M}))$ with $\bar{v} = \bar{v}^i \partial_i$, $\bar{w} = \bar{w}^i \partial_i$, $\bar{p} = \bar{p}_i \mathrm{d}x^i$. Let $\left[\bar{g}^{ij}\right]$ denote the inverse matrix of $[\bar{g}_{ij}]$, then the map inverse to (6.1) is simply given by

$$\bar{v}^i = \bar{g}^{ij} \bar{p}_j \ , \tag{6.2}$$

which allows us to map a covector p to a tangent vector v.

On a Riemannian manifold we can define a new derivation, called the *covariant derivative*, which is based on a new structure, on the connection.

Definition 6.2. A smooth *connection* on an m-dimensional, smooth Riemannian manifold $\mathcal{M}$ is a smooth mapping $\nabla : \Gamma(\mathcal{T}(\mathcal{M})) \times \Gamma(\mathcal{T}(\mathcal{M})) \to \Gamma(\mathcal{T}(\mathcal{M}))$ denoted by $(v, w) \to \nabla_v(w)$, which has the following properties

1. $\nabla_{f^1 v_1 + f^2 v_2}(v_3) = f^1 \nabla_{v_1}(v_3) + f^2 \nabla_{v_2}(v_3)$
2. $\nabla_{v_3}\left(f^1 v_1 + f^2 v_2\right) = v_3\left(f^1\right) v_1 + f^1 \nabla_{v_3}(v_1) + v_3\left(f^2\right) v_2 + f^2 \nabla_{v_3}(v_2)$
   for all $v_1, v_2, v_3 \in \Gamma(\mathcal{T}(\mathcal{M}))$, $f^1, f^2 \in C^\infty(\mathcal{M})$.

If the following properties

3. $[v_1, v_2] = \nabla_{v_1}(v_2) - \nabla_{v_2}(v_1)$
4. $v_3((v_1, v_2)) = (\nabla_{v_3}(v_1), v_2) + (v_1, \nabla_{v_3}(v_2))$

are met in addition, then the connection is called a Riemannian connection.

Until now, we do not know whether a connection exists. The following theorem, the fundamental theorem of Riemannian geometry, gives an answer to this problem.

Theorem 6.3. *Let $\mathcal{M}$ be an m-dimensional, smooth Riemannian manifold, then there exists a uniquely defined Riemannian connection.*

Again, we omit the proof. The derivative $\nabla_{v_1}(v_2)$ is also called the *covariant derivative.*

The properties 1 to 4 of Definition (6.2) uniquely determine the connection and therefore, the covariant derivative. The next theorem tells us its coordinate expressions.

Theorem 6.4. *Let $x = \left(x^1, \ldots, x^m\right)$ be local coordinates for the m-dimensional, smooth Riemannian manifold $\mathcal{M}$ with metric g. Let $\bar{v} = \bar{v}^i \partial_i$, $\bar{w} = \bar{w}^i \partial_i$ be vector fields $v, w \in \Gamma(\mathcal{T}(\mathcal{M}))$ in these coordinates, then the covariant derivative of w along v is given by*

$$\nabla_{\bar{v}}(\bar{w}) = \left(\bar{v}\left(\bar{w}^i\right) + \Gamma^i_{jk} \bar{v}^j \bar{w}^k\right) \partial_i$$

with the Christoffel symbols *(of the second kind)*

$$\Gamma^i_{jk} = \frac{1}{2} \bar{g}^{il} \left(\partial_j \bar{g}_{kl} + \partial_k \bar{g}_{jl} - \partial_l \bar{g}_{jk}\right) \ ,$$

which follow from the metric $\bar{g} = \bar{g}_{ij} \mathrm{d}x^i \otimes \mathrm{d}x^j$.

The reader is asked to prove this theorem.

Analogously to the Lie derivative one can extend the covariant derivative to higher order tensors. Let $f \in C^\infty(\mathcal{M})$ be a function defined on $\mathcal{M}$, then we set

$$\nabla_v(f) = v(f)$$

for $v \in \Gamma(\mathcal{T}(\mathcal{M}))$. The covariant derivative of a covector $\omega \in \Gamma(\mathcal{T}^*(\mathcal{M}))$ follows from the requirement that the product rule

$$\nabla_v(\langle \omega, w \rangle) = \langle \nabla_v(\omega), w \rangle + \langle \omega, \nabla_v(w) \rangle \tag{6.3}$$

is met for any field $w \in \Gamma(\mathcal{T}(\mathcal{M}))$. To derive a coordinate expression, we set $\bar{v} = \bar{v}^i \partial_i$, $\bar{w} = \bar{w}^i \partial_i$ and $\bar{\omega} = \bar{\omega}_i \mathrm{d}x^i$. Now, from Theorem 6.4 and (3.15) it follows

$$\bar{v}(\bar{\omega}_i)\bar{w}^i + \bar{\omega}_i \bar{v}(\bar{w}^i) = \langle \nabla_v(\bar{\omega}_i \mathrm{d}x^i), \bar{w}^i \partial_i \rangle + \bar{\omega}_i \left(\bar{v}(\bar{w}^i) + \Gamma^i_{jk} \bar{v}^j \bar{w}^k\right)$$

and finally

$$\nabla_{\bar{v}}(\bar{\omega}_i \mathrm{d}x^i) = \bar{v}(\bar{\omega}_i)\mathrm{d}x^i - \bar{\omega}_k \Gamma^k_{ji} \bar{v}^j \mathrm{d}x^i \ . \tag{6.4}$$

The extension of the covariant derivative to higher order tensors is straightforward and follows from the requirement that the product rule with the tensor product is met. For instance, let us choose $\omega \in \Gamma(\mathcal{T}^1_0(\mathcal{M}))$, $v \in \Gamma(\mathcal{T}^0_1(\mathcal{M}))$, then the covariant derivative of $\omega \otimes v$ along $w \in \Gamma(\mathcal{T}(\mathcal{M}))$ follows from the relations

$$\nabla_w(\omega \otimes v) = \nabla_w(\omega) \otimes v + \omega \otimes \nabla_w(v) \ . \tag{6.5}$$

Due to our knowledge of the coordinate expression of the covariant derivative of the terms ω and v we immediately get a coordinate expression for the left hand side of equation (6.5). Accordingly, this approach can be extended to higher order tensor fields in a straightforward manner.

Before we close this section, we mention that on a Riemannian manifold there exists a distinguished volume form, whose structure is preserved by coordinate changes. Let us choose coordinates $x = (x^1, \ldots, x^m)$, then this volume form is given by

$$\omega = \sqrt{\bar{g}}\mathrm{d}x^1 \wedge \cdots \wedge \mathrm{d}x^m \ , \qquad \sqrt{\bar{g}} = \det[\bar{g}_{ij}] \ ,$$

where $\bar{g}$ denotes the metric tensor in these coordinates.

7 Jet Manifolds

Jet manifolds are manifolds, which play an important role for systems of partial differential equations like they appear in continuum mechanics. Since we are able to give only a very short overview, the interested reader is referred to, e.g., (Olver, 1995), (Saunders, 1989) for a detailed treatment of this subject.

Roughly speaking, a jet manifold is the extension of a given fibred manifold $(\mathcal{E}, \pi, \mathcal{B})$ with local coordinates (x^i), $i = 1, \ldots, p$, for $\mathcal{B}$ and (x^i, u^α), $\alpha = 1, \ldots, q$, for $\mathcal{E}$ by adding all partial derivatives of sections with respect to the independent coordinates up to a certain order such that the new structures are again fibred manifolds. In order to see the basic idea of a jet manifold, let us consider a section $\sigma \in \Gamma(\mathcal{E})$ at a point $p \in \mathcal{B}$ with its equivalence class of all sections $\gamma \in \Gamma(\mathcal{E})$ satisfying

$$\gamma(p) = \sigma(p) \quad \text{and} \quad \partial_i \gamma(p) = \partial_i \sigma(p) . \tag{7.1}$$

We call two sections σ and γ satisfying (7.1) *1-equivalent* at p and the equivalence class defined by (7.1) the *1-jet* of σ, which is denoted by $j^1(\sigma)(p)$ (or $j(\sigma)(p)$). Based on this preliminary conventions we can define the first jet manifold

Definition 7.1. The first jet manifold $J^1(\mathcal{E})$ (or $J(\mathcal{E})$) of $(\mathcal{E}, \pi, \mathcal{B})$ is the set

$$\left\{ j^1(\sigma)(p) : p \in \mathcal{B}, \sigma \in \Gamma(\mathcal{E}) \right\} .$$

We see, that the notion of a jet manifold can be defined without reference to a special coordinate system. However, we can again find so-called adapted coordinates as following. From now on, we use ordered multi-indices to indicate partial derivatives of arbitrary order. Let f be a smooth section of the bundle $(\mathcal{E}, \pi, \mathcal{B})$. The kth order partial derivatives of f will be denoted by

$$\frac{\partial^k}{\partial_1^{j_1} \cdots \partial_p^{j_p}} f = \partial_J f \quad , \quad \partial_i = \frac{\partial}{\partial x^i}$$

with the ordered multi index $J = j_1, \ldots, j_p$, and $k = \#J = \sum_{i=1}^{p} j_i$. The special index $J = j_1, \ldots, j_p$, $j_i = \delta_{ik}$ will be denoted by 1_k and $J + 1_k$ is a shortcut for $j_i + \delta_{ik}$ with the Kronecker symbol δ_{ik}.

Definition 7.2. Given a bundle $(\mathcal{E}, \pi, \mathcal{B})$ and adapted coordinates (x^i, u^α) with $i = 1, \ldots, p$ and $\alpha = 1, \ldots, q$. The induced coordinate system for $J^1(\mathcal{E})$ is defined by $(x^i, u^\alpha, u^\alpha_{1_i})$. Given a section $\sigma \in \Gamma(\mathcal{E})$, i.e., in adapted coordinates $u^\alpha = \sigma^\alpha(x)$ holds, the pq new, so-called derivative coordinates $u^\alpha_{1_i}$ satisfy $u^\alpha_{1_i} = \partial_i \sigma^\alpha(x)$.

On the first jet manifold $J^1(\mathcal{E})$ there exist two natural projections denoted by $\pi^1 : J^1(\mathcal{E}) \to \mathcal{B}$ and $\pi^1_0 : J^1(\mathcal{E}) \to \mathcal{E}$ defined as $\pi^1(x^i, u^\alpha, u^\alpha_{1_i}) = x$ and $\pi^1_0(x^i, u^\alpha, u^\alpha_{1_i}) = (x^i, u^\alpha)$. One can show that $(J^1(\mathcal{E}), \pi^1_0, \mathcal{E})$ is an affine bundle, whereas $(J^1(\mathcal{E}), \pi^1, \mathcal{B})$ is a bundle only if $\mathcal{E}$ is one, see (Saunders, 1989). It is worth mentioning that sections of these fibred manifolds are not necessarily jets of sections of $\mathcal{E}$. E.g., this is true for a section σ of $(J^1(\mathcal{E}), \pi^1, \mathcal{B})$ with $u^\alpha = \sigma^\alpha$, $u^\alpha_{1_i} = \sigma^\alpha_{1_i}$, iff $\partial_i \sigma^\alpha = \sigma^\alpha_{1_i}$ is met.

Analogously to the first jet manifold, the nth-jet manifold $J^n(\mathcal{E})$ of $\mathcal{E}$ can be constructed. Two sections σ,$\gamma \in \Gamma(\mathcal{E})$ are said to be *n-equivalent* at p, if they meet $\partial_I \sigma(p) = \partial_I \gamma(p)$ with $0 \leq \#J \leq n$. This equivalence class is called the *n-jet* of σ. It is denoted by $j^n(\sigma)(p)$. For $J^n(\mathcal{E})$ there exist the natural projections $\pi^n : J^n(\mathcal{E}) \to \mathcal{B}$ and $\pi^n_k : J^n(\mathcal{E}) \to J^k(\mathcal{E})$ with $n > k \geq 0$ and $J^0(\mathcal{E}) = \mathcal{E}$.

Jet manifolds are equipped with a very rich geometric structure. An important one is given by special vector fields d_i, the so-called total derivatives with respect to the independent coordinates x^i of nth-order. In adapted coordinates (x^i, u^α) the $d_i \in \Gamma\left(\pi^{n+1,*}_n\left(\mathcal{T}\left(J^n(\mathcal{E})\right)\right)\right)$ take the form

$$d_i = \partial_i + u^\alpha_{J+1_i} \partial^J_\alpha \quad , \text{with} \quad \partial^J_\alpha = \frac{\partial}{\partial u^\alpha_J} \quad , \quad \#J = 0, \ldots, n . \tag{7.2}$$

Given a section $\sigma \in \Gamma(\mathcal{E})$ and a smooth real valued function $f \in C^\infty(J^n(\mathcal{E}))$, then d_i is the map $J^n(\mathcal{E}) \to J^{n+1}(\mathcal{E})$ which meets

$$d_i(f) \circ j^{n+1}(\sigma) = \partial_i(f \circ j^n(\sigma))$$

for any f. The dual objects to the fields d_i are the 1-forms $\omega_J^\alpha \in \Gamma\left(\pi_n^{n+1,*}(\mathcal{T}^*(J^n\mathcal{E}))\right)$, called contact forms,

$$\omega_J^\alpha = \mathrm{d}u_J^\alpha - u_{J+1_i}^\alpha \mathrm{d}x^i \ , \quad \#J = 0, \ldots, n \, . \tag{7.3}$$

Total derivatives and contact forms are related by $d_i \rfloor \omega_J^\alpha = 0$.

Now, let us take a short look at the bundle $\left(J^1(\mathcal{E}), \pi_0^1, \mathcal{E}\right)$ of the fibred manifold $(\mathcal{E}, \pi, \mathcal{B})$. There is a unique bundle monomorphism $\Lambda : J^1(\mathcal{E}) \hookrightarrow \mathcal{T}^*(\mathcal{B}) \otimes_{\mathcal{E}} \mathcal{T}(\mathcal{E})$ given by

$$\Lambda = \mathrm{d}x^i \otimes \left(\partial_i + u_{1_i}^\alpha \partial_\alpha\right) \tag{7.4}$$

such that $\Lambda \circ j(\sigma) = T(\sigma)$ is met with the tangent map $T(\sigma) : \mathcal{T}(\mathcal{B}) \to \mathcal{T}(\mathcal{E})$ induced by $\sigma : \mathcal{B} \to \mathcal{E}$. Roughly speaking, $\Lambda \circ j(\sigma)$ describes the Jacobian of σ. Furthermore, the fields $\partial_i + u_{1_i}^\alpha \partial_\alpha$ are nothing else then the first order total derivatives. This relation shows, that the first jet bundle and the tangent bundle are connected with each other.

7.1 Systems of Differential Equations

Jet manifolds are a powerful tool to represent differential equations. They allow to separate the properties of the equations from the properties of their solution in a clear manner. To see this, let us consider a bundle $(\mathcal{E}, \pi, \mathcal{B})$ with adapted coordinates (x^i, u^α), $i = 1, \ldots, p$, $\alpha = 1, \ldots, q$ and its nth-order jet bundle $J^n(\mathcal{E})$. Because of Definition 2.3, a regular submanifold $\mathcal{S}_p^n \subset J^n(\mathcal{E})$, $p \in \mathcal{S}^n$ corresponds locally to a system of (algebraic) equations

$$f^l\left(x, u^{(n)}\right) = 0 \ , \quad l = 1, \ldots, n_e \tag{7.5}$$

involving jet coordinates $u^{(n)}$ of u up to the order n. If we substitute the nth-jet $j^n(\sigma)$ of a smooth section σ into (7.5), then we get a system of ordinary $(p = 1)$ or partial differential equations $(p > 1)$ of the type

$$f^l \circ j^n(\sigma(x)) = 0 \ , \quad l = 1, \ldots, n_e \, . \tag{7.6}$$

In this context, σ denotes a solution of (7.6).

From now on, we will deal exclusively with the submanifold $\mathcal{S}_p^n$ instead of the differential equations (7.5). Moreover in order to state the following operations we identify the submanifold $\mathcal{S}_p^n$ with its ideal $I\left(\mathcal{S}_p^n\right)$, given by the set of functions $f \in C^\infty(J^n(\mathcal{E}))$ such that $f\left(x^i, u^{(n)}\right) = 0$ for all $\left(x^i, u^{(n)}\right) \in \mathcal{S}_p^n$. Now, there exist two natural operations for $\mathcal{S}_p^n$ and $I\left(\mathcal{S}_p^n\right)$, their projection and their prolongation, induced by the operations defined in this section (Pommaret, 1978). The projection of $I\left(\mathcal{S}_p^n\right)$ is simply given by $I\left(\mathcal{S}_p^n\right) \cap C^\infty(J^m(\mathcal{E}))$ and will be denoted by $\pi_m^n\left(I\left(\mathcal{S}_p^n\right)\right)$. We assume that the ideal

$\pi_m^n\left(I\left(\mathcal{S}_p^n\right)\right)$ describes a regular submanifold of $J^m(\mathcal{E})$ denoted by $\pi_m^n\left(\mathcal{S}_p^n\right)$, the projection of $\mathcal{S}_p^n$. To define the prolongation of $\mathcal{S}_p^n$, we use the total derivative d_i to define the ideal

$$J\left(I\left(\mathcal{S}_p^n\right)\right) = I\left(\mathcal{S}_p^n\right) + \left\langle d_i f,\ i = 1, \ldots, p,\ f \in I\left(\mathcal{S}_p^n\right)\right\rangle$$

which denotes the first prolongation of $I\left(\mathcal{S}_p^n\right)$. Obviously, the ideal $J^1\left(I\left(\mathcal{S}_p^n\right)\right)$ contains all functions of $I\left(\mathcal{S}_p^n\right)$, as well as their total derivatives with respect to the independent variables. Again, we assume that $J\left(I\left(\mathcal{S}_p^n\right)\right)$ defines a regular submanifold, which we denote by $J\left(\mathcal{S}_p^n\right)$ by some abuse of notation. Accordingly, the kth prolongation of $I\left(\mathcal{S}_p^n\right)$ is defined by $J^k\left(I\left(\mathcal{S}_p^n\right)\right) = J\left(J^{k-1}\left(I\left(\mathcal{S}_p^n\right)\right)\right)$ and $J^0\left(I\left(\mathcal{S}_p^n\right)\right) = I\left(\mathcal{S}_p^n\right)$. Given a minimal basis $B\left(\mathcal{S}_p^n\right) = \left\{f^1, \ldots, f^k\right\}$ of $I\left(\mathcal{S}_p^n\right)$, which always exists as long as $\mathcal{S}_p^n$ is a regular submanifold, one can construct a minimal basis of $J\left(I\left(\mathcal{S}_p^n\right)\right)$ in a straightforward manner. We just determine the set $B \cup \left\{d_i\left(f^j\right) : j = 1, \ldots, k,\ i = 1, \ldots, p\right\}$ and take out a minimal number of functions such that the members of the remaining set $B\left(J\left(\mathcal{S}_p^n\right)\right)$ are functionally independent.

Let σ be a section of $\mathcal{E}$ such that $j^n(\sigma)(p) \in \mathcal{S}_p^n$ is met. By construction any $\sigma \in \Gamma(\mathcal{E})$, which solves (7.5) is a solution of the prolonged and projected systems. Unfortunately, this is not true for points in general, since a point $j^{n+r}\sigma(p)$ with $\pi_n^{n+r}\left(j^{n+r}\sigma\right)(p) \in J^n(\mathcal{E})$ may fail to meet $j^{n+r}\sigma(p) \in J^r\left(\mathcal{S}_p^n\right)$. This is a result of the fact that in general $I\left(\mathcal{S}_p^n\right) \subset \pi_n^{n+r}\left(J^r\left(I\left(\mathcal{S}_p^n\right)\right)\right)$ or $\pi_n^{n+r}\left(J^r\left(\mathcal{S}_p^n\right)\right) \subset \mathcal{S}_p^n$ is met only. But formally integrable systems do not show this unpleasant behavior.

Definition 7.3. Let $\mathcal{S}_p^n$ be a locally regular submanifold of the nth jet manifold $J^n(\mathcal{E})$ of the bundle $\mathcal{E}$. We call the system $\mathcal{S}_p^n$ *formally integrable*, iff its rth prolongations $J^r\left(\mathcal{S}_p^n\right)$ are regular submanifolds of $J^{n+r}(\mathcal{E})$ and

$$\pi_{n+r}^{n+r+s}\left(J^{r+s}\left(\mathcal{S}_p^n\right)\right) = J^r\left(\mathcal{S}_p^n\right)$$

is met for all $r, s \geq 0$.

Roughly speaking, formally integrable systems do not contain hidden constraints. These constraints are often called integrability conditions. The task of finding all the hidden constraints for a given system is called completion. It is an extremely hard job to complete a general system of the form (7.5). Fortunately, this is not true for systems of ordinary differential equations and systems of linear partial differential equations of the Frobenius type. In Section 4 we have already seen, how completion is carried out for Frobenius type systems. The reader is invited to consider the problem to derive the closure for a given distribution formed by vector fields in this context.

Due to lack of space we cannot go into the details and many aspects of jet theory here. Based on the basic concept of jet manifolds and formal integrability many issues in control theory and mechanics can be treated successfully, as the reader will verify by studying other chapters of this book. For a more detailed treatment of this topic in view of explicit ordinary dynamic control systems the reader is referred to, e.g., (Zehetleitner and Schlacher, 2003).

8 Conclusion

This contribution has the goal to present the reader an overview of methods and ideas from differential geometry, which are used in other parts of this book. Of course, this overview is far from being complete. Therefore, several text books on these topics are cited and the interested reader is invited to study them for deeper information. Nevertheless, this contribution is supposed to offer some support and can be used as an essential reference for the study of the other parts of this book.

Bibliography

W. M. Boothby. *An Introduction to Differentiable Manifolds and Riemanmnian Geometry.* Academic Press, Inc., Orlando, USA, 1986.

Y. Choquet-Bruhat and C. DeWitt-Morette. *Analysis, Manifolds and Physics.* North Holland, Amsterdam, 1982.

Th. Frankel. *The Geometry of Physics, An Introduction.* Cambridge University Press, Cambridge, UK, 1998.

P. J. Olver. *Equivalence, Invariants, and Symmetry.* Cambridge University Press, Cambridge, UK, 1995.

J. F. Pommaret. *Systems of Partial Differential Equations and Lie Pseudogroups.* Gordon and Breach Science Publishers, New York, USA, 1978.

D. J. Saunders. *The Geometry of Jet Bundles.* Cambridge University Press, Cambridge, UK, 1989.

M. Spivak. *Differential Geometry, Vol. 1 to 5.* Publish or Perish, Inc., Houston, Texas, 1979.

K. Zehetleitner and K. Schlacher. Computer algebra methods for implicit dynamic systems and applications. *Mathematics and Computers in Simulation*, to appear, 2003.

Port-Hamiltonian Systems: Network Modeling and Control of Nonlinear Physical Systems

A.J. van der Schaft
Department of Applied Mathematics, University of Twente, Enschede, The Netherlands

Abstract It is shown how *port-based modeling* of lumped-parameter complex physical systems (multi-body systems, electrical circuits, electromechanical systems, ..) naturally leads to a geometrically defined class of systems, called *port-Hamiltonian systems*. These are Hamiltonian systems defined with respect to a power-conserving geometric structure capturing the basic interconnection laws, and a Hamiltonian function given by the total stored energy. The structural properties of port-Hamiltonian systems are discussed, in particular the existence of Casimir functions and its implications for stability and stabilization. Furthermore it is shown how passivity-based control results from interconnecting the plant port-Hamiltonian system with a controller port-Hamiltonian system, leading to a closed-loop port-Hamiltonian system. Finally, extensions to the distributed-parameter case are provided by formulating boundary control systems as infinite-dimensional port-Hamiltonian systems.

1 Introduction

Nonlinear systems and control theory has witnessed tremendous developments over the last three decades, see for example the textbooks (Isidori, 1995; Nijmeijer and van der Schaft, 1990). Especially the introduction of geometric tools like Lie brackets of vector fields on manifolds has greatly advanced the theory, and has enabled the proper generalization of many fundamental concepts known for *linear control systems* to the nonlinear world. While the emphasis in the seventies and the eighties has been primarily on the *structural* analysis of smooth nonlinear dynamical control systems, in the nineties this has been combined with analytic techniques for stability, stabilization and robust control, leading e.g. to backstepping techniques and nonlinear H_∞- control. Moreover, in the last decade the theory of *passive* systems, and its implications for regulation and tracking, has undergone a remarkable revival. This last development was also spurred by work in robotics on the possibilities of *shaping* by feedback the *physical energy* in such a way that it can be used as a suitable Lyapunov function for the control purpose at hand, see e.g. the influential paper (Takegaki and Arimoto, 1981). This has led to what is sometimes called *passivity-based control*, see e.g. (Ortega et al., 1998).

In the present paper we want to further stress the importance of *modeling* for nonlinear control theory. Of course, this is well-known for (nonlinear) control applications, but in our opinion also the development of nonlinear control *theory* for physical systems should be integrated with a theoretical framework for modelling. We discuss how *network modeling* of (lumped-parameter) physical systems naturally leads to a geometrically defined class of systems, called *port-Hamiltonian systems*[1]. This provides a unified mathematical framework for the description of physical systems stemming from different physical domains, such as mechanical, electrical, thermal, as well as mixtures of them.

Historically, the Hamiltonian approach has its roots in analytical mechanics and starts from the principle of least action, via the Euler-Lagrange equations and the Legendre transform, towards the Hamiltonian equations of motion. On the other hand, the network approach stems from electrical engineering, and constitutes a cornerstone of systems theory. While most of the *analysis* of physical systems has been performed within the Lagrangian and Hamiltonian framework, the network modelling point of view is prevailing in *modelling* and *simulation* of (complex) physical systems. The framework of port-Hamiltonian systems *combines* both points of view, by associating with the interconnection structure ("generalized junction structure" in bond graph terminology) of the network model a *geometric structure* given by a *Poisson structure*, or more generally a *Dirac structure*. The Hamiltonian dynamics is then defined with respect to this Poisson (or Dirac) structure *and* the Hamiltonian given by the total stored energy, as well as the energy-dissipating elements and the ports of the system. This is discussed in Section 2 for the case of Poisson structures (no algebraic constraints), and in Section 4 for the general case of Dirac structures. Dirac structures encompass the "canonical" structures which are classically being used in the geometrization of mechanics, since they also allow to describe the geometric structure of systems with *constraints* as arising from the interconnection of sub-systems. Furthermore, Dirac structures allow to extend the Hamiltonian description of *distributed-parameter systems* to include variable boundary conditions, leading to distributed-parameter port-Hamiltonian systems with boundary ports. This will be briefly explained in Section 5 based on (van der Schaft and Maschke, 2002; Maschke and van der Schaft, 2000b,a).

The structural properties of lumped-parameter port-Hamiltonian systems are investigated in Section 3 through geometric tools stemming from the theory of Hamiltonian systems. It is indicated how the *interconnection* of port-Hamiltonian systems again leads to a port-Hamiltonian system, and how this may be exploited for control and design. In particular, we investigate the existence of Casimir functions for the feedback interconnection of a plant port-Hamiltonian system and a controller port-Hamiltonian system, leading to a reduced port-Hamiltonian system on invariant manifolds with *shaped* energy. We thus provide an interpretation of *passivity-based control* from an *interconnection* point of view.

Acknowledgements This paper is based on joint work with several co-authors. In particular we thank Bernhard Maschke and Romeo Ortega for fruitful collaborations. Some of the material covered in this paper has appeared in (van der Schaft, 1999b).

[1]In previous publications we also used the terminology "Port-Controlled Hamiltonian (PCH) systems", or "generalized port-controlled Hamiltonian systems".

2 Port-Hamiltonian systems

2.1 From the Euler-Lagrange and Hamiltonian equations to port-Hamiltonian systems

Let us briefly recall the standard Euler-Lagrange and Hamiltonian equations of motion. The standard *Euler-Lagrange equations* are given as

$$\frac{d}{dt}\left(\frac{\partial L}{\partial \dot{q}}(q,\dot{q})\right) - \frac{\partial L}{\partial q}(q,\dot{q}) = \tau, \tag{2.1}$$

where $q = (q_1, \ldots, q_k)^T$ are generalized configuration coordinates for the system with k degrees of freedom, the Lagrangian L equals the difference $K - P$ between kinetic energy K and potential energy P, and $\tau = (\tau_1, \ldots, \tau_k)^T$ is the vector of generalized forces acting on the system. Furthermore, $\frac{\partial L}{\partial \dot{q}}$ denotes the column-vector of partial derivatives of $L(q, \dot{q})$ with respect to the generalized velocities $\dot{q}_1, \ldots, \dot{q}_k$, and similarly for $\frac{\partial L}{\partial q}$. In standard mechanical systems the kinetic energy K is of the form

$$K(q,\dot{q}) = \frac{1}{2}\dot{q}^T M(q)\dot{q} \tag{2.2}$$

where the $k \times k$ inertia (generalized mass) matrix $M(q)$ is symmetric and positive definite for all q. In this case the vector of generalized *momenta* $p = (p_1, \ldots, p_k)^T$, defined for any Lagrangian L as $p = \frac{\partial L}{\partial \dot{q}}$, is simply given by

$$p = M(q)\dot{q}, \tag{2.3}$$

and by defining the state vector $(q_1, \ldots, q_k, p_1, \ldots, p_k)^T$ the k second-order equations (2.1) transform into $2k$ first-order equations

$$\begin{aligned} \dot{q} &= \tfrac{\partial H}{\partial p}(q,p) \qquad (= M^{-1}(q)p) \\ \dot{p} &= -\tfrac{\partial H}{\partial q}(q,p) + \tau \end{aligned} \tag{2.4}$$

where

$$H(q,p) = \frac{1}{2}p^T M^{-1}(q)p + P(q) \quad (= \frac{1}{2}\dot{q}^T M(q)\dot{q} + P(q)) \tag{2.5}$$

is the total energy of the system. The equations (2.4) are called the *Hamiltonian equations* of motion, and H is called the *Hamiltonian.* The following *energy balance* immediately follows from (2.4):

$$\frac{d}{dt}H = \frac{\partial^T H}{\partial q}(q,p)\dot{q} + \frac{\partial^T H}{\partial p}(q,p)\dot{p} = \frac{\partial^T H}{\partial p}(q,p)\tau \ = \ \dot{q}^T \tau, \tag{2.6}$$

expressing that the increase in energy of the system is equal to the supplied work (*conservation of energy*).

If the Hamiltonian $H(q,p)$ is assumed to be the sum of a positive kinetic energy and a potential energy which is *bounded from below*, that is

$$\begin{aligned} H(q,p) &= \frac{1}{2}p^T M^{-1}(q)p + P(q) \\ M(q) = M^T(q) &> 0, \qquad \exists C > -\infty \quad \text{such that } P(q) \geq C. \end{aligned} \tag{2.7}$$

then it follows that (2.4) with inputs $u = \tau$ and outputs $y = \dot{q}$ is a *passive* (in fact, *lossless*) state space system with storage function $H(q,p) - C \geq 0$ (see e.g. (Willems, 1972; Hill and Moylan, 1976; van der Schaft, 1999b) for the general theory of passive and dissipative systems). Since the energy is only defined up to a constant, we may as well as take as potential energy the function $P(q) - C \geq 0$, in which case the total energy $H(q,p)$ becomes nonnegative and thus itself is the storage function.

System (2.4) is an example of a *Hamiltonian system* with collocated inputs and outputs, which more generally is given in the following form

$$\begin{aligned} \dot{q} &= \frac{\partial H}{\partial p}(q,p) \quad , \quad (q,p) = (q_1,\ldots,q_k,p_1,\ldots,p_k) \\ \dot{p} &= -\frac{\partial H}{\partial q}(q,p) + B(q)u, \qquad u \in \mathbb{R}^m, \\ y &= B^T(q)\frac{\partial H}{\partial p}(q,p) \qquad (= B^T(q)\dot{q}), \qquad y \in \mathbb{R}^m, \end{aligned} \tag{2.8}$$

Here $B(q)$ is the input force matrix, with $B(q)u$ denoting the generalized forces resulting from the control inputs $u \in \mathbb{R}^m$. The state space of (2.8) with local coordinates (q,p) is usually called the *phase space*. In case $m < k$ we speak of an *underactuated* system. If $m = k$ and the matrix $B(q)$ is everywhere invertible, then the Hamiltonian system is called fully actuated.

Because of the form of the output equations $y = B^T(q)\dot{q}$ we again obtain the energy balance

$$\frac{dH}{dt}(q(t),p(t)) = u^T(t)y(t) \tag{2.9}$$

Hence if H is non-negative (or, *bounded from below*), any Hamiltonian system (2.8) is a lossless state space system. For a system-theoretic treatment of the Hamiltonian systems (2.8), especially if the output y can be written as the time-derivative of a vector of generalized configuration coordinates, we refer to e.g. (Brockett, 1977; van der Schaft, 1984, 1986; Crouch and van der Schaft, 1987; Nijmeijer and van der Schaft, 1990).

A major generalization of the class of Hamiltonian systems (2.8) is to consider systems which are described in local coordinates as

$$\begin{aligned} \dot{x} &= J(x)\tfrac{\partial H}{\partial x}(x) + g(x)u, \quad x \in \mathcal{X}, u \in \mathbb{R}^m \\ y &= g^T(x)\tfrac{\partial H}{\partial x}(x), \qquad y \in \mathbb{R}^m \end{aligned} \tag{2.10}$$

Here $J(x)$ is an $n \times n$ matrix with entries depending smoothly on x, which is assumed to be *skew-symmetric*

$$J(x) = -J^T(x), \tag{2.11}$$

and $x = (x_1, \ldots, x_n)$ are local coordinates for an n-dimensional state space manifold $\mathcal{X}$. Because of (2.11) we easily recover the energy-balance $\frac{dH}{dt}(x(t)) = u^T(t)y(t)$, showing that (2.10) is lossless if $H \geq 0$. We call (2.10) with J satisfying (2.11) a *port-Hamiltonian system* with *structure matrix* $J(x)$ and *Hamiltonian* H ((Maschke and van der Schaft, 1992; Maschke et al., 1992; Maschke and van der Schaft, 1994)). Note that (2.8) (and hence (2.4)) is a particular case of (2.10) with $x = (q,p)$, and $J(x)$ being given by the constant skew-symmetric matrix $J = \begin{bmatrix} 0 & I_k \\ -I_k & 0 \end{bmatrix}$, and $g(q,p) = \begin{bmatrix} 0 \\ B(q) \end{bmatrix}$.

As an important mathematical note, we remark that in many examples the structure matrix J will satisfy the "*integrability*" conditions

$$\sum_{l=1}^{n} \left[J_{lj}(x)\frac{\partial J_{ik}}{\partial x_l}(x) + J_{li}(x)\frac{\partial J_{kj}}{\partial x_l}(x) + J_{lk}(x)\frac{\partial J_{ji}}{\partial x_l}(x) \right] = 0, i,j,k = 1,\ldots,n \qquad (2.12)$$

In this case we may find, by Darboux's theorem (see e.g. (Weinstein, 1983)) around any point x_0 where the rank of the matrix $J(x)$ is constant, local coordinates $\tilde{x} = (q,p,s) = (q_1,\ldots,q_k,p_1,\ldots,p_k,s_1,\ldots s_l)$, with $2k$ the rank of J and $n = 2k + l$, such that J in these coordinates takes the form

$$J = \begin{bmatrix} 0 & I_k & 0 \\ -I_k & 0 & 0 \\ 0 & 0 & 0 \end{bmatrix} \qquad (2.13)$$

The coordinates (q,p,s) are called *canonical* coordinates, and J satisfying (2.11) and (2.12) is called a *Poisson structure matrix*. In such canonical coordinates the equations (2.10) take the form

$$\begin{aligned} \dot{q} &= \frac{\partial H}{\partial p}(q,p,s) + g_q(q,p,s)u \\ \dot{p} &= -\frac{\partial H}{\partial q}(q,p,s) + g_p(q,p,s)u \\ \dot{s} &= g_s(q,p,s)u \\ y &= g_q^T(q,p,s)\frac{\partial H}{\partial q}(q,p,s) + g_p^T(q,p,s)\frac{\partial H}{\partial p}(q,p,s) + g_s^T(q,p,s)\frac{\partial H}{\partial s}(q,p,s) \end{aligned} \qquad (2.14)$$

which is, apart from the appearance of the variables s, very close to the standard Hamiltonian form (2.8). In particular, if $g_s = 0$, then the variables s are merely an additional set of *constant* parameters.

Although traditionally Hamiltonian systems arise from the Euler-Lagrange equations of motion (which are usually derived from variational principles) the point of departure for the theory of port-Hamiltonian systems is different. Indeed, port-Hamiltonian systems arise systematically from *network models* of physical systems. In network models of complex physical systems the overall system is seen as the interconnection of energy-storing elements via basic interconnection (balance) laws as Newton's third law or Kirchhoff's laws, as well as power-conserving elements like transformers, kinematic pairs and ideal constraints, together with energy-dissipating elements. The basic point of departure for the theory of port-Hamiltonian systems is to formalize the basic interconnection laws

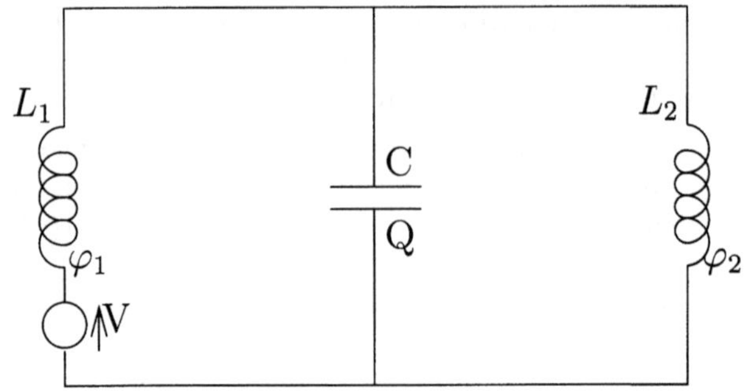

Figure 1. Controlled LC-circuit

together with the power-conserving elements by a *g*eometric structure, and to define the Hamiltonian as the total energy stored in the system. Indeed, for the (restricted) form of port-Hamiltonian systems given above the structure matrix $J(x)$ and the input matrix $g(x)$ may be directly associated with the network interconnection structure, while the Hamiltonian H is just the sum of the energies of all the energy-storing elements; see our papers (Maschke et al., 1992; Maschke and van der Schaft, 1992; Maschke et al., 1994, 1995; van der Schaft and Maschke, 1995a, 1997; Maschke and van der Schaft, 1997; van der Schaft, 1999a; Stramigioli et al., 2000). In particular, network models of complex physical systems formalized within the (generalized) bond graph language (Paynter, 1960; Breedveld, 1984) can be shown to immediately lead to port-Hamiltonian systems; see e.g. (Golo et al., 2003a).

Example 2.1 (LCTG circuits). Consider a controlled LC-circuit (see Figure 1) consisting of two inductors with magnetic energies $H_1(\varphi_1), H_2(\varphi_2)$ (φ_1 and φ_2 being the magnetic flux linkages), and a capacitor with electric energy $H_3(Q)$ (Q being the charge). If the elements are linear then $H_1(\varphi_1) = \frac{1}{2L_1}\varphi_1^2$, $H_2(\varphi_2) = \frac{1}{2L_2}\varphi_2^2$ and $H_3(Q) = \frac{1}{2C}Q^2$. Furthermore let $V = u$ denote a voltage source. Using Kirchhoff's laws one immediately arrives at the dynamical equations

$$\begin{bmatrix} \dot{Q} \\ \dot{\varphi}_1 \\ \dot{\varphi}_2 \end{bmatrix} = \underbrace{\begin{bmatrix} 0 & 1 & -1 \\ -1 & 0 & 0 \\ 1 & 0 & 0 \end{bmatrix}}_{J} \begin{bmatrix} \frac{\partial H}{\partial Q} \\ \frac{\partial H}{\partial \varphi_1} \\ \frac{\partial H}{\partial \varphi_2} \end{bmatrix} + \begin{bmatrix} 0 \\ 1 \\ 0 \end{bmatrix} u \tag{2.15}$$

$$y = \frac{\partial H}{\partial \varphi_1} \qquad (= \text{current through first inductor})$$

with $H(Q, \varphi_1, \varphi_2) := H_1(\varphi_1) + H_2(\varphi_2) + H_3(Q)$ the total energy. Clearly the matrix J is skew-symmetric, and since J is constant it trivially satisfies (2.12). In (Maschke et al., 1995) it has been shown that in this way every LC-circuit with independent elements can be modelled as a port-Hamiltonian system. Furthermore, also any LCTG-circuit with independent elements can be modelled as a port-Hamiltonian system, with J determined by Kirchhoff's laws *and* the constitutive relations of the transformers T and gyrators G. □

Example 2.2 (Actuated rigid body). Consider a rigid body spinning around its center of mass in the absence of gravity. The energy variables are the three components of the body angular momentum p along the three principal axes: $p = (p_x, p_y, p_z)$, and the energy is the kinetic energy

$$H(p) = \frac{1}{2}\left(\frac{p_x^2}{I_x} + \frac{p_y^2}{I_y} + \frac{p_z^2}{I_z}\right),$$

where I_x, I_y, I_z are the principal moments of inertia. Euler's equations describing the dynamics are

$$\begin{bmatrix} \dot{p}_x \\ \dot{p}_y \\ \dot{p}_z \end{bmatrix} = \underbrace{\begin{bmatrix} 0 & -p_z & p_y \\ p_z & 0 & -p_x \\ -p_y & p_x & 0 \end{bmatrix}}_{J(p)} \begin{bmatrix} \frac{\partial H}{\partial p_x} \\ \frac{\partial H}{\partial p_y} \\ \frac{\partial H}{\partial p_z} \end{bmatrix} + g(p)u \tag{2.16}$$

It can be checked that the skew-symmetric matrix $J(p)$ satisfies (2.12). (In fact, $J(p)$ is the canonical Lie-Poisson structure matrix on the dual of the Lie algebra $so(3)$ corresponding to the configuration space $SO(3)$ of the rigid body.) In the scalar input case the term $g(p)u$ denotes the torque around an axis with coordinates $g = (b_x \; b_y \; b_z)^T$, with corresponding collocated output given as

$$y = b_x \frac{p_x}{I_x} + b_y \frac{p_y}{I_y} + b_z \frac{p_z}{I_z}, \tag{2.17}$$

which is the velocity around the same axis $(b_x \; b_y \; b_z)^T$. □

Example 2.3. A third important class of systems that naturally can be written as port-Hamiltonian systems, is constituted by mechanical systems with *kinematic constraints*. Consider as before a mechanical system with k degrees of freedom, locally described by k configuration variables $q = (q_1, \ldots, q_k)$. Suppose that there are constraints on the generalized velocities $\dot{q}$, described as

$$A^T(q)\dot{q} = 0, \tag{2.18}$$

with $A(q)$ a $r \times k$ matrix of rank r everywhere (that is, there are r independent kinematic constraints). Classically, the constraints (2.18) are called *holonomic* if it is possible to find new configuration coordinates $\overline{q} = (\overline{q}_1, \ldots, \overline{q}_k)$ such that the constraints are equivalently expressed as

$$\dot{\overline{q}}_{k-r+1} = \dot{\overline{q}}_{n-r+2} = \cdots = \dot{\overline{q}}_k = 0 , \tag{2.19}$$

in which case one can eliminate the configuration variables $\overline{q}_{k-r+1}, \ldots, \overline{q}_k$, since the kinematic constraints (2.19) are equivalent to the *geometric* constraints

$$\overline{q}_{k-r+1} = c_{k-r+1}, \ldots, \overline{q}_k = c_k , \tag{2.20}$$

for certain constants $c_{k-r+1}, \dots, c_k$ determined by the initial conditions. Then the system reduces to an *unconstrained* system in the remaining configuration coordinates $(\overline{q}_1, \dots, \overline{q}_{k-r})$. If it is *not* possible to find coordinates $\overline{q}$ such that (2.19) holds (that is, if we are not able to *integrate* the kinematic constraints as above), then the constraints are called *nonholonomic*.

The equations of motion for the mechanical system with Lagrangian $L(q, \dot{q})$ and constraints (2.18) are given by the Euler-Lagrange equations (Neimark and Fufaev, 1972)

$$\begin{aligned} \frac{d}{dt}\left(\frac{\partial L}{\partial \dot{q}}\right) - \frac{\partial L}{\partial q} &= A(q)\lambda + B(q)u, \qquad \lambda \in \mathbb{R}^r, u \in \mathbb{R}^m \\ A^T(q)\dot{q} &= 0 \end{aligned} \tag{2.21}$$

where $B(q)u$ are the external forces (controls) applied to the system, for some $k \times m$ matrix $B(q)$, while $A(q)\lambda$ are the *constraint forces.* The Lagrange multipliers $\lambda(t)$ are uniquely determined by the requirement that the constraints $A^T(q(t))\dot{q}(t) = 0$ have to be satisfied for all t.

Defining as before (cf. (2.3)) the generalized momenta the constrained Euler-Lagrange equations (2.21) transform into *constrained Hamiltonian equations* (compare with (2.8)),

$$\begin{aligned} \dot{q} &= \frac{\partial H}{\partial p}(q,p) \\ \dot{p} &= -\frac{\partial H}{\partial q}(q,p) + A(q)\lambda + B(q)u \\ y &= B^T(q)\frac{\partial H}{\partial p}(q,p) \\ 0 &= A^T(q)\frac{\partial H}{\partial p}(q,p) \end{aligned} \tag{2.22}$$

with $H(q,p) = \frac{1}{2}p^T M^{-1}(q)p + P(q)$ the total energy. The *constrained* state space is therefore given as the following subset of the phase space:

$$\mathcal{X}_c = \{(q,p) \mid A^T(q)\frac{\partial H}{\partial p}(q,p) = 0\} \tag{2.23}$$

One way of proceeding with these equations is to *eliminate* the constraint forces, and to *reduce* the equations of motion to the constrained state space. In (van der Schaft and Maschke, 1994) it has been shown that this leads to a port-Hamiltonian system (2.10). Furthermore, the structure matrix J_c of the port-Hamiltonian system satisfies the integrability conditions (2.12) if and only if the constraints (2.18) are *holonomic.* (In fact, if the constraints are holonomic then the coordinates s as in (2.13) can be taken to be equal to the "integrated constraint functions" $\overline{q}_{k-r+1}, \dots, \overline{q}_k$ of (2.20), and the matrix g_s as in (2.14) is zero.)

An alternative way of approaching the system (2.22) is to formalize it directly as an *implicit* port-Hamiltonian system, as will be sketched in Section 4.

2.2 Basic properties of port-Hamiltonian systems

As allude to above, port-Hamiltonian systems naturally arise from a *network modeling* of physical systems without dissipative elements, see our papers (Maschke and van der Schaft, 1992; Maschke et al., 1992; Maschke and van der Schaft, 1994; Maschke et al., 1994, 1995; Maschke and van der Schaft, 1996; van der Schaft and Maschke, 1995a; van der Schaft et al., 1996; van der Schaft and Maschke, 1997; Maschke and van der Schaft, 1997; van der Schaft, 1999a). Recall that a port-Hamiltonian system is defined by a state space manifold $\mathcal{X}$ endowed with a *triple* (J, g, H). The pair $(J(x), g(x)), x \in \mathcal{X}$, captures the *interconnection structure* of the system, with $g(x)$ modeling in particular the *ports* of the system. This is very clear in Example 2.1, where the pair $(J(x), g(x))$ is determined by Kirchhoff's laws, the paradigmatic example of a power-conserving interconnection structure, but it naturally holds for other physical systems without dissipation as well. Independently from the interconnection structure, the function $H : \mathcal{X} \to \mathbb{R}$ defines the total stored *energy* of the system. Furthermore, port-Hamiltonian systems are intrinsically *modular* in the sense that a power-conserving interconnection of a number of port-Hamiltonian systems again defines a port-Hamiltonian system, with its overall interconnection structure determined by the interconnection structures of the composing individual systems together with their power-conserving interconnection, and the Hamiltonian just the sum of the individual Hamiltonians (see (van der Schaft and Maschke, 1997; van der Schaft, 1999a; Dalsmo and van der Schaft, 1999)).

As we have seen before, a basic property of port-Hamiltonian systems is the energy-balancing property $\frac{dH}{dt}(x(t)) = u^T(t)y(t)$. Physically this corresponds to the fact that the *internal* interconnection structure is power-conserving (because of skew-symmetry of $J(x)$), while u and y are the *power-variables* of the ports defined by $g(x)$, and thus $u^T y$ is the externally supplied power.

From the structure matrix $J(x)$ of a port-Hamiltonian system one can directly extract useful information about the dynamical properties of the system. Since the structure matrix is directly related to the modeling of the system (capturing the interconnection structure) this information usually has a direct physical interpretation.

A very important property which may be directly inferred from the structure matrix is the existence of dynamical invariants *independent* of the Hamiltonian H, called *Casimir functions.* Consider the set of p.d.e.'s

$$\frac{\partial^T C}{\partial x}(x) J(x) = 0, \qquad x \in \mathcal{X}, \tag{2.24}$$

in the unknown (smooth) function $C : \mathcal{X} \to \mathbb{R}$. If (2.24) has a solution C then it follows that the time-derivative of C along the port-controlled Hamiltonian system (2.10) satisfies

$$\begin{aligned} \frac{dC}{dt} &= \frac{\partial^T C}{\partial x}(x) J(x) \frac{\partial H}{\partial x}(x) + \frac{\partial^T C}{\partial x}(x) g(x) u \\ &= \frac{\partial^T C}{\partial x}(x) g(x) u \end{aligned} \tag{2.25}$$

Hence, for the input $u = 0$, or for *arbitrary* input functions if additionally $\frac{\partial^T C}{\partial x}(x) g(x) = 0$, the function $C(x)$ *remains constant* along the trajectories of the port-Hamiltonian system, *irrespective* of the precise form of the Hamiltonian H. A function $C : \mathcal{X} \to \mathbb{R}$ satisfying (2.24) is called a *Casimir function* (of the structure matrix $J(x)$).

The existence of non-trivial solutions C to (2.24) clearly assumes that rank $J(x) < \dim \mathcal{X}$, but is also related to the integrability conditions (2.12). In fact, if canonical coordinates (q, p, s) as in (2.13) have been found, then the Casimir functions are precisely all functions $C : \mathcal{X} \to \mathbb{R}$ depending *only* on the s-coordinates.

From (2.25) it follows that the level sets $L_C := \{x \in \mathcal{X} | C(x) = c\}, c \in \mathbb{R}$, of a Casimir function C are *invariant* sets for the autonomous Hamiltonian system $\dot{x} = J(x)\frac{\partial H}{\partial x}(x)$. Furthermore, the dynamics $\dot{x} = J(x)\frac{\partial H}{\partial x}(x)$ *restricted* to any level set L_C is given as the *reduced* Hamiltonian dynamics

$$\dot{x}_C = J_C(x_C)\frac{\partial H_C}{\partial x}(x_C) \tag{2.26}$$

with H_C and J_C the *restriction* of H, respectively J, to L_C. More generally, if $C = (C_1, \ldots, C_r)$ are independent Casimir functions, then in any set of local coordinates $(z_1, \ldots, z_l, C_1, \ldots, C_r)$ for $\mathcal{X}$ the Hamiltonian dynamics $\dot{x} = J(x)\frac{\partial H}{\partial x}(x)$ takes the form

$$\begin{bmatrix} \dot{z} \\ \dot{C} \end{bmatrix} = \begin{bmatrix} \tilde{J}(z, C) & 0 \\ 0 & 0 \end{bmatrix} \begin{bmatrix} \frac{\partial H}{\partial z} \\ \frac{\partial H}{\partial C} \end{bmatrix},$$

leading to the *reduced* Hamiltonian dynamics

$$\dot{z} = \tilde{J}(z, C = c)\frac{\partial H}{\partial z}$$

on any multi-level set $\{x \in \mathcal{X} | \, (C_1(x), \ldots, C_r(x)) = c \in \mathbb{R}^r\}$.

The existence of Casimir functions has immediate consequences for stability analysis of (2.10) for $u = 0$. Indeed, if $C_1, \cdots, C_r$ are Casimirs, then by (2.24) not only $\frac{dH}{dt} = 0$ for $u = 0$, but

$$\frac{d}{dt}\left(H + H_a(C_1, \cdots, C_r)\right)(x(t)) = 0 \tag{2.27}$$

for *any* function $H_a : \mathbb{R}^r \to \mathbb{R}$. Hence, even if H is not positive definite at an equilibrium $x^* \in \mathcal{X}$, then $H + H_a(C_1, \cdots, C_r)$ may be positive definite at x^* by a proper choice of H_a, and thus may serve as a Lyapunov function. This method for stability analysis is called the *Energy-Casimir method*, see e.g. (Marsden and Ratiu, 1994).

Example 2.4 (Example 2.1 continued). The quantity $\phi_1 + \phi_2$ is a Casimir function.

Example 2.5 (Example 2.2 continued). The quantity $\frac{1}{2}p_x^2 + \frac{1}{2}p_y^2 + \frac{1}{2}p_z^2$ (total angular momentum) is a Casimir function.

For a further discussion of the dynamical properties of Hamiltonian systems (especially if J satisfies the integrability conditions (2.12)) we refer to the extensive literature on this topic, see e.g. (Abraham and Marsden, 1978; Marsden and Ratiu, 1994).

2.3 Port-Hamiltonian systems with dissipation

Energy-dissipation is included in the framework of port-Hamiltonian systems (2.10) by terminating some of the ports by resistive elements. Indeed, consider instead of $g(x)u$ in (2.10) a term

$$\begin{bmatrix} g(x) & g_R(x) \end{bmatrix} \begin{bmatrix} u \\ u_R \end{bmatrix} = g(x)u + g_R(x)u_R \tag{2.28}$$

and extend correspondingly the output equations $y = g^T(x)\frac{\partial H}{\partial x}(x)$ to

$$\begin{bmatrix} y \\ y_R \end{bmatrix} = \begin{bmatrix} g^T(x)\frac{\partial H}{\partial x}(x) \\ \\ g_R^T(x)\frac{\partial H}{\partial x}(x) \end{bmatrix} \tag{2.29}$$

Here $u_R, y_R \in \mathbb{R}^{m_r}$ denote the power variables at the ports which are terminated by static resistive elements

$$u_R = -F(y_R) \tag{2.30}$$

where the resistive characteristic $F : \mathbb{R}^{m_r} \to \mathbb{R}^{m_r}$ satisfies

$$y_R^T F(y_R) \geq 0, \quad y_R \in \mathbb{R}^{m_r} \tag{2.31}$$

(In many cases, F will be derivable from a so-called *Rayleigh dissipation function* $R : \mathbb{R}^{m_r} \to \mathbb{R}$ in the sense that $F(y_R) = \frac{\partial R}{\partial y_R}(y_R)$.) In the sequel we concentrate on port-Hamiltonian systems with ports terminated by *linear* resistive elements

$$u_R = -Sy_R \tag{2.32}$$

for some positive semi-definite symmetric matric $S = S^T \geq 0$. Substitution of (2.32) into (2.28) leads to a model of the form

$$\begin{aligned} \dot{x} &= [J(x) - R(x)]\,\frac{\partial H}{\partial x}(x) + g(x)u \\ y &= g^T(x)\frac{\partial H}{\partial x}(x) \end{aligned} \tag{2.33}$$

where $R(x) := g_R(x)Sg_R^T(x)$ is a positive semi-definite symmetric matrix, depending smoothly on x. In this case the energy-balancing property (2.8) takes the form

$$\begin{aligned} \frac{dH}{dt}(x(t)) &= u^T(t)y(t) - \frac{\partial^T H}{\partial x}(x(t))R(x(t))\frac{\partial H}{\partial x}(x(t)) \\ &\leq u^T(t)y(t). \end{aligned} \tag{2.34}$$

showing that a port-Hamiltonian system is passive if the Hamiltonian H is bounded from below. We call (2.33) a *port-Hamiltonian system with dissipation.* Note that in this case *two* geometric structures play a role: the internal interconnection structure given by $J(x)$, and an additional resistive structure given by $R(x)$, which is determined by the port structure $g_R(x)$ and the linear constitutive relations $u_R = -Sy_R$ of the resistive elements.

Regarding Casimir functions for a port-Hamiltonian system with dissipation (2.33) we consider functions $C : \mathcal{X} \to \mathbb{R}$ satisfying the set of p.d.e.'s

$$\frac{\partial^T C}{\partial x}(x)\,[J(x) - R(x)] = 0, \quad x \in \mathcal{X}, \tag{2.35}$$

implying that the time-derivative of C along solutions of the system (2.33) for $u = 0$ is zero (irrespective of the Hamiltonian H).
A stronger notion of Casimir functions is obtained by considering functions $C : \mathcal{X} \to \mathbb{R}$ which are Casimir functions for *both* geometric structures defined by $J(x)$ and $R(x)$, that is

$$\begin{aligned} \tfrac{\partial^T C}{\partial x}(x)J(x) &= 0 \\ \tfrac{\partial^T C}{\partial x}(x)R(x) &= 0 \end{aligned} \tag{2.36}$$

If (2.36) holds for independent functions $C_1, \ldots, C_r$, then in any set of local coordinates $(z, C) = (z_1, \ldots z_l, C_1, \ldots C_r)$ the dynamics (2.33) for $u = 0$ takes the form

$$\begin{bmatrix} \dot{z} \\ \dot{C} \end{bmatrix} = \left(\begin{bmatrix} \tilde{J}(z,C) & 0 \\ 0 & 0 \end{bmatrix} - \begin{bmatrix} \tilde{R}(z,C) & 0 \\ 0 & 0 \end{bmatrix} \right) \begin{bmatrix} \frac{\partial H}{\partial z} \\ \frac{\partial H}{\partial C} \end{bmatrix}, \tag{2.37}$$

which can be restricted on any multi-level set $\{x \in \mathcal{X} | (C_1(x), \ldots, C_r(x)) = c \in \mathbb{R}^r\}$ to

$$\dot{z} = \left[\tilde{J}(z, C = c) - \tilde{R}(z, C = c)\right] \frac{\partial H}{\partial z}(z, C = c) \tag{2.38}$$

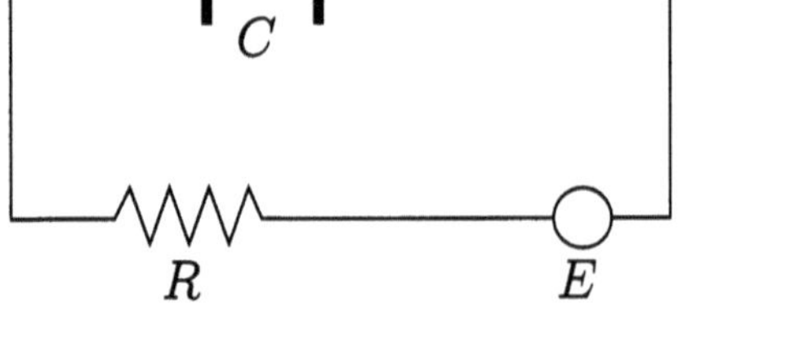

Figure 2. Capacitor microphone

Example 2.6. (Neimark and Fufaev, 1972) Consider the capacitor microphone depicted in Figure 2. Here the capacitance $C(q)$ of the capacitor is varying as a function of the displacement q of the right plate (with mass m), which is attached to a spring (with spring constant $k > 0$) and a damper (with constant $c > 0$), and affected by a mechanical force F (air pressure arising from sound). Furthermore, E is a voltage source. The dynamical equations of motion can be written as the port-Hamiltonian system with dissipation

$$\begin{bmatrix} \dot{q} \\ \dot{p} \\ \dot{Q} \end{bmatrix} = \left(\begin{bmatrix} 0 & 1 & 0 \\ -1 & 0 & 0 \\ 0 & 0 & 0 \end{bmatrix} - \begin{bmatrix} 0 & 0 & 0 \\ 0 & c & 0 \\ 0 & 0 & 1/R \end{bmatrix} \right) \begin{bmatrix} \frac{\partial H}{\partial q} \\ \frac{\partial H}{\partial p} \\ \frac{\partial H}{\partial Q} \end{bmatrix} + \begin{bmatrix} 0 \\ 1 \\ 0 \end{bmatrix} F + \begin{bmatrix} 0 \\ 0 \\ 1/R \end{bmatrix} E \tag{2.39}$$

$$y_1 = \frac{\partial H}{\partial p} = \dot{q}$$

$$y_2 = \frac{1}{R}\frac{\partial H}{\partial Q} = I$$

with p the momentum, R the resistance of the resistor, I the current through the voltage source, and the Hamiltonian H being the total energy

$$H(q,p,Q) = \frac{1}{2m}p^2 + \frac{1}{2}k(q-\bar{q})^2 + \frac{1}{2C(q)}Q^2, \tag{2.40}$$

with $\bar{q}$ denoting the equilibrium position of the spring. Note that $F\dot{q}$ is the mechanical power, and EI the electrical power applied to the system. In the application as a microphone the voltage over the resistor will be used (after amplification) as a measure for the mechanical force F.

Example 2.7. (Ortega et al., 2002) A permanent magnet synchronous motor can be written as a port-Hamiltonian system with dissipation (in a rotating reference, i.e. the dq frame) for the state vector

$$x = M \begin{bmatrix} i_d \\ i_q \\ \omega \end{bmatrix}, \qquad M = \begin{bmatrix} L_d & 0 & 0 \\ 0 & L_q & 0 \\ 0 & 0 & \frac{j}{n_p} \end{bmatrix} \tag{2.41}$$

the magnetic flux linkages and mechanical momentum (i_d, i_q being the currents, and ω the angular velocity), L_d, L_q stator inductances, j the moment of inertia, and n_p the number of pole pairs. The Hamiltonian $H(x)$ is given as $H(x) = \frac{1}{2}x^T M^{-1} x$ (total energy), while furthermore $J(x), R(x)$ and $g(x)$ are determined as

$$J(x) = \begin{bmatrix} 0 & L_0 x_3 & 0 \\ -L_0 x_3 & 0 & -\Phi_{q0} \\ 0 & \Phi_{q0} & 0 \end{bmatrix}, \qquad (2.42)$$
$$R(x) = \begin{bmatrix} R_S & 0 & 0 \\ 0 & R_S & 0 \\ 0 & 0 & 0 \end{bmatrix}, \quad g(x) = \begin{bmatrix} 1 & 0 & 0 \\ 0 & 1 & 0 \\ 0 & 0 & -\frac{1}{n_p} \end{bmatrix}$$

with R_S the stator winding resistance, Φ_{q0} a constant term due to interaction of the permanent magnet and the magnetic material in the stator, and $L_0 := L_d n_p / j$. The three inputs are the stator voltage $(v_d, v_q)^T$ and the (constant) load torque. Outputs are i_d, i_q and ω.

In some cases the interconnection structure $J(x)$ may be actually *varying*, depending on the mode of operation of the system, as exemplified by the following simple dc-to-dc power converter with a single switch. See for a further treatment of power converters in this context (Ortega et al., 2002).

Example 2.8. Consider the ideal boost converter given in Figure 3.

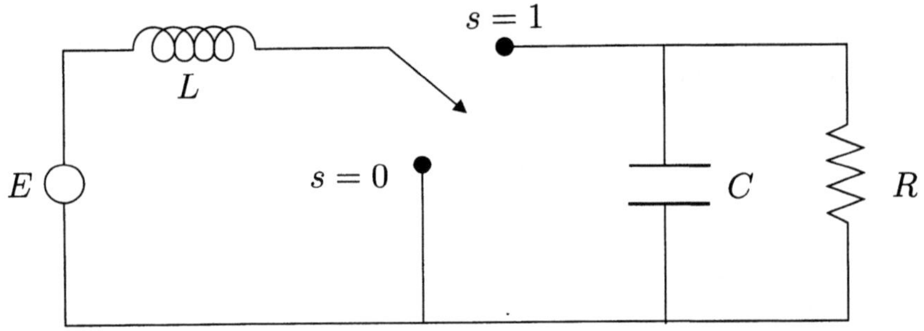

Figure 3. Ideal boost converter

The system equations are given as

$$\begin{aligned} \begin{bmatrix} \dot{x}_1 \\ \dot{x}_2 \end{bmatrix} &= \left(\begin{bmatrix} 0 & -s \\ s & 0 \end{bmatrix} - \begin{bmatrix} 0 & 0 \\ 0 & 1/R \end{bmatrix} \right) \begin{bmatrix} \frac{\partial H}{\partial x_1} \\ \frac{\partial H}{\partial x_2} \end{bmatrix} + \begin{bmatrix} 1 \\ 0 \end{bmatrix} E \\ y &= \frac{\partial H}{\partial x_1} \end{aligned} \qquad (2.43)$$

with x_1 the magnetic flux linkage of the inductor, x_2 the charge of the capacitor, and $H(x_1, x_2) = \frac{1}{2L}x_1^2 + \frac{1}{2C}x_2^2$ the total stored energy. The internal interconnection structure matrix J is either $\begin{bmatrix} 0 & 0 \\ 0 & 0 \end{bmatrix}$ or $\begin{bmatrix} 0 & -1 \\ 1 & 0 \end{bmatrix}$, depending on the ideal switch being in position $s = 0$ or $s = 1$.

3 Control of port-Hamiltonian systems

The aim of this section is to discuss a general methodology for port-Hamiltonian systems (with or without dissipation) which exploits their Hamiltonian properties in an intrinsic way, see e.g. (Maschke et al., 2000, 1999; Ortega et al., 2002; van der Schaft, 1999b). An expected benefit of such a methodology is that it leads to physically interpretable controllers, which possess inherent robustness properties. Future research is aimed at corroborating these claims.

We have already seen that port-Hamiltonian systems are *passive* if the Hamiltonian H is bounded from below. Hence in this case we can use all the results from the theory of passive systems, such as asymptotic stabilization by the insertion of *damping* by negative output feedback, see e.g. (van der Schaft, 1999b). The emphasis in this section is however on the somewhat complementary aspect of *shaping the energy* of the system, which directly involves the Hamiltonian structure of the system, as opposed to the more general passivity structure.

3.1 Control by interconnection

Consider a port-Hamiltonian system (2.33)

$$P: \quad \begin{aligned} \dot{x} &= [J(x) - R(x)]\tfrac{\partial H}{\partial x}(x) + g(x)u \\ y &= g^T(x)\tfrac{\partial H}{\partial x}(x) \end{aligned} \qquad x \in \mathcal{X} \tag{3.1}$$

regarded as a plant system to be controlled. In the previous section we have seen that many physical systems can be modelled in this way, and that the defining entities $J(x), g(x), R(x)$ and H have a concrete physical interpretation. Furthermore, if H is bounded from below then (3.1) is a *passive* system.
Recall the well-known result that the standard feedback interconnection of two passive systems again is a passive system; a basic fact which can be used for various stability and control purposes. In the same vein we can consider the interconnection of the plant (3.1) with *another* port-Hamiltonian system

$$C: \quad \begin{aligned} \dot{\xi} &= [J_C(\xi) - R_C(\xi)]\tfrac{\partial H_C}{\partial \xi}(\xi) + g_C(\xi)u_C \\ y_C &= g_C^T(\xi)\tfrac{\partial H_C}{\partial \xi}(\xi) \end{aligned} \qquad \xi \in \mathcal{X}_C \tag{3.2}$$

regarded as the *controller* system, via the standard feedback interconnection

$$\begin{aligned} u &= -y_C + e \\ u_C &= y + e_C \end{aligned} \tag{3.3}$$

with e, e_C external signals inserted in the feedback loop. The closed-loop system takes the form

$$\begin{aligned}
\begin{bmatrix} \dot{x} \\ \dot{\xi} \end{bmatrix} &= (\underbrace{\begin{bmatrix} J(x) & -g(x)g_C^T(\xi) \\ g_C(\xi)g^T(x) & J_C(\xi) \end{bmatrix}}_{J_{cl}(x,\xi)} - \underbrace{\begin{bmatrix} R(x) & 0 \\ 0 & R_C(\xi) \end{bmatrix}}_{R_{cl}(x,\xi)}) \\
&\begin{bmatrix} \frac{\partial H}{\partial x}(x) \\ \frac{\partial H_C}{\partial \xi}(\xi) \end{bmatrix} + \begin{bmatrix} g(x) & 0 \\ 0 & g_C(\xi) \end{bmatrix} \begin{bmatrix} e \\ e_C \end{bmatrix} \\
\begin{bmatrix} y \\ y_C \end{bmatrix} &= \begin{bmatrix} g(x) & 0 \\ 0 & g_C(\xi) \end{bmatrix} \begin{bmatrix} \frac{\partial H}{\partial x}(x) \\ \frac{\partial H_C}{\partial \xi}(\xi) \end{bmatrix}
\end{aligned} \tag{3.4}$$

which again is a port-Hamiltonian system, with state space given by the product space $\mathcal{X} \times \mathcal{X}_C$, total Hamiltonian $H(x) + H_C(\xi)$, inputs (e, e_C) and outputs (y, y_C). Hence the feedback interconnection of any two port-Hamiltonian systems results in another port-Hamiltonian system; just as in the case of passivity.
It is of interest to investigate the *Casimir functions* of the closed-loop system, especially those relating the state variables ξ of the controller system to the state variables x of the plant system. Indeed, from a control point of view the Hamiltonian H is *given* while H_C can be *assigned*. Thus if we can find Casimir functions $C_i(\xi, x), i = 1, \cdots, r$, relating ξ to x then by the Energy-Casimir method the Hamiltonian $H + H_C$ of the closed-loop system may be replaced by the Hamiltonian $H + H_C + H_a(C_1, \cdots, C_r)$, thus creating the possibility of obtaining a suitable Lyapunov function for the closed-loop system.
In particular, let us consider Casimir functions of the form

$$\xi_i - G_i(x) \quad , \quad i = 1, \ldots, \dim \mathcal{X}_C = n_C \tag{3.5}$$

That means (see (2.35)) that we are looking for solutions of the p.d.e.'s (with e_i denoting the i-th basis vector)

$$\begin{bmatrix} -\frac{\partial^T G_i}{\partial x}(x) & e_i^T \end{bmatrix} \begin{bmatrix} J(x) - R(x) & -g(x)g_C^T(\xi) \\ g_C(\xi)g^T(x) & J_C(\xi) - R_C(\xi) \end{bmatrix} = 0$$

or written out

$$\begin{aligned}
&\frac{\partial^T G_i}{\partial x}(x)\,[J(x) - R(x)] - g_C^i(\xi)g^T(x) = 0 \\
&\frac{\partial^T G_i}{\partial x}(x)g(x)g_C^T(\xi) + J_C^i(\xi) - R_C^i(\xi) = 0
\end{aligned} \tag{3.6}$$

with $\frac{\partial^T G_i}{\partial x}$ denoting as before the gradient vector $\left(\frac{\partial G_i}{\partial x_1}, \ldots, \frac{\partial G_i}{\partial x_n}\right)$, and g_C^i, J_C^i, R_C^i denoting the i-th row of g_C, J_C, respectively R_C.

Suppose we want to solve (3.6) for $i = 1, \ldots, \bar{n}$, with $\bar{n} \leq n_C$ (possibly after permutation of $\xi_1, \ldots, \xi_{n_C}$). with $\bar{J}_C(\xi), \bar{R}_C(\xi)$ the $\bar{n} \times \bar{n}$ left-upper submatrices of J_C, respectively R_C. The following proposition has been shown in (van der Schaft, 1999b).

Proposition 3.1. *The functions $\xi_i - G_i(x)$, $i = 1, \dots, \bar{n} \leq n_C$, satisfy (3.6) (and thus are Casimirs of the closed-loop port-controlled Hamiltonian system (3.4) for $e = 0, e_c = 0$) if and only if $G = (G_1, \dots, G_{\bar{n}})^T$ satisfies*

$$\begin{aligned}
&\tfrac{\partial^T G}{\partial x}(x) J(x) \tfrac{\partial G}{\partial x}(x) = \bar{J}_C(\xi) \\
&R(x) \tfrac{\partial G}{\partial x}(x) = 0 \\
&\bar{R}_C(\xi) = 0 \\
&\tfrac{\partial^T G}{\partial x}(x) J(x) = \tilde{g}_C(\xi)\, g^T(x)
\end{aligned} \tag{3.7}$$

with $\bar{J}_C(\xi), \bar{R}_C(\xi)$ the $\bar{n} \times \bar{n}$ left-upper submatrices of J_C, respectively R_C.

In particular, we conclude that the functions $\xi_i - G_i(x), i = 1, \dots \bar{n}$, are Casimirs of (3.4) for $e = 0, e_C = 0$, if and only if they are Casimirs for *both* the internal interconnection structure $J_{cl}(x, \xi)$ as well as for the dissipation structure $R_{cl}(x, \xi)$. Hence, as in (2.38), it follows directly how the closed-loop port-controlled Hamiltonian system with dissipation (3.4) for $e = 0, e_C = 0$ reduces to a system any multi-level set $\{(x, \xi) \,|\xi_i = G_i(x) + c_i,\ i = 1, \dots, \bar{n}\}$, by restricting both J_{cl} and R_{cl} to this multi-level set.

Example 3.2. (Stramigioli, 1998) Consider the "plant" system

$$\begin{aligned}
\begin{bmatrix} \dot{q} \\ \dot{p} \end{bmatrix} &= \begin{bmatrix} 0 & 1 \\ -1 & 0 \end{bmatrix} \begin{bmatrix} \frac{\partial H}{\partial q} \\ \frac{\partial H}{\partial p} \end{bmatrix} + \begin{bmatrix} 0 \\ 1 \end{bmatrix} u \\
y &= \begin{bmatrix} 0 & 1 \end{bmatrix} \begin{bmatrix} \frac{\partial H}{\partial q} \\ \frac{\partial H}{\partial p} \end{bmatrix}
\end{aligned} \tag{3.8}$$

with q the position and p being the momentum of the mass m, in feedback interconnection $(u = -y_C + e, u_C = y)$ with the controller system (see Figure 4)

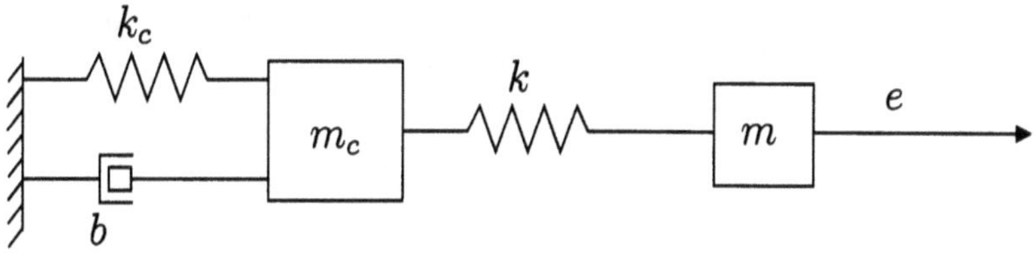

Figure 4. Controlled mass

$$\begin{bmatrix} \dot{\Delta q_c} \\ \dot{p}_c \\ \dot{\Delta q} \end{bmatrix} = \begin{bmatrix} 0 & 1 & 0 \\ -1 & -b & 1 \\ 0 & -1 & 0 \end{bmatrix} \begin{bmatrix} \frac{\partial H_C}{\partial \Delta q_c} \\ \frac{\partial H_C}{\partial p_c} \\ \frac{\partial H_C}{\partial \Delta q} \end{bmatrix} + \begin{bmatrix} 0 \\ 0 \\ 1 \end{bmatrix} u_C \qquad (3.9)$$
$$y_C = \frac{\partial H_C}{\partial \Delta q}$$

where Δq_c is the displacement of the spring k_c, Δq is the displacement of the spring k, and p_c is the momentum of the mass m_c. The plant Hamiltonian is $H(p) = \frac{1}{2m}p^2$, and the controller Hamiltonian is given as $H_C(\Delta q_c, p_c, \Delta q) = \frac{1}{2}(\frac{p_c^2}{m_c} + k(\Delta q)^2 + k_c(\Delta q_c)^2)$. The variable $b > 0$ is the damping constant, and e is an external force. The closed-loop system possesses the Casimir function

$$C(q, \Delta q_c, \Delta q) = \Delta q - (q - \Delta q_c), \qquad (3.10)$$

implying that along the solutions of the closed-loop system

$$\Delta q = q - \Delta q_c + c \qquad (3.11)$$

with c a constant depending on the initial conditions. With the help of LaSalle's Invariance principle it can be shown that *restricted* to the invariant manifolds (3.11) the system is asymptotically stable for the equilibria $q = \Delta q_c = p = p_c = 0$. □

Let us next consider the special case $\bar{n} = n_C$, in which case we wish to relate *all* the controller state variables $\xi_1, \ldots, \xi_{n_C}$ to the plant state variables x via Casimir functions $\xi_1 - G_1(x), \ldots, \xi_{n_C} - G_{n_C}(x)$. Denoting $G = (G_1, \ldots, G_{n_C})^T$ this means that G should satisfy (see (3.7))

$$\begin{aligned} &\tfrac{\partial^T G}{\partial x}(x)J(x)\tfrac{\partial G}{\partial x}(x) = J_C(\xi) \\ &R(x)\tfrac{\partial G}{\partial x}(x) = 0 = R_C(\xi) \\ &\tfrac{\partial^T G}{\partial x}(x)J(x) = g_C(\xi)\, g^T(x) \end{aligned} \qquad (3.12)$$

In this case the reduced dynamics on any multi-level set

$$L_C = \{(x,\xi) | \xi_i = G_i(x) + c_i, i = 1, \ldots n_C\} \qquad (3.13)$$

can be immediately recognized. Indeed, the x-coordinates also serve as coordinates for L_C. Furthermore, the x-dynamics of (3.4) with $e = 0, e_C = 0$ is given as

$$\dot{x} = [J(x) - R(x)]\frac{\partial H}{\partial x}(x) - g(x)g_C^T(\xi)\frac{\partial H_C}{\partial \xi}(\xi). \qquad (3.14)$$

Using the second and the third equality of (3.12) this can be rewritten as

$$\dot{x} = [J(x) - R(x)]\left(\frac{\partial H}{\partial x}(x) + \frac{\partial G}{\partial x}(x)\frac{\partial H_C}{\partial \xi}(\xi)\right). \qquad (3.15)$$

and by the chain-rule property for differentiation this reduces to the port-Hamiltonian system

$$\dot{x} = [J(x) - R(x)] \frac{\partial H_s}{\partial x}(x), \tag{3.16}$$

with the same interconnection and dissipation structure as before, but with *shaped* Hamiltonian H_s given by

$$H_s(x) = H(x) + H_C(G(x) + c). \tag{3.17}$$

An interpretation of the shaped Hamiltonian H_s in terms of *energy-balancing* is the following. Since $R_C(\xi) = 0$ by (3.12) the controller Hamiltonian H_C satisfies $\frac{dH_C}{dt} = u_C^T y_C$. Hence along any multi-level set L_C given by (3.13), invariant for the closed loop port-Hamiltonian system (3.4) for $e = 0, e_C = 0$

$$\frac{dH_s}{dt} = \frac{dH}{dt} + \frac{dH_C}{dt} = \frac{dH}{dt} - u^T y \tag{3.18}$$

since $u = -y_C$ and $u_C = y$. Therefore, up to a constant,

$$H_s(x(t)) = H(x(t)) - \int_0^t u^T(\tau) y(\tau) d\tau, \tag{3.19}$$

and the shaped Hamiltonian H_s is the original Hamiltonian H *minus the energy* supplied to the plant system (3.1) by the controller system (3.2) (modulo a constant; depending on the initial states of the plant and controller).

Remark 3.3. Note that from a stability analysis point of view (3.19) can be regarded as an effective way of *generating* candidate Lyapunov functions H_s from the Hamiltonian H. (Compare with the classical construction of Lur'e functions.)

Example 3.4. A mechanical system with damping and actuated by external forces $u \in \mathbb{R}^m$ is described as a port-Hamiltonian system

$$\begin{aligned} \begin{bmatrix} \dot{q} \\ \dot{p} \end{bmatrix} &= \left(\begin{bmatrix} 0 & I_k \\ -I_k & 0 \end{bmatrix} - \begin{bmatrix} 0 & 0 \\ 0 & D(q) \end{bmatrix} \right) \begin{bmatrix} \frac{\partial H}{\partial q} \\ \frac{\partial H}{\partial p} \end{bmatrix} + \begin{bmatrix} 0 \\ B(q) \end{bmatrix} u \\ y &= B^T(q) \frac{\partial H}{\partial p} \end{aligned} \tag{3.20}$$

with $x = \left[\begin{smallmatrix} q \\ p \end{smallmatrix}\right]$, where $q \in \mathbb{R}^k$ are the generalized configuration coordinates, $p \in \mathbb{R}^k$ the generalized momenta, and $D(q) = D^T(q) \geq 0$ is the damping matrix. In most cases the Hamiltonian $H(q,p)$ takes the form

$$H(q,p) = \frac{1}{2} p^T M^{-1}(q) p + P(q) \tag{3.21}$$

where $M(q) = M^T(q) > 0$ is the generalized inertia matrix, $\frac{1}{2} p^T M^{-1}(q) p = \frac{1}{2} \dot{q}^T M(q) \dot{q}$ is the kinetic energy, and $P(q)$ is the potential energy of the system. Now consider a general port-Hamiltonian controller system (3.2), with state space $\mathbb{R}^m$. Then the equations (3.12) for $G = (G_1(q,p), \ldots, G_m(q,p))^T$ take the form

$$\frac{\partial^T G}{\partial q}\frac{\partial G}{\partial p} - \frac{\partial^T G}{\partial p}\frac{\partial G}{\partial q} = J_C(\xi)$$

$$D(q)\frac{\partial G}{\partial p} = 0 \tag{3.22}$$

$$\frac{\partial^T G}{\partial p} = 0, \frac{\partial^T G}{\partial q} = g_C(\xi)B^T(q)$$

or equivalently

$$J_C = 0, \qquad \frac{\partial G}{\partial p} = 0, \qquad g_C^T(\xi)B(q) = \frac{\partial G}{\partial q}(q) \tag{3.23}$$

Now let $g_C(\xi)$ be the $m \times m$ identity matrix. Then there exists a solution $G = (G_1(q), \ldots, G_m(q))$ to (3.23) if and only if the columns of the input force matrix $B(q)$ satisfy the integrability conditions

$$\frac{\partial B_{il}}{\partial q_j}(q) = \frac{\partial B_{jl}}{\partial q_i}(q), \qquad i,j = 1,\ldots k, \quad l = 1,\ldots m \tag{3.24}$$

Hence, if $B(q)$ satisfies (3.24), then the closed-loop port-Hamiltonian system (3.4) for the controller system (3.2) with $J_C = 0$ admits Casimirs $\xi_i - G_i(q), i = 1, \ldots, m$, leading to a reduced port-Hamiltonian system

$$\begin{bmatrix} \dot{q} \\ \dot{p} \end{bmatrix} = \left(\begin{bmatrix} 0 & I_k \\ -I_k & 0 \end{bmatrix} - \begin{bmatrix} 0 & 0 \\ 0 & D(q) \end{bmatrix} \right) \begin{bmatrix} \frac{\partial H_s}{\partial q} \\ \frac{\partial H_s}{\partial p} \end{bmatrix} + \begin{bmatrix} 0 \\ B(q) \end{bmatrix} e \tag{3.25}$$

$$y = B^T(q)\frac{\partial H_s}{\partial p}$$

for the shaped Hamiltonian

$$H_s(q,p) = H(q,p) + H_C(G_1(q) + c_1, \ldots, G_m(q) + c_m) \tag{3.26}$$

If $H(q,p)$ is as given in (3.21), then

$$H_s(q,p) = \frac{1}{2}p^T M^{-1}(q)p + [P(q) + H_C(G_1(q) + c_1, \cdots, G_m(q) + c_m)] \tag{3.27}$$

and the control amounts to shaping the *potential energy* of the system, see (Takegaki and Arimoto, 1981; Ortega et al., 1998). □

3.2 Passivity-based control of port-Hamiltonian systems

In the previous section we have seen how under certain conditions the feedback interconnection of a port-Hamiltonian system having Hamiltonian H (the "plant") with another port-Hamiltonian system with Hamiltonian H_C (the "controller") leads to a reduced dynamics given by (see (3.16))

$$\dot{x} = [J(x) - R(x)]\frac{\partial H_s}{\partial x}(x) \tag{3.28}$$

for the shaped Hamiltonian $H_s(x) = H(x)+H_C(G(x)+c)$, with $G(x)$ a solution of (3.12). From a *state feedback* point of view the dynamics (3.28) could have been directly obtained by a state feedback $u = \alpha(x)$ such that

$$g(x)\alpha(x) = [J(x) - R(x)]\frac{\partial H_C}{\partial x}(G(x)+c) \tag{3.29}$$

Indeed, such an $\alpha(x)$ is given in explicit form as

$$\alpha(x) = -g_C^T(G(x)+c)\frac{\partial H_C}{\partial \xi}(G(x)+c) \tag{3.30}$$

A state feedback $u = \alpha(x)$ satisfying (3.29) is customarily called a *passivity-based control law*, since it is based on the passivity properties of the original plant system (3.1) and transforms (3.1) into *another* passive system with *shaped* storage function (in this case H_s).

Seen from this perspective we have shown in the previous section that the passivity-based state feedback $u = \alpha(x)$ satisfying (3.29) *can be derived* from the interconnection of the port-Hamiltonian plant system (3.1) with a port-Hamiltonian controller system (3.2). This fact has some favorable consequences. Indeed, it implies that the passivity-based control law defined by (3.29) can be equivalently *generated* as the feedback interconnection of the passive system (3.1) with another passive system (3.2). In particular, this implies an inherent *invariance* property of the controlled system: the plant system (3.28), the controller system (3.17), as well as any other passive system interconnected to (3.28) in a power-conserving fashion, may change in any way as long as they remain passive, and for any perturbation of this kind the controlled system will remain stable.

The *implementation* of the resulting passivity-based control $u = \alpha(x)$ is a somewhat complex issue. In cases of *analog* controller design the interconnection of the plant port-Hamiltonian system (3.1) with the port-Hamiltonian controller system (3.2) seems to be the logical option. Furthermore, in general it may be favorable to avoid an explicit state feedback, but instead to use the dynamic output feedback controller (3.2). On the other hand, in some applications the measurement of the passive output y may pose some problems, while the state feedback $u = \alpha(x)$ is in fact easier to implement, as illustrated in the next example.

Example 3.5 (Example 3.4 continued). The passivity-based control $u = \alpha(x)$ resulting from (3.23) is given by (assuming g_C to be the identity matrix and $B(q)$ to satisfy (3.24))

$$u_i = -\frac{\partial H_C}{\partial \xi_i}(G(q)+c), \qquad i = 1,\dots m \tag{3.31}$$

This follows from (3.30), and can be directly checked by substituting (3.31) into (3.20) and using the equality $B(q) = \frac{\partial G}{\partial q}(q)$. Comparing the implementation of the state feedback controller (3.31) with the implementation of the port-Hamiltonian controller system based on the measurement of

$$y = B^T(q)\frac{\partial H}{\partial p}(q,p) = B^T(q)\dot{q} \tag{3.32}$$

one may note that the measurement of the generalized velocities (3.32) is in some cases (for example in a robotics context) more problematic than the measurement of the generalized positions (3.31). □

Remark 3.6. On the other hand, the control of the port-Hamiltonian plant system (3.1) by interconnection with the port-Hamiltonian controller system (3.2) allows for the possibility of inserting an asymptotically stabilizing *damping* not directly in the plant but instead in the controller system, cf. Example 3.2.

In the rest of this section we concentrate on the passivity-based (state feedback) control $u = \alpha(x)$. The purpose is to more systematically indicate how a port-Hamiltonian system with dissipation (3.1) may be asymptotically stabilized around a desired equilibrium x^* in two steps:

I Shape by passivity-based control the Hamiltonian in such a way that it has a strict minimum at $x = x^*$. Then x^* is a (marginally) stable equilibrium of the controlled system.

II Add damping to the system in such a way that x^* becomes an *asymptotically* stable equilibrium of the controlled system.

As before, we shall concentrate on Step I. Therefore, let us consider a port-Hamiltonian system with dissipation (3.1) with $\mathcal{X}$ the n-dimensional state space manifold. Suppose we wish to stabilize the system around a desired equilibrium x^*, assigning a closed-loop energy function $H_d(x)$ to the system which has a strict minimum at x^* (that is, $H_d(x) > H_d(x^*)$ for all $x \neq x^*$ in a neighbourhood of x^*). Denote

$$H_d(x) = H(x) + H_a(x), \tag{3.33}$$

where the to be defined function H_a is the energy *added* to the system (by the control action). We have the following

Proposition 3.7. *(Ortega et al., 2002; Maschke et al., 1999) Assume we can find a feedback $u = \alpha(x)$ and a vector function $K(x)$ satisfying*

$$[J(x) - R(x)]\, K(x) = g(x)\alpha(x) \tag{3.34}$$

such that

$$\begin{array}{lrcll} (i) & \frac{\partial K_i}{\partial x_j}(x) & = & \frac{\partial K_j}{\partial x_i}(x), & i,j = 1,\ldots,n \\ (ii) & K(x^*) & = & -\frac{\partial H}{\partial x}(x^*) & \\ (iii) & \frac{\partial K}{\partial x}(x^*) & > & -\frac{\partial^2 H}{\partial x^2}(x^*) & \end{array} \tag{3.35}$$

with $\frac{\partial K}{\partial x}$ the $n \times n$ matrix with i-th column given by $\frac{\partial K_i}{\partial x}(x)$, and $\frac{\partial^2 H}{\partial x^2}(x^*)$ denoting the Hessian matrix of H at x^*. Then the closed-loop system is a Hamiltonian system with *dissipation*

$$\dot{x} = [J(x) - R(x)] \frac{\partial H_d}{\partial x}(x) \tag{3.36}$$

where H_d is given by (3.33), with H_a such that

$$K(x) = \frac{\partial H_a}{\partial x}(x) \tag{3.37}$$

Furthermore, x^* is a stable equilibrium of (3.36).

A further generalization is to use state feedback in order to *change* the interconnection structure and the resistive structure of the plant system, and thereby to create more flexibility to shape the storage function for the (modified) port-controlled Hamiltonian system to a desired form. This methodology has been called Interconnection-Damping Assignment Passivity-Based Control (IDA-PBC) in (Ortega et al., 2002, 2001), and has been succesfully applied to a number of applications. The method is especially attractive if the newly assigned interconnection and resistive structures are judiciously chosen on the basis of physical considerations, and represent some "ideal" interconnection and resistive structures for the physical plant. For an extensive treatment of IDA-PBC we refer to (Ortega et al., 2002, 2001).

4 Implicit port-Hamiltonian systems

From a general modeling point of view physical systems are, at least in first instance, often described as DAE's, that is, a mixed set of differential and *algebraic* equations. This stems from the fact that in many modeling approaches the system under consideration is naturally regarded as obtained from interconnecting simpler sub-systems. These interconnections in general, give rise to algebraic constraints between the state space variables of the sub-systems; thus leading to implicit systems. While in the linear case one may argue that it is often relatively straightforward to eliminate the algebraic constraints, and thus to reduce the system to an *explicit* form, in the nonlinear case such a conversion from implicit to explicit form is usually fraught with difficulties. Indeed, if the algebraic constraints are nonlinear they need not be analytically solvable (locally or globally). More importantly perhaps, even if they are analytically solvable, then often one would prefer *not* to eliminate the algebraic constraints, because of the complicated and physically not easily interpretable expressions for the reduced system which may arise.

Therefore it is important to extend the framework of port-Hamiltonian systems, as sketched in the previous sections, to the context of implicit systems. In order to give the definition of an implicit port-Hamiltonian system (with dissipation) we first consider the notion of a Dirac structure, formalizing the concept of a power-conserving interconnection, and generalizing the notion of a structure matrix $J(x)$ as encountered before.

4.1 Power-conserving interconnections

Let us return to the basic setting of passivity, starting with a finite-dimensional linear space and its dual, in order to define *power*. Thus, let $\mathcal{F}$ be an ℓ-dimensional linear space, and denote its dual (the space of linear functions on $\mathcal{F}$) by $\mathcal{F}^*$. The product space $\mathcal{F} \times \mathcal{F}^*$ is considered to be the space of power variables, with power defined by

$$P = < f^* | f >, \quad (f, f^*) \in \mathcal{F} \times \mathcal{F}^*, \tag{4.1}$$

where $< f^* | f >$ denotes the duality product, that is, the linear function $f^* \in \mathcal{F}^*$ acting on $f \in \mathcal{F}$. Often we call $\mathcal{F}$ the space of *flows* f, and $\mathcal{F}^*$ the space of *efforts* e, with the power of an element $(f, e) \in \mathcal{F} \times \mathcal{F}^*$ denoted as $< e | f >$.

Remark 4.1. If $\mathcal{F}$ is endowed with an *inner product* structure $<,>$, then $\mathcal{F}^*$ can be naturally *identified* with $\mathcal{F}$ in such a way that $< e | f > = < e, f >$, $f \in \mathcal{F}$, $e \in \mathcal{F}^* \simeq \mathcal{F}$.

Example 4.2. Let $\mathcal{F}$ be the space of generalized *velocities*, and $\mathcal{F}^*$ be the space of generalized *forces*, then $< e | f >$ is mechanical power. Similarly, let $\mathcal{F}$ be the space of *currents*, and $\mathcal{F}^*$ be the space of *voltages*, then $< e | f >$ is electrical power.

There exists on $\mathcal{F} \times \mathcal{F}^*$ a canonically defined symmetric bilinear form

$$< (f_1, e_1), (f_2, e_2) >_{\mathcal{F} \times \mathcal{F}^*} := < e_1 | f_2 > + < e_2 | f_1 > \tag{4.2}$$

for $f_i \in \mathcal{F}$, $e_i \in \mathcal{F}^*, i = 1, 2$. Now consider a linear subspace

$$S \subset \mathcal{F} \times \mathcal{F}^* \tag{4.3}$$

and its orthogonal complement with respect to the bilinear form $<,>_{\mathcal{F} \times \mathcal{F}^*}$ on $\mathcal{F} \times \mathcal{F}^*$, denoted as

$$S^{\perp} \subset \mathcal{F} \times \mathcal{F}^*. \tag{4.4}$$

Clearly, if S has dimension d, then the subspace $S^{\perp}$ has dimension $2\ell - d$. (Since dim $(\mathcal{F} \times \mathcal{F}^*) = 2\ell$, and $<,>_{\mathcal{F} \times \mathcal{F}^*}$ is a non-degenerate form.)

Definition 4.3. (Courant, 1990; Dorfman, 1993; Dalsmo and van der Schaft, 1999) A constant Dirac structure on $\mathcal{F}$ is a linear subspace $\mathcal{D} \subset \mathcal{F} \times \mathcal{F}^*$ such that

$$\mathcal{D} = \mathcal{D}^{\perp} \tag{4.5}$$

It immediately follows that the dimension of any Dirac structure $\mathcal{D}$ on an ℓ-dimensional linear space is equal to ℓ. Furthermore, let $(f, e) \in \mathcal{D} = \mathcal{D}^{\perp}$. Then by (4.2)

$$0 = < (f, e), (f, e) >_{\mathcal{F} \times \mathcal{F}^*} = 2 < e | f > . \tag{4.6}$$

Thus for all $(f, e) \in \mathcal{D}$ we obtain

$$< e \mid f > = 0. \tag{4.7}$$

Hence a Dirac structure $\mathcal{D}$ on $\mathcal{F}$ defines a power-conserving relation between the power variables $(f, e) \in \mathcal{F} \times \mathcal{F}^*$.

Remark 4.4. The condition dim $\mathcal{D} =$ dim $\mathcal{F}$ is intimately related to the usually expressed statement that a physical interconnection can *not* determine at the same time both the flow and effort (e.g. current *and* voltage, or velocity *and* force).

Constant Dirac structures admit different *matrix representations.* Here we just list a number of them, without giving proofs and algorithms to convert one representation into another, see e.g. (Dalsmo and van der Schaft, 1999).
Let $\mathcal{D} \subset \mathcal{F} \times \mathcal{F}^*$, with dim $\mathcal{F} = \ell$, be a constant Dirac structure. Then $\mathcal{D}$ can be represented as

1. (*Kernel and Image representation,* (Dalsmo and van der Schaft, 1999; van der Schaft and Maschke, 1995a)).

$$\mathcal{D} = \{(f, e) \in \mathcal{F} \times \mathcal{F}^* | Ff + Ee = 0\} \tag{4.8}$$

for $\ell \times \ell$ matrices F and E satisfying

$$\begin{aligned} (i) \quad & EF^T + FE^T = 0 \\ (ii) \quad & \text{rank } [F \vdots E] = \ell \end{aligned} \tag{4.9}$$

Equivalently,

$$\mathcal{D} = \{(f, e) \in \mathcal{F} \times \mathcal{F}^* | f = E^T \lambda, \ e = F^T \lambda, \ \lambda \in \mathbb{R}^\ell\} \tag{4.10}$$

2. (*Constrained input-output representation,* (Dalsmo and van der Schaft, 1999)).

$$\mathcal{D} = \{(f, e) \in \mathcal{F} \times \mathcal{F}^* | f = Je + G\lambda, \ G^T e = 0\} \tag{4.11}$$

for an $\ell \times \ell$ skew-symmetric matrix J, and a matrix G such that ImG = $\{f | (f, 0) \in \mathcal{D}\}$. Furthermore, Ker$J = \{e | (0, e) \in \mathcal{D}\}$.

3. (*Hybrid input-output representation,* (Bloch and Crouch, 1999)).
Let $\mathcal{D}$ be given as in (4.8). Suppose rank $F = \ell^1 (\leq \ell)$. Select ℓ^1 independent columns of F, and group them into a matrix F^1. Write (possibly after permutations) $F = [F^1 \vdots F^2]$ and, correspondingly $E = [E^1 \vdots E^2]$, $f = \begin{bmatrix} f^1 \\ f^2 \end{bmatrix}, e = \begin{bmatrix} e^1 \\ e^2 \end{bmatrix}$.

Then the matrix $[F^1 \vdots E^2]$ can be shown to be invertible, and

$$\mathcal{D} = \left\{ \left(\begin{pmatrix} f^1 \\ f^2 \end{pmatrix}, \begin{pmatrix} e^1 \\ e^2 \end{pmatrix} \right) \middle| \begin{pmatrix} f^1 \\ e^2 \end{pmatrix} = J \begin{pmatrix} e^1 \\ f^2 \end{pmatrix} \right\} \tag{4.12}$$

with $J := -\left[F^1 \vdots E^2\right]^{-1} \left[F^2 \vdots E^1\right]$ skew-symmetric.

4. (*Canonical coordinate representation,* (Courant, 1990)).
There exist linear coordinates (q, p, r, s) for $\mathcal{F}$ such in these coordinates and dual coordinates for $\mathcal{F}^*$, $(f, e) = (f_q, f_p, f_r, f_s, e_q, e_p, e_r, e_s) \in \mathcal{D}$ if and only if

$$\left\{ \begin{array}{rclrcl} f_q & = & e_p, & f_p & = & -e_q \\ f_r & = & 0, & e_s & = & 0 \end{array} \right. \tag{4.13}$$

Example 4.5. Kirchhoff's laws are an example of (4.8), taking $\mathcal{F}$ the space of currents and $\mathcal{F}^*$ the space of voltages.

Given a Dirac structure $\mathcal{D}$ on $\mathcal{F}$, the following subspaces of $\mathcal{F}$, respectively $\mathcal{F}^*$, are of importance

$$\begin{array}{rcl} G_1 & := & \{f \in \mathcal{F} \mid \exists e \in \mathcal{F}^* \text{ s.t. } (f,e) \in \mathcal{D}\} \\ P_1 & := & \{e \in \mathcal{F}^* \mid \exists f \in \mathcal{F} \text{ s.t. } (f,e) \in \mathcal{D}\} \end{array} \tag{4.14}$$

The subspace G_1 expresses the set of admissible flows, and P_1 the set of admissible efforts. It follows from the image representation (4.10) that

$$\begin{array}{rcl} G_1 & = & \text{Im } E^T \\ P_1 & = & \text{Im } F^T \end{array} \tag{4.15}$$

4.2 Implicit port-Hamiltonian systems

From a network modeling perspective a (lumped-parameter) physical system is naturally described by a set of (possibly multi-dimensional) *energy-storing* elements, a set of *energy-dissipating* or *resistive* elements, and a set of *ports* (by which interaction with the environment can take place), interconnected to each other by a *power-conserving interconnection*, see Figure 5.

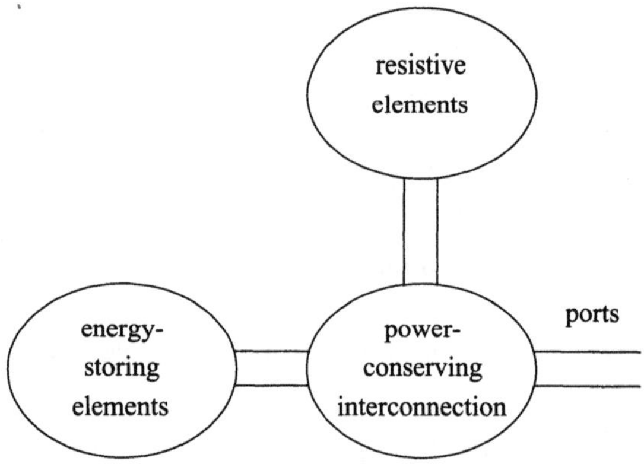

Figure 5. Implicit port-Hamiltonian system with dissipation

Here the power-conserving interconnection also includes power-conserving elements like (in the electrical domain) transformers, gyrators, or (in the mechanical domain) transformers, kinematic pairs and kinematic constraints.

Associated with the energy-storing elements are energy-variables $x_1, \cdots, x_n$, being coordinates for some n-dimensional state space manifold $\mathcal{X}$, and a total energy $H : \mathcal{X} \to \mathbb{R}$. The power-conserving interconnection is formalized in first instance (see later on for the non-constant case) by a constant Dirac structure $\mathcal{D}$ on the finite-dimensional linear space $\mathcal{F} := \mathcal{F}_S \times \mathcal{F}_R \times \mathcal{F}_P$, with $\mathcal{F}_S$ denoting the space of flows f_S connected to the energy-storing elements, $\mathcal{F}_R$ denoting the space of flows f_R connected to the dissipative (resistive) elements, and $\mathcal{F}_P$ the space of external flows f_P which can be connected to the environment. Dually, we write $\mathcal{F}^* = \mathcal{F}_S^* \times \mathcal{F}_R^* \times \mathcal{F}_P^*$, with $e_S \in \mathcal{F}_S^*$ the efforts connected to the energy-storing elements, $e_R \in \mathcal{F}_R^*$ the efforts connected to the resistive elements, and $e_P \in \mathcal{F}_P^*$ the efforts to be connected to the environment of the system.

The flow variables of the energy-storing elements are given as $\dot{x}(t) = \frac{dx}{dt}(t), t \in \mathbb{R}$, and the effort variables of the energy-storing elements as $\frac{\partial H}{\partial x}(x(t))$ (implying that $< \frac{\partial H}{\partial x}(x(t)) | \dot{x}(t) >= \frac{dH}{dt}(x(t))$ is the increase in energy). In order to have a consistent sign convention for energy flow we put

$$\begin{aligned} f_S &= -\dot{x} \\ e_S &= \frac{\partial H}{\partial x}(x) \end{aligned} \tag{4.16}$$

Similarly, restricting to *linear* resistive elements as in (2.32), the flow and effort variables connected to the resistive elements are related as

$$f_R = -Re_R \tag{4.17}$$

for some matrix $R = R^T \geq 0$.

Substitution of (4.16) and (4.17) into the Dirac structure $\mathcal{D}$ leads to the following geometric description of the dynamics

$$(f_S = -\dot{x}, f_R = -Re_R, f_P, e_S = \frac{\partial H}{\partial x}(x), e_R, e_P) \in \mathcal{D} \tag{4.18}$$

We call (4.18) an *implicit port-Hamiltonian system* (with dissipation), defined with respect to the constant Dirac structure $\mathcal{D}$, the Hamiltonian H and the resistive structure R.

An equational representation of an implicit port-Hamiltonian system is obtained by taking a matrix representation of the Dirac structure $\mathcal{D}$ as discussed in the previous subsection. For example, in kernel representation the Dirac structure on $\mathcal{F} = \mathcal{F}_S \times \mathcal{F}_R \times \mathcal{F}_P$ may be given as

$$\begin{aligned} \mathcal{D} = \{ & (f_S, f_R, f_P, e_S, e_R, e_P) \mid \\ & F_S f_S + E_S e_S + F_R f_R + E_R e_R + F_P f_P + E_P e_P = 0 \} \end{aligned} \tag{4.19}$$

for certain matrices $F_S, E_S, F_R, E_R, F_P, E_P$ satisfying

$$\begin{aligned} &(i) \quad E_S F_S^T + F_S E_S^T + E_R F_R^T + F_R E_R^T + E_P F_P^T + F_P E_P^T = 0 \\ &(ii) \quad \text{rank} \left[F_S \vdots F_R \vdots F_P \vdots E_S \vdots E_R \vdots E_P \right] = \dim \mathcal{F} \end{aligned} \tag{4.20}$$

Then substitution of (4.16) and (4.17) into (4.19) yields the following set of differential-algebraic equations for the implicit port-Hamiltonian system

$$F_S \dot{x}(t) = E_S \frac{\partial H}{\partial x}(x(t)) - F_R R e_R + E_R e_R + F_P f_P + E_P e_P, \tag{4.21}$$

Different representations of the Dirac structure $\mathcal{D}$ lead to different representations of the implicit port-Hamiltonian system, and this freedom may be exploited for simulation and analysis.

Actually, for many purposes this definition of port-Hamiltonian system is not general enough, since often the Dirac structure is not constant, but *modulated* by the state variables x. In this case the matrices $F_S, E_S, F_R, E_R, F_P, E_P$ in the kernel representation depend (smoothly) on x, leading to the implicit port-Hamiltonian system

$$\begin{aligned} F_S(x(t))\dot{x}(t) &= E_S(x(t))\tfrac{\partial H}{\partial x}(x(t)) - F_R(x(t))Re_R(t) \\ &+E_R(x(t))e_R(t) + F_P(x(t))f_P(t) + E_P(x(t))e_P(t), \quad t \in \mathbb{R} \end{aligned} \tag{4.22}$$

with

$$\begin{aligned} & E_S(x)F_S^T(x) + F_S(x)E_S^T(x) + E_R(x)F_R^T(x) + F_R(x)E_R^T(x) \\ & + E_P(x)F_P^T(x) + F_P(x)E_P^T(x) = 0, \quad \forall x \in \mathcal{X} \\ & \text{rank} \left[F_S(x) \vdots F_R(x) \vdots F_P(x) \vdots E_S(x) \vdots E_R(x) \vdots E_P(x) \right] = \dim \mathcal{F} \end{aligned} \tag{4.23}$$

Remark 4.6. Strictly speaking the flow and effort variables $\dot{x}(t) = -f_S(t)$, respectively $\frac{\partial H}{\partial x}(x(t)) = e_S(t)$, are not living in a constant linear space $\mathcal{F}_S$, respectively $\mathcal{F}_S^*$, but instead in the tangent spaces $T_{x(t)}\mathcal{X}$, respectively co-tangent spaces $T^*_{x(t)}\mathcal{X}$, to the state space manifold $\mathcal{X}$. This is formalized in the definition of a *non-constant Dirac structure on a manifold*; see the references (Courant, 1990; Dorfman, 1993; Dalsmo and van der Schaft, 1999; van der Schaft, 1999b).

It can be checked that the definition of a port-Hamiltonian system as given in (2.33) is a special case of (4.22), see (van der Schaft, 1999b). By the power-conservation property of a Dirac structure (cf. (4.7)) it follows directly that any implicit port-Hamiltonian system satisfies the energy-balance

$$\begin{aligned} \tfrac{dH}{dt}(x(t)) &= < \tfrac{\partial H}{\partial x}(x(t)) | \dot{x}(t) >= \\ &= -e_R^T(t)Re_R(t) + e_P^T(t)f_P(t), \end{aligned} \tag{4.24}$$

as was the case for an (explicit) port-Hamiltonian system (2.33).

The *algebraic constraints* that are present in the implicit system (4.22) are expressed by the subspace P_1, and the Hamiltonian H. In fact, since the Dirac structure $\mathcal{D}$ is modulated by the x-variables, also the subspace P_1 is modulated by the x-variables, and thus the effort variables e_S, e_R and e_P necessarily satisfy

$$(e_S, e_R, e_P) \in P_1(x), \quad x \in \mathcal{X}, \tag{4.25}$$

or, because of (4.15),

$$e_S \in \text{ Im } F_S^T(x), e_R \in \text{ Im } F_R^T(x), e_P \in \text{ Im } F_P^T(x). \tag{4.26}$$

The second and third inclusions entail the expression of e_R and e_P in terms of the other variables, while the first inclusion determines, since $e_S = \frac{\partial H}{\partial x}(x)$, the following algebraic constraints on the state variables

$$\frac{\partial H}{\partial x}(x) \in \text{ Im } F_S^T(x). \tag{4.27}$$

Remark 4.7. Under certain non-degeneracy conditions the elimination of the algebraic constraints (4.27) for an implicit port-Hamiltonian system (4.18) can be shown to result in an *explicit* port-Hamiltonian system.

The *Casimir functions* $C : \mathcal{X} \to \mathbb{R}$ of the implicit system (4.22) are determined by the subspace $G_1(x)$. Indeed, necessarily $(f_S, f_R, f_P) \in G_1(x)$, and thus by (4.15)

$$f_S \in \text{ Im } E_S^T(x), f_R \in \text{ Im } E_R^T(x), f_P \in \text{ Im } E_P^T(x). \tag{4.28}$$

Since $f_S = \dot{x}(t)$, the first inclusion yields the *flow constraints*

$$\dot{x}(t) \in \text{ Im } E_S^T(x(t)), \quad t \in \mathbb{R}. \tag{4.29}$$

Thus $C : \mathcal{X} \to \mathbb{R}$ is a Casimir function if $\frac{dC}{dt}(x(t)) = \frac{\partial^T C}{\partial x}(x(t))\dot{x}(t) = 0$ for all $\dot{x}(t) \in$ Im $E_S^T(x(t))$. Hence $C : \mathcal{X} \to \mathbb{R}$ is a Casimir of the implicit port-Hamiltonian system (4.18) if it satisfies the set of p.d.e.'s

$$\frac{\partial C}{\partial x}(x) \in \text{ Ker } E_S(x) \tag{4.30}$$

Remark 4.8. Note that $C : \mathcal{X} \to \mathbb{R}$ satisfying (4.30) is a Casimir function of (4.18) in a *strong sense*: it is a dynamical invariant $\left(\frac{dC}{dt}(x(t)) = 0\right)$ for *every* port behavior and every resistive relation (4.17).

Example 4.9. (Dalsmo and van der Schaft, 1999; van der Schaft and Maschke, 1995b) The constrained Hamiltonian equations (2.22) can be viewed as an implicit port-Hamiltonian system, with respect to the Dirac structure $\mathcal{D}$, given in constrained input-output representation (4.11) by

$$\begin{aligned} \mathcal{D} = \{(f_S, f_P, e_S, e_P) | 0 = A^T(q)e_S,\ e_P = B^T(q)e_S, \\ -f_S = \begin{bmatrix} 0 & I_k \\ -I_k & 0 \end{bmatrix} e_S + \begin{bmatrix} 0 \\ A(q) \end{bmatrix} \lambda + \begin{bmatrix} 0 \\ B(q) \end{bmatrix} f_P,\ \lambda \in \mathbb{R}^r\} \end{aligned} \tag{4.31}$$

In this case, the algebraic constraints on the state variables (q, p) are given as

$$0 = A^T(q)\frac{\partial H}{\partial p}(q, p) \tag{4.32}$$

while the Casimir functions C are determined by the equations

$$\frac{\partial^T C}{\partial q}(q)\dot{q} = 0, \qquad \text{for all } \dot{q} \text{ satisfying } A^T(q)\dot{q} = 0. \tag{4.33}$$

Hence, finding Casimir functions amounts to *integrating* the kinematic constraints $A^T(q)\dot{q} = 0$. In particular, if the kinematic constraints are *holonomic*, and thus can be expressed as (2.19), then $\bar{q}_{k-r+1}, \cdots, \bar{q}_k$ generate all the Casimir functions. □

Remark 4.10. For a proper notion of *integrability* of non-constant Dirac structures, generalizing the integrability conditions (2.12) of the structure matrix $J(x)$, we refer e.g. to (Dalsmo and van der Schaft, 1999).

In principle, the theory presented before for *explicit* port-Hamiltonian systems can be directly extended, mutatis mutandis, to *implicit* port-Hamiltonian system. In particular, the standard feedback interconnection of an implicit port-Hamiltonian system P with port variables f_P, e_P (the "plant") with another implicit port-Hamiltonian system with port variables f_P^C, e_P^C (the "controller") is readily seen to result in a closed-loop implicit port-Hamiltonian system with port variables. Furthermore, as in the explicit case, the Hamiltonian of this closed-loop system is just the *sum* of the Hamiltonian of the plant port-Hamiltonian system and the Hamiltonian of the controller port-Hamiltonian system. Finally, the Casimir analysis for the closed-loop system can be performed along the same lines as before.

5 Distributed-parameter port-Hamiltonian systems

From a modeling and control point of view it is very desirable to be able to include *distributed-parameter* components into the Hamiltonian description of complex physical systems. However, in extending the Hamiltonian theory as for instance exposed in (Olver, 1993) to distributed-parameter control systems a fundamental difficulty arises in the treatment of *boundary conditions.* Indeed, the treatment of infinite-dimensional Hamiltonian systems in the literature seems mostly focussed on systems with infinite spatial domain, where the variables go to zero for the spatial variables tending to infinity, or on systems with boundary conditions such that the energy exchange through the boundary is *zero.* On the other hand, from a control and interconnection point of view it is essential to be able to describe a distributed-parameter system with varying boundary conditions inducing *energy exchange through the boundary,* since in many applications interaction with the environment (e.g. actuation or measurement) takes place through the boundary of the system. Clear examples are the telegraph equations (describing the dynamics of a transmission line), where the boundary of the system is described by the behavior of the voltages and currents at both ends of the transmission line, or a vibrating string (or, more generally, a flexible beam), where it is natural to consider the evolution of the forces and velocities at the ends of the string. Furthermore, in both examples it is obvious that in general the boundary exchange of power (voltage times current in the transmission line example, and force times velocity for the vibrating string) will be non-zero, and that in fact one would like to consider the voltages and currents or forces and velocities as additional *boundary variables* of the system, which can be interconnected to other systems. Also for numerical integration and *simulation* of complex distibuted-parameter systems it is essential to be able to describe the complex system as the interconnection or coupling of its subsystems via their boundary variables; for example in the case of coupled fluid-solid dynamics.

From a mathematical point of view, it is not obvious how to incorporate non-zero energy flow through the boundary in the existing Hamiltonian framework for distributed-parameter systems. The problem is already illustrated by the Hamiltonian formulation of e.g. the Korteweg-de Vries equation (see e.g. (Olver, 1993)). Here for zero boundary conditions a *Poisson bracket* can be formulated with the use of the differential operator $\frac{d}{dx}$, since by integration by parts this operator is obviously skew-symmetric. However, for boundary conditions corresponding to non-zero energy flow the differential operator is not skew-symmetric anymore (since after integrating by parts the remainders are not zero).

In (van der Schaft and Maschke, 2002; Maschke and van der Schaft, 2000b,a) we have provided a framework to overcome this fundamental problem by using the notion of an *infinite-dimensional Dirac structure.* The infinite-dimensional Dirac structure employed in these papers has a specific form by being defined on certain spaces of differential forms on the spatial domain of the system and its boundary, and making use of Stokes' theorem. Its construction emphasizes the geometrical content of the physical variables involved, by identifying them as differential k-forms, for appropriate k. This framework has been used (van der Schaft and Maschke, 2002) for a port-Hamiltonian representation of the telegrapher's equations of an ideal transmission line, Maxwell's equations on a bounded domain with non-zero Poynting vector at its boundary, a vibrating string with traction forces at its ends, and planar beam models (Golo et al., 2002b, 2003b). Furthermore the framework has been extended to cover Euler's equations for an ideal fluid on a domain with permeable boundary, see also (van der Schaft and Maschke, 2001).

Throughout, let Z be an n-dimensional smooth manifold with smooth $(n-1)$-dimensional boundary ∂Z, representing the space of *spatial variables.*

Denote by $\Omega^k(Z)$, $k = 0, 1, \cdots, n$, the space of exterior k-forms on Z, and by $\Omega^k(\partial Z)$, $k = 0, 1, \cdots, n-1$, the space of k-forms on ∂Z. (Note that $\Omega^0(Z)$, respectively $\Omega^0(\partial Z)$, is the space of smooth functions on Z, respectively ∂Z.) Clearly, $\Omega^k(Z)$ and $\Omega^k(\partial Z)$ are (infinite-dimensional) linear spaces (over $\mathbb{R}$). Furthermore, there is a natural pairing between $\Omega^k(Z)$ and $\Omega^{n-k}(Z)$ given by

$$< \beta | \alpha > := \int_Z \beta \wedge \alpha \quad (\in \mathbb{R}) \tag{5.1}$$

with $\alpha \in \Omega^k(Z), \beta \in \Omega^{n-k}(Z)$, where $\wedge$ is the usual wedge product of differential forms yielding the n-form $\beta \wedge \alpha$. In fact, the pairing (5.1) is *non-degenerate* in the sense that if $< \beta | \alpha > = 0$ for all α, respectively for all β, then $\beta = 0$, respectively $\alpha = 0$.

Similarly, there is a pairing between $\Omega^k(\partial Z)$ and $\Omega^{n-1-k}(\partial Z)$ given by

$$< \beta | \alpha > := \int_{\partial Z} \beta \wedge \alpha \tag{5.2}$$

with $\alpha \in \Omega^k(\partial Z), \beta \in \Omega^{n-1-k}(\partial Z)$. Now let us define the linear space

$$\mathcal{F}_{p,q} := \Omega^p(Z) \times \Omega^q(Z) \times \Omega^{n-p}(\partial Z), \tag{5.3}$$

for any pair p, q of positive integers satisfying

$$p + q = n + 1, \tag{5.4}$$

and correspondingly let us define

$$\mathcal{E}_{p,q} := \Omega^{n-p}(Z) \times \Omega^{n-q}(Z) \times \Omega^{n-q}(\partial Z). \tag{5.5}$$

Then the pairing (5.1) and (5.2) yields a (non-degenerate) pairing between $\mathcal{F}_{p,q}$ and $\mathcal{E}_{p,q}$ (note that by (5.4) $(n-p)+(n-q)=n-1$). As before, symmetrization of this pairing yields the following bilinear form on $\mathcal{F}_{p,q} \times \mathcal{E}_{p,q}$ with values in $\mathbb{R}$:

$$\begin{aligned} &\ll \left(f_p^1, f_q^1, f_b^1, e_p^1, e_q^1, e_b^1\right), \left(f_p^2, f_q^2, f_b^2, e_p^2, e_q^2, e_b^2\right) \gg := \\ &\int_Z \left[e_p^1 \wedge f_p^2 + e_q^1 \wedge f_q^2 + e_p^2 \wedge f_p^1 + e_q^2 \wedge f_q^1\right] + \int_{\partial Z} \left[e_b^1 \wedge f_b^2 + e_b^2 \wedge f_b^1\right] \end{aligned} \tag{5.6}$$

where for $i = 1, 2$

$$\begin{aligned} &f_p^i \in \Omega^p(Z),\ f_q^i \in \Omega^q(Z) \\ &e_p^i \in \Omega^{n-p}(Z),\ e_p^i \in \Omega^{n-q}(Z) \\ &f_b^i \in \Omega^{n-p}(\partial Z),\ e_b^i \in \Omega^{n-q}(\partial Z) \end{aligned} \tag{5.7}$$

The spaces of differential forms $\Omega^p(Z)$ and $\Omega^q(Z)$ will represent the energy variables of two different physical energy domains interacting with each other, while $\Omega^{n-p}(\partial Z)$ and $\Omega^{n-q}(\partial Z)$ will denote the boundary variables whose (wedge) product represents the boundary energy flow. For example, in Maxwell's equations (Example 5.4) we will have $n = 3$ and $p = q = 2$; with $\Omega^p(Z) = \Omega^2(Z)$, respectively $\Omega^q(Z) = \Omega^2(Z)$, being the space of electric field inductions, respectively magnetic field inductions, and $\Omega^{n-p}(\partial Z) = \Omega^1(\partial Z)$ denoting the electric and magnetic field intensities at the boundary, with product the Poynting vector.

Theorem 5.1. *Consider $\mathcal{F}_{p,q}$ and $\mathcal{E}_{p,q}$ given in (5.3), (5.5) with p, q satisfying (5.4), and bilinear form $\ll, \gg$ given by (5.6). Define the following linear subspace D of $\mathcal{F}_{p,q} \times \mathcal{E}_{p,q}$*

$$\begin{aligned} D \;=\; &\{(f_p, f_q, f_b, e_p, e_q, e_b) \in \mathcal{F}_{p,q} \times \mathcal{E}_{p,q} | \\ &\begin{bmatrix} f_p \\ f_q \end{bmatrix} = \begin{bmatrix} 0 & (-1)^r d \\ d & 0 \end{bmatrix} \begin{bmatrix} e_p \\ e_q \end{bmatrix}, \\ &\begin{bmatrix} f_b \\ e_b \end{bmatrix} = \begin{bmatrix} 1 & 0 \\ 0 & -(-1)^{n-q} \end{bmatrix} \begin{bmatrix} e_{p|\partial Z} \\ e_{q|\partial Z} \end{bmatrix} \} \end{aligned} \tag{5.8}$$

where $|_{\partial Z}$ denotes restriction to the boundary ∂Z, and $r := pq + 1$. Then $D = D^\perp$, that is, D is a Dirac structure.

The proof of Theorem 5.1 relies on Stokes' theorem, and the infinite-dimensional Dirac structure defined in Theorem 5.1 is therefore called a Stokes-Dirac structure.

The definition of a distributed-parameter Hamiltonian system with respect to a Stokes-Dirac structure can now be stated as follows. Let Z be an n-dimensional manifold with boundary ∂Z, and let D be a Stokes-Dirac structure. Consider furthermore a *Hamiltonian density* (energy per volume element)

$$\mathcal{H} : \Omega^p(Z) \times \Omega^q(Z) \times Z \to \Omega^n(Z) \tag{5.9}$$

resulting in the total energy

$$H := \int_Z \mathcal{H} \in \mathbb{R} \tag{5.10}$$

Recall, see (5.1), that there exists a non-degenerate pairing between $\Omega^p(Z)$ and $\Omega^{n-p}(Z)$, respectively between $\Omega^q(Z)$ and $\Omega^{n-q}(Z)$. This means that $\Omega^{n-p}(Z)$ and $\Omega^{n-q}(Z)$ can be regarded as *dual spaces* to $\Omega^p(Z)$, respectively $\Omega^q(Z)$ (although strictly contained in their functional analytic duals). Let now $\alpha_p, \partial\alpha_p \in \Omega^p(Z)$, $\alpha_q, \partial\alpha_q \in \Omega^q(Z)$. Then under weak smoothness conditions on $\mathcal{H}$

$$H(\alpha_p + \partial\alpha_p, \alpha_q + \partial\alpha_q) = \int_Z \mathcal{H}\left(\alpha_p + \partial\alpha_p, \alpha_q + \partial\alpha_q, z\right) =$$

$$\int_Z \mathcal{H}\left(\alpha_p, \alpha_q, z\right) + \int_Z \left[\delta_p H \wedge \partial\alpha_p + \delta_q H \wedge \partial\alpha_q\right] + \text{ higher order terms in } \partial\alpha_p, \partial\alpha_q \tag{5.11}$$

for certain differential forms

$$\begin{aligned} \delta_p H &\in \Omega^{n-p}(Z) \\ \delta_q H &\in \Omega^{n-q}(Z) \end{aligned} \tag{5.12}$$

Furthermore, from the non-degeneracity of the pairing between $\Omega^p(Z)$ and $\Omega^{n-p}(Z)$, respectively between $\Omega^q(Z)$ and $\Omega^{n-q}(Z)$, it immediately follows that these differential forms are uniquely determined. This means that $(\delta_p H, \delta_q H) \in \Omega^{n-p}(Z) \times \Omega^{n-q}(Z)$ can be regarded as the (partial) *variational derivatives* (see e.g. (Olver, 1993)) of H at $(\alpha_p, \alpha_q) \in \Omega^p(Z) \times \Omega^q(Z)$. Throughout this paper we shall assume that the Hamiltonian H admits variational derivatives satisfying (5.11).

Now consider a time-function

$$(\alpha_p(t), \alpha_q(t)) \in \Omega^p(Z) \times \Omega^q(Z), \quad t \in \mathbb{R}, \tag{5.13}$$

and the Hamiltonian $H(\alpha_p(t), \alpha_q(t))$ evaluated along this trajectory. It follows that at any time t

$$\frac{dH}{dt} = \int_Z \left[\delta_p H \wedge \frac{\partial\alpha_p}{\partial t} + \delta_q H \wedge \frac{\partial\alpha_q}{\partial t}\right] \tag{5.14}$$

The differential forms $\frac{\partial\alpha_p}{\partial t}, \frac{\partial\alpha_q}{\partial t}$ represent the generalized velocities of the energy variables α_p, α_q. They are connected to the Stokes-Dirac structure D by setting

$$\begin{aligned} f_p &= -\tfrac{\partial\alpha_p}{\partial t} \\ f_q &= -\tfrac{\partial\alpha_q}{\partial t} \end{aligned} \tag{5.15}$$

(again the minus sign is included to have a consistent energy flow description). Since the right-hand side of (5.14) is the rate of increase of the stored energy H, we set

$$\begin{aligned} e_p &= \delta_p H \\ e_q &= \delta_q H \end{aligned} \tag{5.16}$$

Definition 5.2. The *distributed-parameter port-Hamiltonian system* with n-dimensional manifold of spatial variables Z, state space $\Omega^p(Z) \times \Omega^q(Z)$ (with $p+q=n+1$), Stokes-Dirac structure D given by (5.8), and Hamiltonian H, is given as (with $r = pq+1$)

$$\begin{bmatrix} -\frac{\partial \alpha_p}{\partial t} \\ -\frac{\partial \alpha_q}{\partial t} \end{bmatrix} = \begin{bmatrix} 0 & (-1)^r d \\ d & 0 \end{bmatrix} \begin{bmatrix} \delta_p H \\ \delta_q H \end{bmatrix}$$
$$\begin{bmatrix} f_b \\ e_b \end{bmatrix} = \begin{bmatrix} 1 & 0 \\ 0 & -(-1)^{n-q} \end{bmatrix} \begin{bmatrix} \delta_p H|_{\partial Z} \\ \delta_q H|_{\partial Z} \end{bmatrix} \tag{5.17}$$

By the power-conserving property of any Dirac structure it immediately follows that for any $(f_p, f_q, f_b, e_p, e_q, e_b)$ in the Stokes-Dirac structure D

$$\int_Z [e_p \wedge f_p + e_q \wedge f_q] + \int_{\partial Z} e_b \wedge f_b = 0 \tag{5.18}$$

Hence by substitution of (5.15), (5.16) and using (5.14) we obtain

Proposition 5.3. *Consider the distributed parameter port-Hamiltonian system (5.17). Then*

$$\frac{dH}{dt} = \int_{\partial Z} e_b \wedge f_b, \tag{5.19}$$

expressing that the increase in energy on the domain Z is equal to the power supplied to the system through the boundary ∂Z.

The system (5.17) can be called a (nonlinear) *boundary control* system in the sense of e.g. (Fattorini, 1968). Indeed, we could interpret f_b as the boundary control inputs to the system, and e_b as the measured outputs (or the other way around).

Example 5.4 (Maxwell's equations). Let $Z \subset \mathbb{R}^3$ be a 3-dimensional manifold with boundary ∂Z, defining the spatial domain, and consider the electromagnetic field in Z. The energy variables are the *electric field induction* 2-form $\alpha_p = \mathcal{D} \in \Omega^2(Z)$:

$$\mathcal{D} = \frac{1}{2} D_{ij}(t,z) dz^i \wedge dz^j \tag{5.20}$$

and the *magnetic field induction* 2-form $\alpha_q = \mathcal{B} \in \Omega^2(Z)$:

$$\mathcal{B} = \frac{1}{2} B_{ij}(t,z) dz^i \wedge dz^j \tag{5.21}$$

The corresponding Stokes-Dirac structure ($n=3, p=2, q=2$) is given as (cf. (5.8))

$$\begin{bmatrix} f_p \\ f_q \end{bmatrix} = \begin{bmatrix} 0 & -d \\ d & 0 \end{bmatrix} \begin{bmatrix} e_p \\ e_q \end{bmatrix}, \begin{bmatrix} f_b \\ e_b \end{bmatrix} = \begin{bmatrix} 1 & 0 \\ 0 & 1 \end{bmatrix} \begin{bmatrix} e_{p|\partial Z} \\ e_{q|\partial Z} \end{bmatrix} \tag{5.22}$$

Usually in this case one does *not* start with the definition of the total energy (Hamiltonian) H, but instead with the co-energy variables $\delta_p H, \delta_q H$, given, respectively, as the electric field intensity $\mathcal{E} \in \Omega^1(Z)$:

$$\mathcal{E} = E_i(t,z) dz^i \tag{5.23}$$

and the magnetic field intensity $\mathcal{H} \in \Omega^1(Z)$:

$$\mathcal{H} = H_i(t,z)dz^i \tag{5.24}$$

They are related to the energy variables through the constitutive relations of the medium (or material equations)

$$\begin{array}{rcl} *\mathcal{D} & = & \epsilon \mathcal{E} \\ *\mathcal{B} & = & \mu H \end{array} \tag{5.25}$$

with the scalar functions $\epsilon(t,z)$ and $\mu(t,z)$ denoting the electric permittivity, respectively magnetic permeability, and $*$ denoting the Hodge star operator (corresponding to a Riemannian metric on Z), converting 2-forms into 1-forms. Then one *defines* the Hamiltonian H as

$$H = \int_Z \frac{1}{2}(\mathcal{E} \wedge \mathcal{D} + \mathcal{H} \wedge \mathcal{B}), \tag{5.26}$$

and one immediately verifies that $\delta_p H = \mathcal{E}, \delta_q H = \mathcal{H}$.
Nevertheless there are other cases (corresponding to a *nonlinear* theory of the electromagnetic field, such as the Born-Infeld theory, see e.g. (Ingarden and Jamiolkowski, 1985)) where one starts with a more general Hamiltonian $H = \int_Z h$, with the energy density $h(\mathcal{D}, B)$ being a more general expression than $\frac{1}{2}(\epsilon^{-1} * \mathcal{D} \wedge \mathcal{D} + \mu^{-1} * \mathcal{B} \wedge \mathcal{B})$.

Assuming that there is no current in the medium Maxwell's equations can now be written as (see (Ingarden and Jamiolkowski, 1985))

$$\begin{array}{rcl} \frac{\partial \mathcal{D}}{\partial t} & = & d\mathcal{H} \\ \frac{\partial \mathcal{B}}{\partial t} & = & -d\mathcal{E} \end{array} \tag{5.27}$$

Explicitly taking into account the behavior at the boundary, Maxwell's equations on a domain $Z \subset \mathbb{R}^3$ are then represented as the port-Hamiltonian system with respect to the Stokes-Dirac structure given by (5.22), as

$$\begin{array}{rcl} \begin{bmatrix} -\frac{\partial \mathcal{D}}{\partial t} \\ -\frac{\partial \mathcal{B}}{\partial t} \end{bmatrix} & = & \begin{bmatrix} 0 & -d \\ d & 0 \end{bmatrix} \begin{bmatrix} \delta_D H \\ \delta_B H \end{bmatrix} \\ \begin{bmatrix} f_b \\ e_b \end{bmatrix} & = & \begin{bmatrix} \delta_D H|_{\partial Z} \\ \delta_B H|_{\partial Z} \end{bmatrix} \end{array} \tag{5.28}$$

Note that the first line of (5.27) is nothing else than (the differential version of) Ampère's law, while the second line of (5.27) is Faraday's law. Hence the Stokes-Dirac structure in (5.27), (5.28) expresses the basic physical laws connecting $\mathcal{D}, \mathcal{B}, \mathcal{H}$ and $\mathcal{E}$.
The energy-balance (5.19) in the case of Maxwell's equations takes the form

$$\frac{dH}{dt} = \int_{\partial Z} \delta_B H \wedge \delta_D H = \int_{\partial Z} \mathcal{H} \wedge \mathcal{E} = -\int_{\partial Z} \mathcal{E} \wedge \mathcal{H} \tag{5.29}$$

with $\mathcal{E} \wedge \mathcal{H}$ a 2-form corresponding to the *Poynting vector* (see (Ingarden and Jamiolkowski, 1985)).

Example 5.5 (Vibrating string). Consider an elastic string subject to traction forces at its ends. The spatial variable z belongs to the interval $Z = [0,1] \subset \mathbb{R}$. Let us denote by $u(t,z)$ the displacement of the string. The elastic potential energy is a function of the *strain* given by the 1-form

$$\alpha_q(t) = \epsilon(t,z)dz = \frac{\partial u}{\partial z}(t,z)dz \tag{5.30}$$

The associated co-energy variable is the *stress* given by the 0-form

$$\sigma = T * \alpha_q \tag{5.31}$$

with T the elasticity modulus and $*$ the Hodge star operator. Hence the potential energy is the quadratic function

$$U(\alpha_q) = \int_0^1 \sigma\alpha_q = \int_0^1 T * \alpha_q \wedge \alpha_q = \int_0^1 T\left(\frac{\partial u}{\partial z}\right)^2 dz \tag{5.32}$$

and $\sigma = \delta_q U$.

The kinetic energy K is a function of the kinetic *momentum* defined as the 1-form

$$\alpha_p(t) = p(t,z)dz \tag{5.33}$$

given by the quadratic function

$$K(\alpha_p) = \int_0^1 \frac{p^2}{\mu} dz \tag{5.34}$$

The associated co-energy variable is the *velocity* given by the 0-form

$$v = \frac{1}{\mu} * \alpha_p = \delta_p K \tag{5.35}$$

In this case the Dirac structure is the Stokes-Dirac structure for $n = p = q = 1$, with an *opposite sign* convention leading to the equations (with $H := U + K$)

$$\begin{aligned} \begin{bmatrix} -\frac{\partial\alpha_p}{\partial t} \\ -\frac{\partial\alpha_q}{\partial t} \end{bmatrix} &= \begin{bmatrix} 0 & -d \\ -d & 0 \end{bmatrix} \begin{bmatrix} \delta_p H \\ \delta_q H \end{bmatrix} \\ \begin{bmatrix} f_b \\ e_b \end{bmatrix} &= \begin{bmatrix} 1 & 0 \\ 0 & 1 \end{bmatrix} \begin{bmatrix} \delta_p H|_{\partial Z} \\ \delta_q H|_{\partial Z} \end{bmatrix} \end{aligned} \tag{5.36}$$

or, in more down-to-earth notation

$$\begin{aligned} \frac{\partial p}{\partial t} &= \frac{\partial\sigma}{\partial z} = \frac{\partial}{\partial z}(T\epsilon) \\ \frac{\partial\epsilon}{\partial t} &= \frac{\partial v}{\partial z} = \frac{\partial}{\partial z}\left(\frac{1}{\mu}p\right) \\ f_b &= v|_{\{0,1\}} \\ e_b &= \sigma|_{\{0,1\}} \end{aligned} \tag{5.37}$$

with boundary variables the velocity and stress at the ends of the string. Of course, by substituting $\epsilon = \frac{\partial u}{\partial z}$ into the 2nd equation of (5.37) one obtains $\frac{\partial}{\partial z}\left(\frac{\partial u}{\partial t} - \frac{p}{\mu}\right) = 0$, implying that

$$p = \mu \frac{\partial u}{\partial t} + \mu f(t) \tag{5.38}$$

for some function f, which may be set to zero. Substitution of (5.38) into the first equation of (5.37) then yields the wave equation

$$\mu \frac{\partial^2 u}{\partial t^2} = \frac{\partial}{\partial z}\left(T \frac{\partial u}{\partial z}\right) \tag{5.39}$$

This framework can be extended to general beam models, see (Golo et al., 2002b, 2003b).

6 Conclusions and future research

The theory presented in this paper opens up the way for many analysis, simulation and control problems. Its potential for set-point regulation has already received some attention (see (Maschke et al., 2000, 1999; Ortega et al., 2002, 2001; van der Schaft, 1999b)), while the extension to *tracking problems* is wide open. In this context we also like to refer to some recent work concerned with the shaping of the *Lagrangian*, see e.g. (Bloch et al., 1998). Also, the control of mechanical systems with nonholonomic kinematic constraints can be fruitfully approached from this point of view, see e.g. (Fujimoto and Sugie, 1999), as well as the modelling and control of multi-body systems, see (Maschke et al., 1994; Maschke and van der Schaft, 1997; Stramigioli et al., 2000). The framework of port-Hamiltonian systems seems perfectly suited to theoretical investigations on the topic of *impedance control*; see already (Stramigioli, 1998) for some initial results in this direction. Furthermore, the connection with multi-modal (*hybrid*) systems, corresponding to port-Hamiltonian systems with varying interconnection structure, needs further investigations. Some applications of the framework of distributed-parameter port-Hamiltonian systems have been already reported in (van der Schaft and Maschke, 2001; Rodriguez et al., 2001); see also (Schlacher and Kugi, 1999) for related work on smart structures. Finally, current research is concerned with the spatial discretization of port-Hamiltonian distributed-parameter systems to finite-dimensional port-Hamiltonian systems (Golo et al., 2002a), with direct applications towards simulation and control of such systems.

Bibliography

R.A. Abraham and J.E. Marsden. *Foundations of Mechanics.* Benjamin/Cummings, Reading, MA, 2nd edition, 1978.

A. Bloch, N. Leonard, and J.E. Marsden. Matching and stabilization by the method of controlled Lagrangians. In *37th IEEE Conf. on Decision and Control*, pages 1446–1451, Tampa, FL, 1998.

A.M. Bloch and P.E. Crouch. Representations of Dirac structures on vector spaces and nonlinear *LC* circuits. In G. Ferreyra, R. Gardner, H. Hermes, and H. Sussmann, editors, *Symposia in Pure Mathematics, Differential Geometry and Control Theory*, volume 64, pages 103–117. AMS, 1999.

P.C. Breedveld. *Physical systems theory in terms of bond graphs.* PhD thesis, University of Twente, Faculty of Electrical Engineering, 1984.

R.W. Brockett. Control theory and analytical mechanics. In C. Martin and R. Hermann, editors, *Geometric Control Theory*, volume VII of *Lie Groups: History, Frontiers and Applications*, pages 1–46. Math. Sci. Press, Brookline, 1977.

T.J. Courant. Dirac manifolds. *Trans. American Math. Soc.*, 319:631–661, 1990.

P.E. Crouch and A.J. van der Schaft. *Variational and Hamiltonian Control Systems.* Number 101 in Lect. Notes in Control and Inf. Sciences. Springer, Berlin, 1987.

M. Dalsmo and A.J. van der Schaft. On representations and integrability of mathematical structures in energy-conserving physical systems. *SIAM J. Control and Optimization*, 37(1):54–91, 1999.

I. Dorfman. *Dirac Structures and Integrability of Nonlinear Evolution Equations.* John Wiley, Chichester, 1993.

H.O. Fattorini. Boundary control systems. *SIAM J. Control*, 6:349–385, 1968.

K. Fujimoto and T. Sugie. Stabilization of a class of Hamiltonian systems with nonholonomic constraints via canonical transformations. In *Proc. European Control Conference '99*, Karlsruhe, Sept. 1999.

G. Golo, V. Talasila, and A.J. van der Schaft. Approximation of the telegrapher's equations. In *Proc. 41st IEEE Conf. Decision and Control*, Las Vegas, Nevada, Dec. 2002a.

G. Golo, V. Talasila, and A.J. van der Schaft. A Hamiltonian formulation of the Timoshenko beam. In *Mechatronics 2002*, pages 838–847, Enschede, June 2002b.

G. Golo, A. van der Schaft, P.C. Breedveld, and B.M. Maschke. Hamiltonian formulation of bond graphs. In R. Johansson and A. Rantzer, editors, *Nonlinear and Hybrid Systems in Automotive Control*, pages 351–372. Springer, London, 2003a.

G. Golo, A.J. van der Schaft, and S.Stramigioli. Hamiltonian formulation of planar beams. In A. Astolfi, F. Gordillo, and A.J. van der Schaft, editors, *Proc. 2nd IFAC Workshop on Lagrangian and Hamiltonian Methods for Nonlinear Control*, pages 169–174, Sevilla, 2003b.

D.J. Hill and P.J. Moylan. Stability of nonlinear dissipative systems. *IEEE Trans. Aut. Contr.*, 21:708–711, 1976.

R.S. Ingarden and A. Jamiolkowski. *Classical Electrodynamics.* PWN-Polish Sc. Publ., Elsevier, Warszawa, 1985.

A. Isidori. *Nonlinear Control Systems.* Communications and Control Engineering Series. Springer, London, 3rd edition, 1995.

J.E. Marsden and T.S. Ratiu. *Introduction to Mechanics and Symmetry.* Number 17 in Texts in Applied Mathematics. Springer, New York, 1994.

B. Maschke, R. Ortega, and A.J. van der Schaft. Energy-based Lyapunov functions for forced Hamiltonian systems with dissipation. *IEEE Transactions on Automatic Control*, 45:1498–1502, 2000.

B.M. Maschke, C. Bidard, and A.J. van der Schaft. Screw-vector bond graphs for the kinestatic and dynamic modeling of multibody systems. In *Proc. ASME Int. Mech. Engg. Congress*, volume 55, pages 637–644, Chicago, U.S.A., 1994.

B.M. Maschke, R. Ortega, A.J. van der Schaft, and G. Escobar. An energy-based derivation of Lyapunov functions for forced systems with application to stabilizing control. In *Proc. 14th IFAC World Congress*, volume E, pages 409–414, Beijing, 1999.

B.M. Maschke and A.J. van der Schaft. Port-controlled Hamiltonian systems: Modelling origins and system-theoretic properties. In *Proc. 2nd IFAC NOLCOS*, pages 282–288, Bordeaux, 1992.

B.M. Maschke and A.J. van der Schaft. System-theoretic properties of port-controlled Hamiltonian systems. In *Systems and Networks: Mathematical Theory and Applications*, volume II, pages 349–352. Akademie-Verlag, Berlin, 1994.

B.M. Maschke and A.J. van der Schaft. Interconnection of systems: the network paradigm. In *Proc. of the IEEE Int. Conf. on Decision and Control, CDC'96*, pages 207–212, Kobe, Japan, Dec. 1996.

B.M. Maschke and A.J. van der Schaft. Interconnected mechanical systems, part II: The dynamics of spatial mechanical networks. In A. Astolfi, D.J.N. Limebeer, C. Melchiorri, A. Tornambe, and R.B. Vinter, editors, *Modelling and Control of Mechanical Systems*, pages 17–30. Imperial College Press, London, 1997.

B.M. Maschke and A.J. van der Schaft. Hamiltonian representation of distributed parameter systems with boundary energy flow. In A. Isidori, F. Lamnabhi-Lagarrigue, and W. Respondek, editors, *Nonlinear Control in the Year 2000*, number 258 in Lect. Notes Control and Inf. Sciences, pages 137–142. Springer, 2000a.

B.M. Maschke and A.J. van der Schaft. Port controlled Hamiltonian representation of distributed parameter systems. In N.E. Leonard and R. Ortega, editors, *IFAC Workshop on Lagrangian and Hamiltonian methods for nonlinear control*, pages 28–38. Princeton University, 2000b.

B.M. Maschke, A.J. van der Schaft, and P.C. Breedveld. An intrinsic Hamiltonian formulation of network dynamics: Non-standard poisson structures and gyrators. *Journal of the Franklin Institute*, 329(5):923–966, 1992.

B.M. Maschke, A.J. van der Schaft, and P.C. Breedveld. An intrinsic Hamiltonian formulation of the dynamics of LC-circuits. *IEEE Trans. Circ. and Syst.*, 42:73–82, 1995.

J.I. Neimark and N.A. Fufaev. Dynamics of nonholonomic systems. *Translations of Mathematical Monographs, American Mathematical Society*, 33, 1972.

H. Nijmeijer and A. J. van der Schaft. *Nonlinear Dynamical Control Systems.* Springer, New York, 1990.

P.J. Olver. *Applications of Lie Groups to Differential Equations.* Springer, 2nd edition, 1993.

R. Ortega, A. Loria, P.J. Nicklasson, and H. Sira-Ramirez. *Passivity-based Control of Euler-Lagrange Systems.* Springer, London, 1998.

R. Ortega, A.J. van der Schaft, I. Mareels, and B.M. Maschke. Putting energy back in control. *Control Systems Magazine*, 21:18–33, 2001.

R. Ortega, A.J. van der Schaft, B.M. Maschke, and G. Escobar. Interconnection and damping assignment passivity-based control of port-controlled Hamiltonian systems. *Automatica*, 38:585–596, 2002.

H. M. Paynter. *Analysis and Design of Engineering Systems.* M.I.T. Press, MA, 1960.

H. Rodriguez, A.J. van der Schaft, and R. Ortega. On stabilization of nonlinear distributed parameter port-controlled Hamiltonian systems via energy-shaping. In *Proc. 40th IEEE Conf. on Decision and Control*, pages 131–136, Orlando, Florida, 2001.

K. Schlacher and A. Kugi. Control of mechanical structures by piezoelectric actuators and sensors. In D. Aeyels, F. Lamnabhi-Lagarrigue, and A.J. van der Schaft, editors, *Stability and Stabilization of Nonlinear Systems*, volume 246 of *Lecture Notes in Control and Information Sciences*, pages 275–292. Springer, London, 1999.

S. Stramigioli. *From Differentiable Manifolds to Interactive Robot Control.* PhD thesis, University of Delft, Dec. 1998.

S. Stramigioli, B.M. Maschke, and C. Bidard. A Hamiltonian formulation of the dynamics of spatial mechanism using Lie groups and screw theory. In *Proc. (to appear) Symposium Commemorating the Legacy, Work and Life of Sir R.S. Ball, J. Duffy and H. Lipkin organizers*, University of Cambridge, Trinity College, Cambridge, U.K., July, 9-11 2000.

M. Takegaki and S. Arimoto. A new feedback method for dynamic control of manipulators. *Trans. ASME, J. Dyn. Systems, Meas. Control*, 103:119–125, 1981.

A.J. van der Schaft. *System Theoretic Properties of Physical Systems.* CWI Tract 3, CWI, Amsterdam, 1984.

A.J. van der Schaft. Stabilization of Hamiltonian systems. *Nonl. An. Th. Math. Appl.*, 10:1021–1035, 1986.

A.J. van der Schaft. Interconnection and geometry. In J.W. Polderman and H.L. Trentelman, editors, *From Intelligent Control to Behavioral Systems*, pages 203–218, Groningen, 1999a.

A.J. van der Schaft. *L_2-Gain and Passivity Techniques in Nonlinear Control.* Communications and Control Engineering Series. Springer, London, 2nd revised and enlarged edition, 1999b.

A.J. van der Schaft, M. Dalsmo, and B.M. Maschke. Mathematical structures in the network representation of energy-conserving physical systems. In *Proc. 35th IEEE Conf. on Decision and Control*, pages 201–206, Kobe, Japan, 1996.

A.J. van der Schaft and B.M. Maschke. On the Hamiltonian formulation of nonholonomic mechanical systems. *Rep. Math. Phys.*, 34:225–233, 1994.

A.J. van der Schaft and B.M. Maschke. The Hamiltonian formulation of energy conserving physical systems with external ports. *Archiv für Elektronik und Übertragungstechnik*, 49:362–371, 1995a.

A.J. van der Schaft and B.M. Maschke. Mathematical modeling of constrained Hamiltonian systems. In *Proc. 3rd IFAC NOLCOS '95*, pages 678–683, Tahoe City, CA, 1995b.

A.J. van der Schaft and B.M Maschke. Interconnected mechanical systems, part I: geometry of interconnection and implicit Hamiltonian systems. In A. Astolfi, D.J.N Limebeer, C. Melchiorri, A. Tornambè, and R.B. Vinter, editors, *Modelling and Control of Mechanical Systems*, pages 1–15. Imperial College Press, 1997.

A.J. van der Schaft and B.M. Maschke. Fluid dynamical systems as Hamiltonian boundary control systems. In *Proc. 40th IEEE Conf. on Decision and Control*, pages 4497–4502, Orlando, Florida, 2001.

A.J. van der Schaft and B.M Maschke. Hamiltonian formulation of distributed-parameter systems with boundary energy flow. *Journal of Geometry and Physics*, 42:166–194, 2002.

A. Weinstein. The local structure of poisson manifolds. *J. Differential Geometry*, 18: 523–557, 1983.

J.C. Willems. Dissipative dynamical systems - part I: General theory. *Archive for Rational Mechanics and Analysis*, 45:321–351, 1972.

Periodic Motion Induced by the Painlevé Paradox

R. I. Leine [a]* B. Brogliato [b] and H. Nijmeijer [a]
[a] Eindhoven University of Technology, Eindhoven, The Netherlands
[b] INRIA Rhône-Alpes, Saint Ismier, France

Abstract. This chapter describes the periodic motion of the Frictional Impact Oscillator, which consists of an object with normal and tangential degrees of freedom that comes in contact with a rigid surface. The Frictional Impact Oscillator contains the basic mechanism for a hopping phenomenon observed in many practical applications. We will show that the hopping or bouncing motion in this type of systems is closely related to the Painlevé paradox. A dynamical system exhibiting the Painlevé paradox has non-uniqueness and non-existence of solutions in certain sliding modes. Furthermore, we will show that this type of systems can exhibit the Painlevé paradox for physically realistic values of the friction coefficient.

1 Introduction

If a finger is pushed over a table, then a hopping motion of the finger can be observed when the finger and table are sufficiently rough. The same phenomenon occurs when a piece of chalk is pushed over the blackboard, often resulting in a dotted line which indicates periodic detachment. Brake systems, consisting of a pin-on-disk mechanism, can exhibit an intermittent motion, where the pin periodically loses contact with the disk, accompanied by a squealing noise (Ibrahim, 1994). Robotic manipulators, which touch a surface of an obstacle, may also show a kind of bouncing motion, normally regarded as undesirable (Brogliato, 1999). In this chapter we will study a Frictional Impact Oscillator, which consists of a rigid object with normal and tangential degrees of freedom that comes in contact with a rigid surface. The Frictional Impact Oscillator contains the basic mechanism for the hopping phenomenon observed in many practical applications. We will show that the hopping or bouncing motion in this type of systems is closely related to the Painlevé paradox (which we explain in the sequel). Furthermore, we will show that this type of systems can exhibit the Painlevé paradox for arbitrary (positive) values of the friction coefficient.

The hopping phenomenon is explained and analyzed using Amontons-Coulomb's law and Newton's restitution law. We acknowledge that these contact laws provide only a crude approximation of the reality of dry friction and impact. However, in most engineering situations, the higher order information needed for a more accurate description is not available and we have to be content with some moderately precise calculation,

*This project was supported by the Dutch Technology Foundation, STW (grant EWO.5647), and the Royal Dutch Academy of Sciences.

accounting at least for the main features of phenomena (Moreau, 1988). The hopping phenomenon occurs generally in complex systems with many degrees of freedom and many contact points. Our aim is to describe the basic mechanism of the hopping motion with a system that is as simple as possible. We will show that a two degree of freedom model (which we call the Frictional Impact Oscillator) with a single contact, assuming Amontons-Coulomb's law and Newton's restitution law, is sufficient to describe the essential features of the phenomenon. This relatively simple model shows a complex dynamical behaviour, which might give insight in the dynamics of more complex systems.

The rigid multibody theory (Brogliato, 1999; Glocker, 1995, 2001; Moreau, 1988; Pfeiffer and Glocker, 1996), which assumes instantaneous impact between rigid bodies and Amontons-Coulomb friction model in tangential direction, provides a good approach to study the dynamics when one is interested in the global motion of the system. Mechanical systems with friction and impact, modelled with a rigid multibody approach, belong to the class of hybrid systems (i.e. systems with a mixed continuous and discrete nature in time). The transitions in time from one hybrid mode to another (e.g. from contact to detachment or stick to slip) demand the choice of the next hybrid mode. A rigorous way to perform this search is to formulate the problem as a Linear Complementarity problem (LCP). The solution of the LCP can be non-unique, giving an undeterminate next hybrid mode, and the solution of the LCP may even not exist, which leads to inconsistency in the model. The occurrence of inconsistency in a mechanical rigid multibody system due to friction is known as the *Painlevé paradox.*

It has been known since the end of the 19th century that the combination of Amontons–Coulomb friction together with the assumption of rigidity can cause inconsistencies for high values of the friction coefficient (Jellet, 1872; Painlevé, 1895a,b, 1905). Painlevé (1895a,b, 1905) considered the sliding motion of rigid objects in contact with the ground, which have inconsistencies in the sliding mode. The problem of a sliding rod studied by Painlevé became the classical example of a hybrid system with inconsistencies. Lecornu (1905) proposed velocity jumps to escape from inconsistent modes, often now addressed as impacts without collisions. A number of studies show that inconsistencies occur in the classical Painlevé example when the friction coefficient μ is larger than $\frac{4}{3}$ (Glocker, 1995; Lötstedt, 1981; Mason and Wang, 1988). The classical Painlevé example was recently studied in more detail in (Génot and Brogliato, 1999). New results on how the solution passes singular points in the sliding mode were presented in Génot and Brogliato (1999).

The critical value of $\frac{4}{3}$ is very large and not likely to occur in physical situations. In (Génot, 1998) an adapted version of the classical Painlevé example was proposed, where the rod has a rounded tip, which lowers the critical friction coefficient to 0.63 (the example with a rounded tip was inspired by Moreau (1986)). The original example studied by Painlevé (1895b), being the contact of a box with the ground, can have a much lower critical friction coefficient for a small radius of gyration. The occurrence of the Painlevé paradox of a rigid body with a friction rotor was studied in (Wilms and Cohen, 1997) and a critical friction coefficient was found which can take arbitrary (positive) values depending on other parameters. A similar conclusion is drawn in the present study.

A model of a hopping finger was studied by Moreau (1986). Stick-slip motion with and without flight phase were reported, as well period-doubled motion, but the occurrence of the Painlevé paradox was not studied. The Frictional Impact Oscillator, presented in this chapter, is slightly different and lends itself better to an analytical investigation of the Painlevé paradox.

Control of the sliding motion of an end-effector of a robot on the surface of an obstacle can be hampered by the same instability phenomena which govern the Frictional Impact Oscillator. It is expected that the results on the Frictional Impact Oscillator, which will be studied in the sequel, can help to design better control schemes for constrained robotic motion.

The main results on the classical Painlevé example will be briefly discussed in Section 3. The Frictional Impact Oscillator will be studied in Section 4. This model can be simplified which leads to the Simplified Frictional Impact Oscillator (Section 5). Critical friction coefficients of the Painlevé paradox will be derived for both models and numerical results for the periodic solutions will be given. The notation as well as the methods of Glocker (1995) and Pfeiffer and Glocker (1996) are used for the dynamics of rigid multibody systems with impact and friction and are reviewed in Section 2.

This chapter is based on the work presented in (Leine et al., 2002); further information and proofs can be found in (Leine et al., 2002).

2 Mathematical Modeling of Impact with Friction

In this section we review some aspects of a mathematical theory for the dynamics of rigid bodies with completely inelastic frictional impact as has been formulated in (Glocker, 1995, 2001; Pfeiffer and Glocker, 1996). A shorter version of this formulation can be found in (Leine et al., 2003).

For the description on the dynamics of multibody systems it is convenient to introduce four contact sets, which describe the kinematic state of the constraints of the system:

$$\begin{aligned} I_G &= \{1, 2, \ldots, n_G\} && \\ I_S &= \{i \in I_G \mid g_{Ni} = 0\} && \text{with } n_S \text{ elements} \\ I_N &= \{i \in I_S \mid \dot{g}_{Ni} = 0\} && \text{with } n_N \text{ elements} \\ I_H &= \{i \in I_N \mid \dot{g}_{Ti} = 0\} && \text{with } n_H \text{ elements,} \end{aligned} \tag{2.1}$$

where g_{Ni} and $\dot{g}_{Ti}$ denote the normal contact distance and tangential relative velocity of constraint i. The set I_G consists of n_G indices of all constraints, which we want to take into account. I_S contains the n_S indices of the constraints with vanishing normal distance but arbitrary relative velocity. In the set I_N are the n_N indices of the potentially active normal constraints, which fulfill the necessary conditions for continuous contact (vanishing normal distance and no relative velocity in normal direction). The set I_N contains therefore all indices of slipping or sticking contacts. The n_H elements of the set I_H correspond to the potentially active constraints in tangential direction (sticking). The sets I_S, I_N, I_H are not constant, because the contact configuration of the dynamical system changes with time due to stick-slip transitions, impact and contact loss.

The dynamics of a multibody system can be expressed for almost all t by the equation of motion

$$\boldsymbol{M}(t, \boldsymbol{q})\ddot{q} - \boldsymbol{h}(t, \boldsymbol{q}, \dot{\boldsymbol{q}}) - \sum_{i \in I_S} (\boldsymbol{w}_N(t, \boldsymbol{q})\lambda_N + \boldsymbol{w}_T(t, \boldsymbol{q})\lambda_T)_i = \boldsymbol{0}, \tag{2.2}$$

where $\boldsymbol{M}$ is the symmetric mass matrix, $\boldsymbol{q}$ the vector of generalized coordinates, $\boldsymbol{h}$ the vector with all smooth elastic, gyroscopic and dissipating generalized forces and $\boldsymbol{\lambda}_N$ and $\boldsymbol{\lambda}_T$ the vectors with normal and tangential contact forces. The vectors $\boldsymbol{w}_N$ and $\boldsymbol{w}_T$ are the normal and tangential force directions.

We assume the generalized coordinates $\boldsymbol{q}(t)$ to be absolute continuous functions in time and the generalized velocities $\dot{\boldsymbol{q}}(t)$ to be functions of locally bounded variations, when a solution exists of course. We therefore can define the left and right limits $\dot{\boldsymbol{q}}(t^-)$ and $\dot{\boldsymbol{q}}(t^+)$ at each time instant $t > t_0$ (Brogliato, 1999; Glocker, 2001). Furthermore, we assume the generalized velocities to be right-continuous, i.e. $\dot{\boldsymbol{q}}(t) = \dot{\boldsymbol{q}}(t^+)$. We specify the initial condition at $t = t_0$ by $(\boldsymbol{q}_0, \dot{\boldsymbol{q}}_0) = (\boldsymbol{q}(t_0), \dot{\boldsymbol{q}}(t_0))$. If we allow a first velocity jump to occur at t_0 then we set $\dot{\boldsymbol{q}}_0 = \dot{\boldsymbol{q}}(t_0^-)$ (Moreau, 1988). Typically, the contact forces λ_N and λ_T become impulsive when an impact occurs and we have $\dot{\boldsymbol{q}}(t^-) \neq \dot{\boldsymbol{q}}(t^+)$.

More conveniently we put (2.2) in the form

$$\boldsymbol{M}\ddot{\boldsymbol{q}} - \boldsymbol{h} - \boldsymbol{W}_N\boldsymbol{\lambda}_N - \boldsymbol{W}_T\boldsymbol{\lambda}_T = \boldsymbol{0} \tag{2.3}$$

where the dependencies on $t, \boldsymbol{q}, \dot{\boldsymbol{q}}$ have been omitted for brevity and where $\boldsymbol{W}_N$ and $\boldsymbol{W}_T$ are matrices containing the generalized force directions in normal and tangential direction. The contact distances g_{Ni} and g_{Ti} are gathered in the vectors $\boldsymbol{g}_N$ and $\boldsymbol{g}_T$. We can express the contact velocities and accelerations in the generalized accelerations by

$$\begin{bmatrix}\dot{\boldsymbol{g}}_N \\ \dot{\boldsymbol{g}}_T\end{bmatrix} = \begin{bmatrix}\boldsymbol{W}_N^{\mathrm{T}} \\ \boldsymbol{W}_T^{\mathrm{T}}\end{bmatrix}\dot{\boldsymbol{q}} + \begin{bmatrix}\tilde{\boldsymbol{w}}_N \\ \tilde{\boldsymbol{w}}_T\end{bmatrix}, \qquad \begin{bmatrix}\ddot{\boldsymbol{g}}_N \\ \ddot{\boldsymbol{g}}_T\end{bmatrix} = \begin{bmatrix}\boldsymbol{W}_N^{\mathrm{T}} \\ \boldsymbol{W}_T^{\mathrm{T}}\end{bmatrix}\ddot{\boldsymbol{q}} + \begin{bmatrix}\bar{\boldsymbol{w}}_N \\ \bar{\boldsymbol{w}}_T\end{bmatrix}. \tag{2.4}$$

Each closed contact $i \in I_N$ is characterized by a vanishing contact distance g_{Ni} and normal relative velocity $\dot{g}_{Ni}$. Under the assumption of impenetrability $g_{Ni} \geq 0$, only two situations may occur:

$$\begin{array}{ll} \ddot{g}_{Ni} = 0 \wedge \lambda_{Ni} \geq 0 & \text{contact is maintained} \\ \ddot{g}_{Ni} > 0 \wedge \lambda_{Ni} = 0 & \text{detachment} \end{array} \tag{2.5}$$

From (2.5) we see that the normal contact law shows a complementary behaviour: the product of the contact force and acceleration is always zero:

$$\ddot{g}_{Ni}\lambda_{Ni} = 0, \qquad i \in I_N. \tag{2.6}$$

The complementary behaviour of the normal contact law is depicted in Figure 1a and shows a corner of admissible combinations of $\ddot{g}_{Ni}$ and λ_{Ni}.

We assume Amontons-Coulomb law to hold in the tangential direction. For a closed contact $i \in I_N$, with friction coefficient μ_i, the following three cases are possible:

$$\begin{array}{llll} \dot{g}_{Ti} = 0 & \Rightarrow & |\lambda_{Ti}| \leq \mu_i\lambda_{Ni} & \text{sticking} \\ \dot{g}_{Ti} < 0 & \Rightarrow & \lambda_{Ti} = +\mu_i\lambda_{Ni} & \text{negative sliding} \\ \dot{g}_{Ti} > 0 & \Rightarrow & \lambda_{Ti} = -\mu_i\lambda_{Ni} & \text{positive sliding} \end{array} \qquad i \in I_N. \tag{2.7}$$

To determine the tangential contact force during sticking, one can formulate unilateral laws for sticking contacts. For a closed sticking contact $i \in I_H$ the following three cases

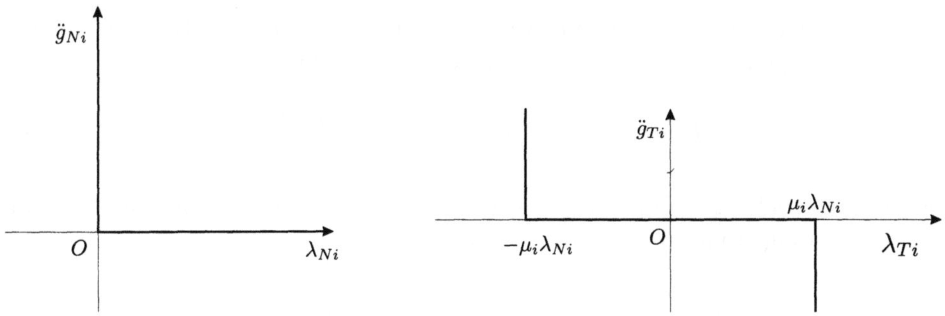

(a) normal direction (b) tangential direction

Figure 1. Complementarity of contacts.

are possible

$$\left.\begin{array}{lll} \ddot{g}_{Ti} = 0 & \Rightarrow \quad |\lambda_{Ti}| \leq \mu_i \lambda_{Ni} & \text{remains sticking} \\ \ddot{g}_{Ti} < 0 & \Rightarrow \quad \lambda_{Ti} = +\mu_i \lambda_{Ni} & \text{commences negative sliding} \\ \ddot{g}_{Ti} > 0 & \Rightarrow \quad \lambda_{Ti} = -\mu_i \lambda_{Ni} & \text{commences positive sliding} \end{array}\right. \qquad i \in I_H. \tag{2.8}$$

The normal and tangential contact law lead, together with the equations of motion, to the coupled normal–tangential contact problem for the stick-slip and detachment transitions of the multibody system

$$\begin{bmatrix} \ddot{\boldsymbol{g}}_N \\ \ddot{\boldsymbol{g}}_H \end{bmatrix} = \begin{bmatrix} \boldsymbol{W}_N^{\mathrm{T}}\boldsymbol{M}^{-1}(\boldsymbol{W}_N + \boldsymbol{W}_G\boldsymbol{\mu}_G) & \boldsymbol{W}_N^{\mathrm{T}}\boldsymbol{M}^{-1}\boldsymbol{W}_H \\ \boldsymbol{W}_H^{\mathrm{T}}\boldsymbol{M}^{-1}(\boldsymbol{W}_N + \boldsymbol{W}_G\boldsymbol{\mu}_G) & \boldsymbol{W}_H^{\mathrm{T}}\boldsymbol{M}^{-1}\boldsymbol{W}_H \end{bmatrix} \begin{bmatrix} \boldsymbol{\lambda}_N \\ \boldsymbol{\lambda}_H \end{bmatrix} + \begin{bmatrix} \boldsymbol{W}_N^{\mathrm{T}}\boldsymbol{M}^{-1}\boldsymbol{h} + \bar{\boldsymbol{w}}_N \\ \boldsymbol{W}_H^{\mathrm{T}}\boldsymbol{M}^{-1}\boldsymbol{h} + \bar{\boldsymbol{w}}_H \end{bmatrix} \tag{2.9}$$

with[1]

$$\begin{array}{ll} 0 \leq \ddot{\boldsymbol{g}}_N \perp \boldsymbol{\lambda}_N \geq 0 & \in I_N \\ \boldsymbol{\lambda}_H \in -\boldsymbol{\mu}_H \boldsymbol{\lambda}_N \mathrm{diag}(\partial|\ddot{\boldsymbol{g}}_H|) & \in I_H \end{array} \tag{2.10}$$

where

$$\begin{array}{c} \boldsymbol{W}_G = \boldsymbol{w}_{Ti}, \qquad i \in I_N \backslash I_H, \\ \boldsymbol{g}_H = g_{Ti}, \quad \boldsymbol{\lambda}_H = \lambda_{Ti}, \quad \boldsymbol{W}_H = \boldsymbol{w}_{Ti}, \quad \bar{\boldsymbol{w}}_H = \bar{w}_{Ti}, \qquad i \in I_H \\ \boldsymbol{\mu}_H \in \mathbb{R}^{n_H \times n_N}, \quad \boldsymbol{\mu}_G \in \mathbb{R}^{n_N - n_H \times n_N} \end{array} \tag{2.11}$$

and where the subdifferential of convex analysis $\partial|x| = \mathrm{Sign}(x)$ has been used, for which $\mathrm{Sign}(0) = [-1, 1]$. Equations (2.9) can be transformed into a LCP as has been formulated in (Glocker, 1995; Pfeiffer and Glocker, 1996).

If an impact occurs, then we generally have a discontinuity of in the generalized velocities $\dot{\boldsymbol{q}}$. The impact is assumed to begin at a time t^- and to end at a time t^+.

[1]The notation $\boldsymbol{a} \perp \boldsymbol{b}$ means that $\boldsymbol{a}$ stands perpendicular to $\boldsymbol{b}$, i.e. $\boldsymbol{a}^{\mathrm{T}}\boldsymbol{b} = 0$. It follows therefore from the complementarity conditions $\boldsymbol{0} \leq \boldsymbol{a} \perp \boldsymbol{b} \geq \boldsymbol{0}$, that if $a_i > 0$ then $b_i = 0$ and vice versa.

The time difference $t^+ - t^-$ is assumed to be "infinitely small"[2] in the rigid multibody approach. The equation of motion is integrated over the impact time

$$\boldsymbol{M}\big(\dot{\boldsymbol{q}}(t^+) - \dot{\boldsymbol{q}}(t^-)\big) = \begin{bmatrix}\boldsymbol{W}_N & \boldsymbol{W}_T\end{bmatrix}\begin{bmatrix}\boldsymbol{\Lambda}_N \\ \boldsymbol{\Lambda}_T\end{bmatrix}, \tag{2.12}$$

which yields the velocity jump $\dot{\boldsymbol{q}}(t^+) - \dot{\boldsymbol{q}}(t^-)$ as a function of the impulses $\boldsymbol{\Lambda}_N$ and $\boldsymbol{\Lambda}_T$ in normal and tangential direction defined by

$$\Lambda_{Ni} = \lim_{t^+ \to t^-} \int_{t^-}^{t^+} \lambda_{Ni} dt, \qquad \Lambda_{Ti} = \lim_{t^+ \to t^-} \int_{t^-}^{t^+} \lambda_{Ti} dt, \qquad i \in I_S. \tag{2.13}$$

Due to the unilateral character of the contact constraint only nonnegative normal contact forces are possible, $\lambda_{Ni} \geq 0$, which results in nonnegative values of the normal impulses $\Lambda_{Ni} \geq 0$. At the end of the completely inelastic impact the approaching process of the bodies has to be completed. Thus negative values of the contact velocities are forbidden, $\dot{g}_{Ni}(t^+) \geq 0$. If an impulse is transferred ($\Lambda_{Ni} > 0$), then the corresponding contact participates in the impact and $\dot{g}_{Ni}(t^+) = 0$. If no impulse is transferred ($\Lambda_{NCi} = 0$), then the corresponding constraint is superfluous and we allow velocities $\dot{g}_{Ni}(t^+) \geq 0$. The impact law in normal direction is therefore expressed by the complementarity condition

$$\Lambda_{Ni} \geq 0, \quad \dot{g}_{Ni}(t^+) \geq 0, \quad \Lambda_{Ni}\dot{g}_{Ni}(t^+) = 0; \quad i \in I_S. \tag{2.14}$$

Possible stick-slip transitions during the collision with reversed sliding prevent an analytical integration of Coulomb's friction law (2.8) over the impact time interval. However, we state the tangential impact law as

$$\begin{array}{lll} \dot{g}_{Ti}(t^+) = 0 & \Rightarrow \quad |\Lambda_{Ti}| \leq \mu_i \Lambda_{Ni} & \text{sticking} \\ \dot{g}_{Ti}(t^+) < 0 & \Rightarrow \quad \Lambda_{Ti} = +\mu_i \Lambda_{Ni} & \text{negative sliding} \\ \dot{g}_{Ti}(t^+) > 0 & \Rightarrow \quad \Lambda_{Ti} = -\mu_i \Lambda_{Ni} & \text{positive sliding} \end{array} \qquad i \in I_S. \tag{2.15}$$

with the remark that (2.15) coincides with Coulomb's friction law (2.8) in the cases of continuous sliding during the impact and of arbitrary transitions to sticking at the end of the impact. Only events of reversed sliding or transitions from sticking to sliding with a sliding phase at the end of the impact are different from Coulomb's law (Glocker, 1995; Moreau, 1988; Pfeiffer and Glocker, 1996).

Evaluating the contact velocities (2.4) at t^+ and t^- gives

$$\begin{bmatrix}\dot{\boldsymbol{g}}_N(t^+) \\ \dot{\boldsymbol{g}}_T(t^+)\end{bmatrix} = \begin{bmatrix}\boldsymbol{W}_N^{\mathrm{T}} \\ \boldsymbol{W}_T^{\mathrm{T}}\end{bmatrix}\big(\dot{\boldsymbol{q}}(t^+) - \dot{\boldsymbol{q}}(t^-)\big) + \begin{bmatrix}\dot{\boldsymbol{g}}_N(t^-) \\ \dot{\boldsymbol{g}}_T(t^-)\end{bmatrix} \qquad \in \mathbb{R}^{2n_s}. \tag{2.16}$$

Substitution of (2.12) in (2.16) together with the contact laws (2.14) and (2.15) gives

$$\begin{bmatrix}\dot{\boldsymbol{g}}_N(t^+) \\ \dot{\boldsymbol{g}}_T(t^+)\end{bmatrix} = \begin{bmatrix}\boldsymbol{W}_N^{\mathrm{T}}\boldsymbol{M}^{-1}\boldsymbol{W}_N & \boldsymbol{W}_N^{\mathrm{T}}\boldsymbol{M}^{-1}\boldsymbol{W}_T \\ \boldsymbol{W}_T^{\mathrm{T}}\boldsymbol{M}^{-1}\boldsymbol{W}_N & \boldsymbol{W}_T^{\mathrm{T}}\boldsymbol{M}^{-1}\boldsymbol{W}_T\end{bmatrix}\begin{bmatrix}\boldsymbol{\Lambda}_N \\ \boldsymbol{\Lambda}_T\end{bmatrix} + \begin{bmatrix}\dot{\boldsymbol{g}}_N(t^-) \\ \dot{\boldsymbol{g}}_T(t^-)\end{bmatrix} \tag{2.17}$$

[2]Mathematically more correct is to consider the impact as a singleton, i.e. a point in time, and the equation of motion as a measure differential equation, which is beyond the scope of this chapter (see (Glocker, 2000; Moreau, 1988)).

with

$$\begin{aligned} 0 \le \dot{\boldsymbol{g}}_N(t^+) \perp \boldsymbol{\Lambda}_N \ge 0 \qquad & \in I_S \\ \boldsymbol{\Lambda}_T \in -\boldsymbol{\mu}_S \boldsymbol{\Lambda}_N \mathrm{diag}(\partial |\dot{\boldsymbol{g}}_T(t^+)|) \quad & \in I_S, \end{aligned} \tag{2.18}$$

where $\boldsymbol{\mu}_S = \mathrm{diag}(\mu_i), i \in I_S$. The formulation of the coupled normal-tangential contact problem for completely inelastic impact is given by (2.17) and is usually solved by reformulating it as a LCP.

In the sequel we will use an event-driven integration scheme to obtain solutions $\boldsymbol{q}(t)$ of unilaterally constrained mechanical systems. The equation of motion (2.3) for given index sets is numerically integrated until an impact, stick-slip or detachment event occurs. If the event is an impact event, then the impact equations (2.17) have to be solved, after which the new generalized velocities $\dot{\boldsymbol{q}}$ are known. Subsequently, the equations (2.9) on acceleration level have to be solved, because the impact might cause stick-slip transitions or detachment of other contacts. The new accelerations $\ddot{\boldsymbol{q}}$ are known after having solved (2.9). The new index sets can then be setup and a new smooth integration phase can start. Basically any ODE-solver can be used for the integration of the smooth phase as long as the solver supports event detection.

The event-driven integration scheme complies with a certain solution concept. The solution remains in a smooth mode as long as some indicator inequalities are met (e.g. $|\lambda_T| \le \lambda_N$ in the stick mode, $g_N \ge 0$ in the unconstrained mode). If an indicator function is violated, then an event occurs, and the equations (2.17) and (2.9) are solved which are basically LCP's. The LCP's derived from (2.17) and (2.9) determine the mode change, i.e. they indicate the next mode. The solution of the LCP's can be non-unique, indicating non-uniqueness of the next mode and therefore non-uniqueness of the solution $\boldsymbol{q}(t)$. The solution of the LCP's can also be non-existent, implying that there is no continuation possible in any mode, and therefore non-existence of the solution $\boldsymbol{q}(t)$.

3 The Classical Painlevé Example

In this section we will briefly summarize some results on the classical Painlevé problem of a sliding rigid rod. The rigid body formulation of the model is taken from (Glocker, 1995) and the main results from (Brogliato, 1999; Génot and Brogliato, 1999).

Consider a rigid homogeneous slender rod with mass m, length $2s$ and inertia $J_S = \frac{1}{3}ms^2$ (see Figure 2). The rod is sliding with one tip over a rigid ground. Amontons-Coulomb friction model is supposed to hold at the contact point with constant friction coefficient $\mu > 0$. The system has three degrees of freedom, which we gather in a vector of generalized coordinates $\boldsymbol{q} = \begin{bmatrix} x & y & \varphi \end{bmatrix}^{\mathrm{T}}$. The normal contact distance and tangential contact velocity are

$$g_N = y - s\sin\varphi, \quad \dot{g}_T = \dot{x} - s\dot{\varphi}\sin\varphi. \tag{3.1}$$

The system matrices of the Painlevé problem become

$$\boldsymbol{M} = \begin{bmatrix} m & 0 & 0 \\ 0 & m & 0 \\ 0 & 0 & J_S \end{bmatrix}, \quad \boldsymbol{h} = \begin{bmatrix} 0 \\ -mg \\ 0 \end{bmatrix}, \quad \boldsymbol{W}_N = \begin{bmatrix} 0 \\ 1 \\ -s\cos\varphi \end{bmatrix}, \quad \boldsymbol{W}_T = \begin{bmatrix} 1 \\ 0 \\ -s\sin\varphi \end{bmatrix}, \tag{3.2}$$

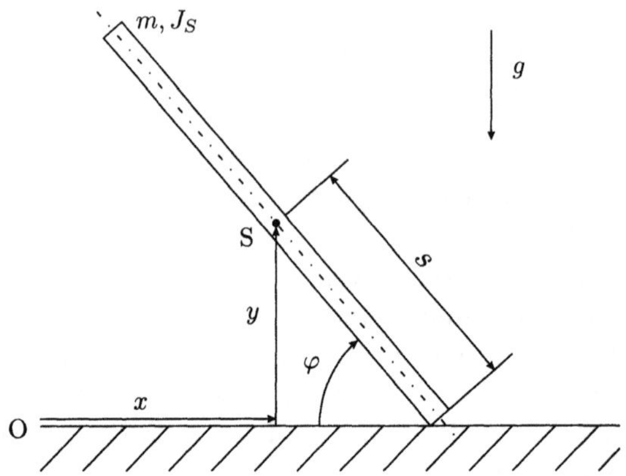

Figure 2. The classical Painlevé example.

$$\hat{\boldsymbol{w}}_N = \hat{\boldsymbol{w}}_T = 0, \quad \bar{\boldsymbol{w}}_N = s\dot{\varphi}^2 \sin\varphi, \quad \bar{\boldsymbol{w}}_T = -s\dot{\varphi}^2 \cos\varphi. \tag{3.3}$$

It is relatively easy to show that a dynamic frictional catastrophe can occur in this simple system. We assume that the rod is sliding from over the ground in forward direction, i.e. $g_N = 0$, $\dot{g}_N = 0$ and $\dot{g}_T > 0$. It therefore holds that $\lambda_T = -\mu\lambda_N$. The equations of motion become in the forward sliding mode

$$\begin{aligned} m\ddot{x} &= -\mu\lambda_N \\ m\ddot{y} &= -mg + \lambda_N \\ J_S\ddot{\varphi} &= s(-\cos\varphi + \mu\sin\varphi)\lambda_N. \end{aligned} \tag{3.4}$$

The rod remains in sliding contact with the ground when $\ddot{g}_N = 0$ and will detach when $\ddot{g}_N > 0$, with

$$\ddot{g}_N = \ddot{y} - s\cos\varphi\ddot{\varphi} + s\dot{\varphi}^2 \sin\varphi. \tag{3.5}$$

The normal contact acceleration becomes in the forward sliding mode (substituting (3.4) in (3.5))

$$\ddot{g}_N = A\lambda_N + b, \tag{3.6}$$

with

$$A(\varphi) = \frac{1}{m}\big(1 + 3\cos\varphi(\cos\varphi - \mu\sin\varphi)\big), \quad b(\varphi, \dot{\varphi}) = s\dot{\varphi}^2 \sin\varphi - g. \tag{3.7}$$

Note that A is a function of φ and μ while b is function of φ and $\dot{\varphi}$. Between the normal contact acceleration $\ddot{g}_N$ and the normal contact force λ_N the following complementarity condition holds

$$0 \leq \ddot{g}_N \perp \lambda_N \geq 0. \tag{3.8}$$

The complementarity condition (3.8) expresses that if the contact detaches, $\ddot{g}_N > 0$, then it must hold that $\lambda_N = 0$, while if the normal contact force is positive $\lambda_N > 0$ the contact remains closed $\ddot{g}_N = 0$. The linear equation (3.6) together with the complementarity condition (3.8) gives a scalar LCP for the detachment of the rod in the forward sliding

Table 1. LCP solutions.

A	b	number of solutions	LCP solutions
$A > 0$	$b > 0$	unique solution	$\lambda_N = 0 \wedge \ddot{g}_N = b$
$A > 0$	$b < 0$	unique solution	$\lambda_N = -\frac{b}{A} \wedge \ddot{g}_N = 0$
$A < 0$	$b > 0$	two solutions	$\lambda_N = -\frac{b}{A} \wedge \ddot{g}_N = 0$ or $\lambda_N = 0 \wedge \ddot{g}_N = b$
$A < 0$	$b < 0$	no solution	$\emptyset$

mode. The scalar LCP with $\ddot{g}_N = A\lambda_N + b$, $0 \leq \ddot{g}_N \perp \lambda_N \geq 0$ has a unique solution for $A > 0$, two solutions for $A < 0 \wedge b > 0$, no solution for $A < 0 \wedge b < 0$ or infinitely many solutions for $A = b = 0$ (hyperstatic case). The solutions of the LCP are indicated in Table 1. The occurrence of inconsistency (the LCP has no solution) and indeterminacy (non-uniqueness of solutions) is called the Painlevé paradox (Génot and Brogliato, 1999).

The Painlevé paradox will occur when $A(\varphi)$ becomes negative, which happens for large enough values of μ. The critical value of μ for which $A(\varphi) = 0$ occurs at $\mu = \mu_c(\varphi)$ with

$$\mu_c(\varphi) = \frac{1 + 3\cos^2\varphi}{3\sin\varphi\cos\varphi}. \tag{3.9}$$

The critical friction $\mu_c(\varphi)$ coefficient is minimal for $\varphi = \arctan 2$, giving $\mu_{\text{cmin}} = \frac{4}{3}$. This minimal critical friction coefficient changes for other ratios of $\frac{ms^2}{J_S}$ but will always be larger than 1 for $m > 0$ and $J_S > 0$. For $\mu \geq \mu_{\text{cmin}}$ there exists an interval of $\varphi_{c1} < \varphi < \varphi_{c2}$ for which $A < 0$ with

$$\varphi_{c1} = \arctan\left(\frac{3\mu - \sqrt{9\mu^2 - 16}}{2}\right), \quad \varphi_{c2} = \arctan\left(\frac{3\mu + \sqrt{9\mu^2 - 16}}{2}\right). \tag{3.10}$$

The $(\varphi, \dot{\varphi})$-plane is depicted in Figure 3 together with the four different LCP modes for detachment in forward sliding. A detailed study of the singular points P_{c1} and P_{c2} (Figure 3) is given in (Génot and Brogliato, 1999). It is proven in (Génot and Brogliato, 1999) that the solution can reach the singular points P_{c1} and P_{c2}, depending on the friction coefficient, either with a finite contact force or with a contact force reaching infinite values.

In the sequel we will present a model, consisting of a body with normal and tangential degrees of freedom, which exhibits the Painlevé paradox. We will prove that a minimal critical friction coefficient for the Painlevé paradox is possible which can be arbitrary small depending on some mass ratio. Furthermore, we will analyze the bifurcations of this model and will show that a figure like Figure 3 can be a very useful tool to analyze bifurcations caused by Painlevé paradoxes.

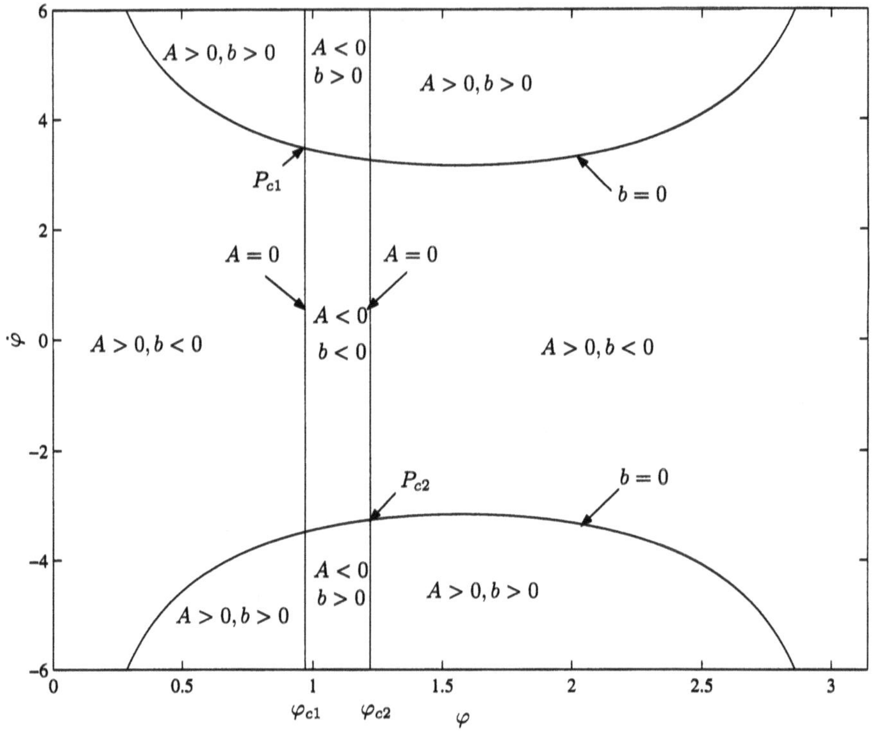

Figure 3. The $(\varphi, \dot{\varphi})$-plane of the classical Painlevé problem (Génot and Brogliato, 1999).

4 The Frictional Impact Oscillator

A human finger which is pushed over a table may exhibit periodic motion with stick and slip phases. When the friction between the finger-tip and the table is sufficiently high, even periodic motion may be observed with phases where the finger is not in contact with the table, i.e. the finger detaches from the table causing a 'flight' phase after which an impact occurs to another phase of the periodic motion. In this section we will study a simple system, called the Frictional Impact Oscillator, which exhibits the same phenomenon. The system is low-dimensional (2 degrees-of-freedom) but shows rich dynamic behaviour, being complicated by the Painlevé paradox.

A similar system was studied by Moreau (1986). The system of Moreau consists of a bar with a rounded tip which is mounted elastically in horizontal and vertical direction, which leads to a 3 degree-of-freedom system. The horizontal compliance is not essential for the phenomenon and is neglected in the present study, which facilitates the analysis.

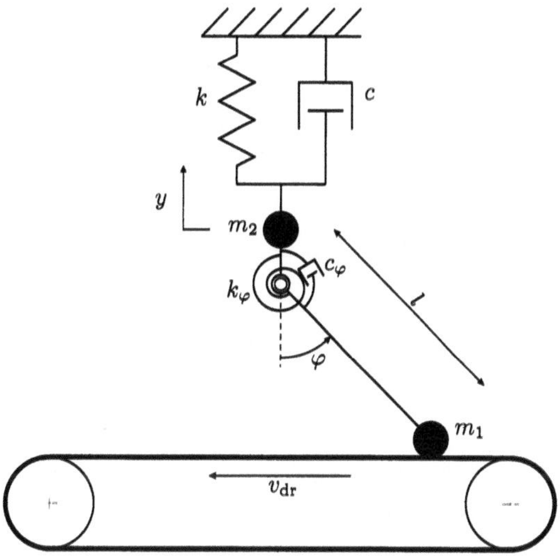

Figure 4. The Frictional Impact Oscillator.

4.1 Model

The Frictional Impact Oscillator is depicted in Figure 4. The system consists of a mass-spring-damper system (with coordinate y and constants m_2, k, c), which can be looked upon as the 'hand', robot arm or supporting structure. A mass-less rigid bar, representing the 'finger' or end-effector of a robot, is attached by a hinge to the 'hand' or robot arm respectively. The hinge can only move in the vertical direction by a displacement y. The bar is mounted at the hinge by a rotational spring and dashpot (k_φ, c_φ). A mass m_1 is located at the tip of the bar. The springs k and k_φ are unstressed when $y = 0$ and $\varphi = \varphi_0$ respectively. The tip of the bar can make contact with a belt, which is moving at constant velocity v_{dr}. The contact is regarded to be completely inelastic but frictional ($\mu > 0$). Pure Amontons-Coulomb friction is assumed without Stribeck effect (μ is not dependent on the relative velocity). The normal contact distance and tangential contact velocity are

$$g_N = l(1 - \cos\varphi) + y, \quad \dot{g}_T = l\dot{\varphi}\cos\varphi + v_{dr}. \tag{4.1}$$

The rotation of the bar φ and the vertical displacement of the hinge y can be gathered in a vector of generalized coordinates $\boldsymbol{q} = \begin{bmatrix}\varphi & y\end{bmatrix}^{\mathrm{T}}$. The kinetic and potential energy of the system can be expressed by

$$\begin{aligned} T &= \frac{1}{2}m_1 l^2\dot{\varphi}^2 + m_1 l \sin\varphi\, \dot{y}\dot{\varphi} + \frac{1}{2}(m_1 + m_2)\dot{y}^2, \\ V &= \frac{1}{2}k_\varphi(\varphi - \varphi_0)^2 + \frac{1}{2}ky^2 + m_1 g(l - l\cos\varphi + y) + m_2 g y. \end{aligned} \tag{4.2}$$

Lagrange's equations of motion can be put in the form of (2.3). The system matrices and vectors of the Frictional Impact Oscillator become

$$\boldsymbol{M} = \begin{bmatrix} m_1 l^2 & m_1 l \sin\varphi \\ m_1 l \sin\varphi & m_1 + m_2 \end{bmatrix}, \qquad \boldsymbol{h} = \begin{bmatrix} -k_\varphi(\varphi - \varphi_0) - c_\varphi\dot{\varphi} - m_1 g l \sin\varphi \\ -ky - c\dot{y} - (m_1 + m_2)g - m_1 l \cos\varphi\, \dot{\varphi}^2 \end{bmatrix}, \tag{4.3}$$

$$\begin{aligned} \boldsymbol{W}_N &= \begin{bmatrix} l \sin\varphi \\ 1 \end{bmatrix}, \quad \tilde{\boldsymbol{w}}_N = 0, \quad \bar{\boldsymbol{w}}_N = l\dot{\varphi}^2 \cos\varphi, \\ \boldsymbol{W}_T &= \begin{bmatrix} l \cos\varphi \\ 0 \end{bmatrix}, \quad \tilde{\boldsymbol{w}}_T = v_{\mathrm{dr}}, \quad \bar{\boldsymbol{w}}_T = -l\dot{\varphi}^2 \sin\varphi. \end{aligned} \tag{4.4}$$

4.2 Forward Slip, $\dot{g}_T > 0$

The motion of the system in the forward slip mode is constraint with $\lambda_N \geq 0$ and $\lambda_T = -\mu\lambda_N$. The equations of motion become

$$\begin{aligned} m_1 l^2 \ddot{\varphi} + m_1 l \sin\varphi\, \ddot{y} + c_\varphi\dot{\varphi} + k_\varphi(\varphi - \varphi_0) &= l(\sin\varphi - \mu\cos\varphi)\lambda_N - m_1 g l \sin\varphi \\ m_1 l \sin\varphi\, \ddot{\varphi} + (m_1 + m_2)\ddot{y} + c\dot{y} + ky + m_1 l \cos\varphi\, \dot{\varphi}^2 &= \lambda_N - (m_1 + m_2)g. \end{aligned} \tag{4.5}$$

Elimination of λ_N together with the conditions $g_N = \dot{g}_N = \ddot{g}_N = 0$ gives a second order differential equation

$$\mathcal{M}(\varphi)\ddot{\varphi} + \mathcal{D}(\varphi)\dot{\varphi}^2 + \mathcal{C}(\varphi)\dot{\varphi} + \mathcal{F}(\varphi) = 0, \tag{4.6}$$

with

$$\begin{aligned} \mathcal{M}(\varphi) =& \big(m_1 \cos^2\varphi + m_2 \sin\varphi(\sin\varphi - \mu\cos\varphi)\big) l^2 \\ \mathcal{D}(\varphi) =& \big((-m_1 + m_2)\sin\varphi\cos\varphi - m_2\mu\cos^2\varphi\big) l^2 \\ \mathcal{C}(\varphi) =& c_\varphi + c l^2 \sin\varphi(\sin\varphi - \mu\cos\varphi) \\ \mathcal{F}(\varphi) =& k_\varphi(\varphi - \varphi_0) + k l^2 (1 - \cos\varphi)(\sin\varphi - \mu\cos\varphi) \\ &+ \big((m_1 + m_2)\mu\cos\varphi - m_2\sin\varphi\big) l g. \end{aligned} \tag{4.7}$$

The system can have (multiple) equilibria φ_{eq} in the forward sliding mode, which obey

$$\mathcal{F}(\varphi_{eq}) = 0 \tag{4.8}$$

and a nonzero contact force

$$\lambda_N = -kl(1 - \cos\varphi_{eq}) + (m_1 + m_2)g \geq 0. \tag{4.9}$$

If the condition

$$\mathcal{K}(\varphi) = \frac{\partial\mathcal{F}(\varphi)}{\partial\varphi} > 0, \quad \forall\varphi \tag{4.10}$$

holds with

$$\begin{aligned} \frac{\partial\mathcal{F}}{\partial\varphi} =& k_\varphi + k l^2 \big((1 - \cos\varphi)(\cos\varphi + \mu\sin\varphi) + \sin\varphi(\sin\varphi - \mu\cos\varphi)\big) \\ &- l g\big((m_1 + m_2)\mu\sin\varphi + m_2\cos\varphi\big) \end{aligned} \tag{4.11}$$

then at most one equilibrium in the forward sliding mode exists. This is a sufficient condition for $\mathcal{F}(\varphi)$ to have at most one zero but not a necessary condition. We assume in the forthcoming that k_φ is chosen large enough that indeed $\mathcal{K}(\varphi) > 0$ for all φ of interest. This means that the unique equilibrium φ_{eq}, if it exists and obeys $\lambda_N > 0$, must be close to φ_0. The equilibrium is approximately located at

$$\varphi_{eq} \approx \varphi_0 - \frac{\mathcal{F}(\varphi_0)}{\mathcal{K}(\varphi_0)}. \tag{4.12}$$

The equation of motion for $\mathcal{M}(\varphi) \neq 0$ can be written in first order form as

$$\begin{aligned} \dot{\varphi} &= \psi \\ \dot{\psi} &= -\mathcal{M}(\varphi)^{-1}\left(\mathcal{D}(\varphi)\psi^2 + \mathcal{C}(\varphi)\psi + \mathcal{F}(\varphi)\right) \end{aligned} . \tag{4.13}$$

Linearization around the equilibrium $(\xi, \psi) = (0, 0)$ with $\xi = \varphi - \varphi_{eq}$ gives

$$\begin{aligned} \dot{\xi} &= \psi \\ \dot{\psi} &= -\left(\mathcal{M}(\varphi_{eq})^{-1} - \mathcal{M}(\varphi_{eq})^{-2}\mathcal{M}'(\varphi_{eq})\xi\right)\left(\mathcal{C}(\varphi_{eq})\psi + \mathcal{F}(\varphi_{eq}) + \mathcal{K}(\varphi_{eq})\xi\right) + H.O.T. \end{aligned} \tag{4.14}$$

with $\mathcal{M}'(\varphi) = \frac{\partial \mathcal{M}}{\partial \varphi}$. It should be noted that $\mathcal{F}(\varphi_{eq}) = 0$. The eigenvalue problem therefore yields

$$\mathcal{M}(\varphi_{eq})\lambda^2 + \mathcal{C}(\varphi_{eq})\lambda + \mathcal{K}(\varphi_{eq}) = 0 \tag{4.15}$$

The equilibrium is stable if $\mathcal{M}(\varphi_{eq}) > 0$, $\mathcal{C}(\varphi_{eq}) > 0$ and $\mathcal{K}(\varphi_{eq}) > 0$ or $\mathcal{M}(\varphi_{eq}) < 0$, $\mathcal{C}(\varphi_{eq}) < 0$ and $\mathcal{K}(\varphi_{eq}) < 0$. Note that the friction coefficient μ can cause a negative feedback of the damping forces. The equilibrium undergoes a Hopf bifurcation for $\mathcal{M}(\varphi_{eq}) > 0$, $\mathcal{K}(\varphi_{eq}) > 0$ when the friction coefficient μ passes the critical value

$$\mu_d = \tan\varphi_{eq} + \frac{c_\varphi}{cl^2 \sin\varphi_{eq} \cos\varphi_{eq}}, \tag{4.16}$$

for which $\mathcal{C}(\varphi_{eq}) = 0$.[3] The value of $\mathcal{M}(\varphi_{eq})$ becomes zero when $\mu = \tan\varphi_{eq} + \frac{m_1}{m_2}\frac{1}{\tan\varphi_{eq}}$, which is in fact the Painlevé paradox.

4.3 Painlevé paradox

In this subsection we will study the Painlevé paradox in the forward and backward sliding mode with similar techniques as presented in (Génot and Brogliato, 1999).

The solution remains in the forward sliding mode if $\dot{g}_T > 0$ and $\ddot{g}_N = 0$ and in the backward sliding mode if $\dot{g}_T < 0$ and $\ddot{g}_N = 0$. The body detaches from the belt when $\ddot{g}_N > 0$. For the normal contact acceleration $\ddot{g}_N$ holds:

$$\ddot{g}_N = l\sin\varphi\,\ddot{\varphi} + \ddot{y} + l\cos\varphi\,\dot{\varphi}^2. \tag{4.17}$$

Substituting (2.3) in (2.4) together with $g_N = \dot{g}_N = 0$ gives

$$\ddot{g}_N = \begin{cases} A^+\lambda_N + b, & \dot{g}_T > 0 \\ A^-\lambda_N + b, & \dot{g}_T < 0 \end{cases}, \tag{4.18}$$

[3] If μ *passes* μ_d, then it follows that $\mathcal{C}(\varphi_{eq})$ changes sign. A pair of complex conjugated eigenvalues cross the imaginary axis, implying a Hopf bifurcation.

with

$$A^+(\varphi) = \boldsymbol{W}_N^{\mathrm{T}}\boldsymbol{M}^{-1}(\boldsymbol{W}_N - \mu\boldsymbol{W}_T) = \frac{\cos^2\varphi}{N(\varphi)}\Big(1 + \frac{m_2}{m_1}\tan\varphi(\tan\varphi - \mu)\Big), \tag{4.19}$$

$$A^-(\varphi) = \boldsymbol{W}_N^{\mathrm{T}}\boldsymbol{M}^{-1}(\boldsymbol{W}_N + \mu\boldsymbol{W}_T) = \frac{\cos^2\varphi}{N(\varphi)}\Big(1 + \frac{m_2}{m_1}\tan\varphi(\tan\varphi + \mu)\Big), \tag{4.20}$$

$$\begin{aligned} b(\varphi,\dot\varphi) &= \boldsymbol{W}_N^{\mathrm{T}}\boldsymbol{M}^{-1}\boldsymbol{h} + \bar{\boldsymbol{w}}_N \\ &= \frac{\cos^2\varphi}{N(\varphi)}\Big(-\frac{m_2}{m_1}\frac{\tan\varphi}{l\cos\varphi}(c_\varphi\dot\varphi + k_\varphi(\varphi-\varphi_0)) + cl\sin\varphi\,\dot\varphi \\ &\qquad + kl(1-\cos\varphi) + \frac{m_2 l}{\cos\varphi}\dot\varphi^2\Big) - g \end{aligned} \tag{4.21}$$

and

$$N(\varphi) = m_1\cos^2\varphi + m_2. \tag{4.22}$$

The notation A^+ is used to denote the value of A in the forward sliding mode, while A^- will be used for the backward sliding mode. Note that $A^\pm$ are functions of φ and μ, and b is a function of φ and $\dot\varphi$. The linear equation (4.18) gives together with the complementarity condition $0 \le \ddot g_N \perp \lambda_N \ge 0$ a scalar Linear Complementarity Problem for the detachment in the forward and backward sliding mode.

The Painlevé paradox occurs in the forward sliding mode when $A^+(\varphi)$ becomes negative, leading to inconsistency or indeterminacy of the forward sliding mode (depending on the sign of $b(\varphi,\dot\varphi)$). The value $A^+(\varphi)$ can become negative for sufficiently large values of the friction coefficient μ. The critical value occurs at $\mu = \mu_c^+(\varphi)$ for which $A^+(\varphi) = 0$,

$$\mu_c^+(\varphi) = \tan\varphi + \frac{m_1}{m_2}\frac{1}{\tan\varphi}. \tag{4.23}$$

The critical value $\mu_c^+(\varphi)$ is minimal when $\varphi = \arctan\sqrt{\frac{m_1}{m_2}}$ giving

$$\mu_{\mathrm{cmin}} = 2\sqrt{\frac{m_1}{m_2}}. \tag{4.24}$$

For $\mu \ge \mu_{\mathrm{cmin}}$ there exists an interval of $\varphi_{c1}^+ < \varphi < \varphi_{c2}^+$ for which $A^+ < 0$ with

$$\varphi_{c1}^+ = \arctan\Big(\frac{1}{2}\mu + \sqrt{\frac{1}{4}\mu^2 - \frac{m_1}{m_2}}\Big), \qquad \varphi_{c2}^+ = \arctan\Big(\frac{1}{2}\mu - \sqrt{\frac{1}{4}\mu^2 - \frac{m_1}{m_2}}\Big). \tag{4.25}$$

With respect to the previous section the following holds (Leine et al., 2002)

Lemma 4.1. *Consider a mechanical system* $\boldsymbol{M}(\boldsymbol{q})\ddot{\boldsymbol{q}} - \boldsymbol{h}(\boldsymbol{q},\dot{\boldsymbol{q}}) - \boldsymbol{W}_N(\boldsymbol{q})\lambda_N - \boldsymbol{W}_T(\boldsymbol{q})\lambda_T = \boldsymbol{0}$ *with n degrees of freedom* $\boldsymbol{q} \in \mathbb{R}^n$ *and with a single unilateral contact point, i.e.* $\lambda_N, \lambda_T \in \mathbb{R}$. *Let* $\boldsymbol{M}(\boldsymbol{q})$ *be nonsingular. Let the system be for some non-zero time-interval $t \in U$ in the forward sliding mode,* $\lambda_T = -\mu\lambda_N$, *and the detachment LCP be described by* $\ddot g_N = A(\boldsymbol{q})\lambda_N + b(\boldsymbol{q},\dot{\boldsymbol{q}})$. *The motion in the forward slip mode can be described in a set of minimal coordinates* $\boldsymbol{a}$ *by the reduced system* $\boldsymbol{\mathcal{M}}(\boldsymbol{a})\ddot{\boldsymbol{a}} + \boldsymbol{\mathcal{H}}(\boldsymbol{a},\dot{\boldsymbol{a}}) = \boldsymbol{0}$, *with* $\boldsymbol{a} \in \mathbb{R}^f$, $f = n-1$. *It holds that* $A(\boldsymbol{q}) = 0$ *if and only if* $\boldsymbol{\mathcal{M}}(\boldsymbol{a}) = \boldsymbol{O}$.

Lemma 4.2. *Let μ and $\boldsymbol{q}$ be such that $A(\boldsymbol{q}) = 0$, with $\boldsymbol{q} = \boldsymbol{q}(t, \boldsymbol{a})$. It holds that $\mathcal{H}(\boldsymbol{a}, \dot{\boldsymbol{a}}) = 0$ if and only if $b(\boldsymbol{q}, \dot{\boldsymbol{q}}) = 0$.*

These lemmas have the following implications for the Frictional Impact Oscillator:

Lemma 4.3. *It holds that $A^+(\varphi) = 0$ if and only if $\mathcal{M}(\varphi) = 0$.*

Lemma 4.4. *Let $\mu = \mu_c^+ > 0$ and be bounded such that $A^+(\varphi) = 0$. Then it holds that $b(\varphi, \dot{\varphi}) = 0$ if and only if $\mathcal{D}(\varphi)\dot{\varphi}^2 + \mathcal{C}(\varphi)\dot{\varphi} + \mathcal{F}(\varphi) = 0$.*

We conclude from Lemmas 4.3 and 4.4 that when φ is such that $A^+ = 0$, then the slip equation (4.6) is fulfilled only when $b = 0$. Note that it must hold in the forward slip mode that $\ddot{g}_N = 0$ from which follows that $b = 0$ when $A^+ = 0$. Let us denote the singular point in the $(\varphi, \dot{\varphi})$-plane for which $A^+ = 0$ and $b = 0$ by P_c. The point P_c is located outside the $(\varphi, \dot{\varphi})$-plane shown in Figure 7, at the intersection of the vertical line A^+ and the curve $b = 0$. It follows that a solution in the forward slip mode can only reach the line $A^+ = 0$ in the $(\varphi, \dot{\varphi})$-plane through the singular point P_c. Singular points of this type were studied in detail in (Génot and Brogliato, 1999) for the classical Painlevé example. The contact force λ_N might become unbounded at P_c. We assume in the sequel that periodic solutions, if they exist, do not pass the critical point P_c (i.e. the critical point is not located on a periodic solution of interest). Indeed, for the hopping phenomenon, which is of interest in the current chapter, critical points like P_c do not play a role (at least not for the parameter set we study).

The Painlevé Paradox can also occur in the backward sliding mode. The friction force λ_T now works in the opposite direction and the Painlevé paradox can therefore only occur for negative values of φ. The value A^- can become negative for sufficiently large values of the friction coefficient μ and negative values of φ. The critical value occurs at $\mu = \mu_c^-$ for which $A^- = 0$. It can easily be verified that $\mu_c^- = -\mu_c^+$, being minimal when $\varphi = -\arctan\sqrt{\frac{m_1}{m_2}}$ giving again the critical value μ_{cmin} (4.24). For $\mu \geq \mu_{\text{cmin}}$ there exists an interval of $\varphi_{c1}^- > \varphi > \varphi_{c2}^-$ for which $A^- < 0$ with $\varphi_{ci}^- = -\varphi_{ci}^+$, $i = 1, 2$. However, we will focus in the sequel on the Painlevé paradox in the forward sliding mode because a possible sliding equilibrium is located in the forward sliding mode for $v_{\text{dr}} > 0$.

4.4 Analysis of Periodic Motion

The Frictional Impact Oscillator can exhibit periodic motion with slip, stick and flight phases. We will explore how the topology of the periodic solution depends on the regions defined by the lines $A = 0$ and $b = 0$. We will consider two cases: case 1 with $\mu = 0.5 < \mu_{\text{cmin}}$ and case 2 with $\mu = 1 > \mu_{\text{cmin}}$, where $\mu_{\text{cmin}} = 0.6325$. The other parameters are $m_1 = 0.1$ kg, $m_2 = 1$ kg, $l = 1$ m, $k = 100$ N/m, $k_\varphi = 100$ Nm, $c = 10$ N/(ms), $c_\varphi = 0$ Nm/s, $\varphi_0 = \frac{\pi}{8}$, $v_{\text{dr}} = 1$ m/s, $g = 10$ Nm/s^2.

Case 1: $\mu = 0.5$. Solutions in the forward slip mode exist and are unique because $A > 0$ $\forall \varphi$ for $\mu < \mu_{\text{cmin}}$. The lines $b = 0$ (4.21) and $\dot{g}_T = 0$ are drawn in the plane $(\varphi, \dot{\varphi})$ of Figure 5. The time history of the contact distance g_N and contact velocity $\dot{g}_T$ are shown in Figure 6. The borders of the friction cone on the line $\dot{g}_T$ in Figure 5 are

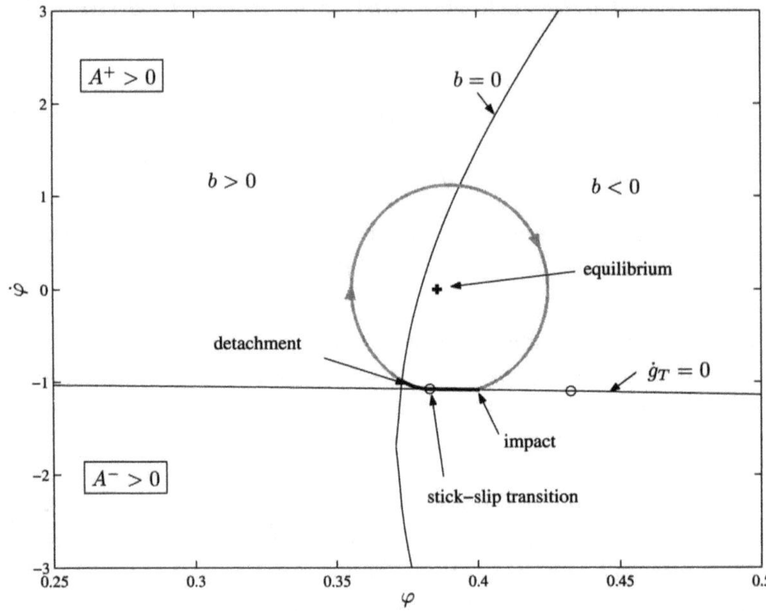

Figure 5. The $(\varphi, \dot{\varphi})$-plane of the Frictional Impact Oscillator, $\mu = 0.5$.

obtained numerically and depicted by a small circle ($\circ$). The part of the line $\dot{g}_T$ between the two $\circ$ signs is therefore the stick mode. Note that the friction cone lies, for this parameter set, totally outside the detachment region $A^+ > 0, b > 0$. A stable periodic solution is found numerically and consists of the following phases: stick–slip–flight. The dynamics of the flight phase is not only dependent on $(\varphi, \dot{\varphi})$ and is therefore depicted in grey. The stick to slip transition occurs at the border of the friction cone. The body remains some time in forward slipping contact with the belt in the space $A^+ > 0, b < 0$ until the line $b = 0$ is hit. The sign of b changes at the line $b = 0$, giving the LCP solution $\lambda_N = 0, \ddot{g}_N > 0$, which means that the body detaches from the belt. The body remains in flight (unconstrained free motion) until an impact occurs ($g_N = 0$). The impact causes the velocities $\dot{\varphi}$ and $\dot{y}$ to jump. However, the jump is so small that the discontinuity cannot be seen at the scale of Figures 5 and 6. The impact results in sticking of the body to the belt. Other coexisting periodic solutions might exist in theory but were not found numerically for this parameter set.

Case 2: $\mu = 1$. All four regions of Table 1 exist for this parameter set and are depicted in Figure 7. The friction cone is now bordered partly by the region $A^+ < 0, b > 0$ (the indeterminate mode of the detachment LCP of the forward slip mode) and is partly bordered by the region $A^+ < 0, b < 0$ (the inconsistent mode of the detachment LCP of the forward slip mode). A periodic solution was found which contains a stick phase and a flight phase (depicted in grey). The velocity jump due to the impact is depicted by a

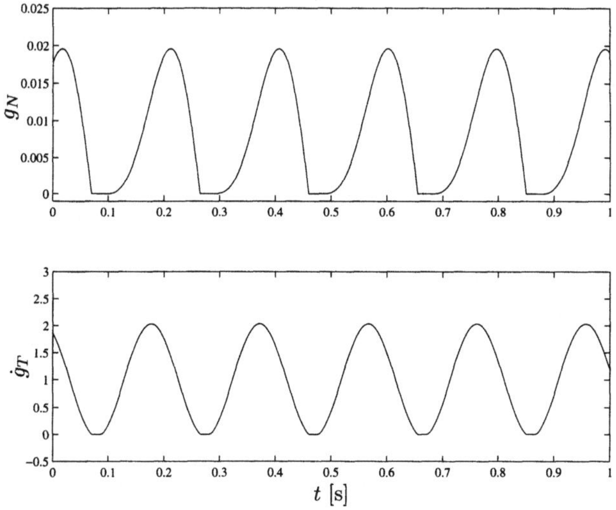

Figure 6. Time history of the Frictional Impact Oscillator, $\mu = 0.5$.

dotted line. The impact causes the post-impact state to be in the stick mode.

A few fundamental questions should be put forward at this point:

1. Is the solution at the end of the stick phase locally unique? The forward slip mode is at this point undeterminate.

2. Can an impact cause a post-impact state which is in the forward sliding mode with non-existence of solution?

These questions are very hard to answer due to the nonlinear terms in the Frictional Impact Oscillator, which make the analysis cumbersome. Instead, it is much more convenient to look at a piecewise linear simplified version of this model, which we call the Simplified Frictional Impact Oscillator. The above questions will be answered for the simplified model. It will be shown that the Simplified Frictional Impact Oscillator is to some extent a good approximation of the Frictional Impact Oscillator.

5 The Simplified Frictional Impact Oscillator

In this section a simplified version of the Frictional Impact Oscillator will be studied. For small values of $\varphi - \varphi_0 \ll 1$ we can approximate the Frictional Impact Oscillator (Figure 4) with the Simplified Frictional Impact Oscillator depicted in Figure 10. Note that the angle γ is constant. Figure 9 might help to understand the relation between the two models. The Frictional Impact Oscillator is pushed over the belt, while the Simplified Impact Oscillator seems to be dragged over the belt.

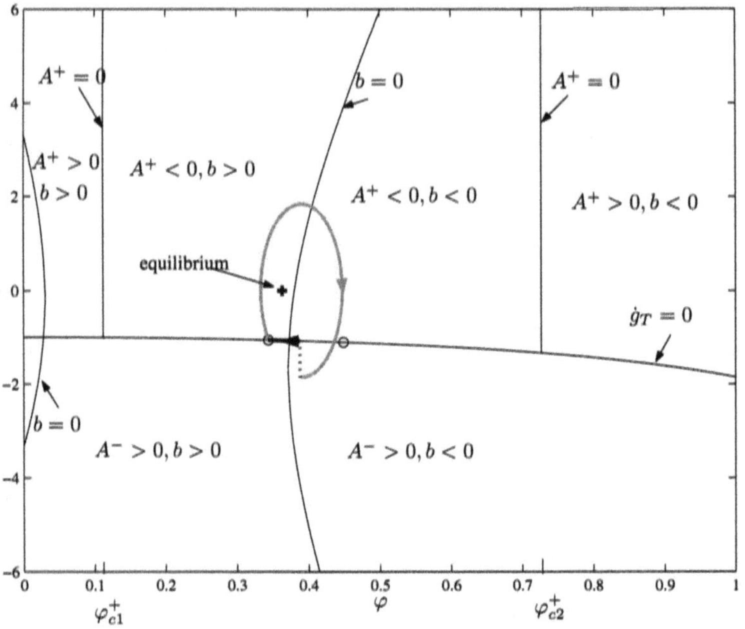

Figure 7. The $(\varphi, \dot{\varphi})$-plane of the Frictional Impact Oscillator, $\mu = 1$.

5.1 Model

The Simplified Frictional Impact Oscillator consists of a mass-spring-damper system (m_1, k_1, c_1) which stands under a fixed angle γ and is attached to a vertically suspended mass-spring-damper system (m_2, k_2, c_2). The springs are unstressed when $u_1 = 0$ and $u_2 = 0$. The correspondence between the constants of the Frictional Impact Oscillator and the Simplified Frictional Impact Oscillator is $u_1 = l(\varphi - \varphi_0)$, $u_2 = y$, $\gamma = \varphi_0$, $k_1 = \frac{k_\varphi}{l^2}$, $c_1 = \frac{c_\varphi}{l^2}$, $k_2 = k$ and $c_2 = c$.

The kinetic and potential energy of the system can be expressed by

$$\begin{aligned} T &= \frac{1}{2} m_1 \dot{u}_1^2 + m_1 \sin\gamma \, \dot{u}_1 \dot{u}_2 + \frac{1}{2}(m_1 + m_2)\dot{u}_2^2, \\ V &= \frac{1}{2} k_1 u_1^2 + \frac{1}{2} k_2 u_2^2 + m_1 g \sin\gamma \, u_1 + (m_1 + m_2) g u_2. \end{aligned} \tag{5.1}$$

The system matrices and vectors of the Simplified Frictional Impact Oscillator become

$$\boldsymbol{M} = \begin{bmatrix} m_1 & m_1 \sin\gamma \\ m_1 \sin\gamma & m_1 + m_2 \end{bmatrix}, \qquad \boldsymbol{h} = \begin{bmatrix} -k_1 u_1 - c_1 \dot{u}_1 - m_1 g \sin\gamma \\ -k_2 u_2 - c_2 \dot{u}_2 - (m_1 + m_2) g \end{bmatrix}, \tag{5.2}$$

$$g_N = u_1 \sin\gamma + u_2, \qquad \dot{g}_T = \dot{u}_1 \cos\gamma + v_{\mathrm{dr}} \tag{5.3}$$

$$\boldsymbol{W}_N = \begin{bmatrix} \sin\gamma \\ 1 \end{bmatrix}, \quad \boldsymbol{W}_T = \begin{bmatrix} \cos\gamma \\ 0 \end{bmatrix}, \quad \tilde{w}_N = 0, \quad \bar{w}_N = 0, \quad \tilde{w}_T = v_{\mathrm{dr}}, \quad \bar{w}_T = 0 \tag{5.4}$$

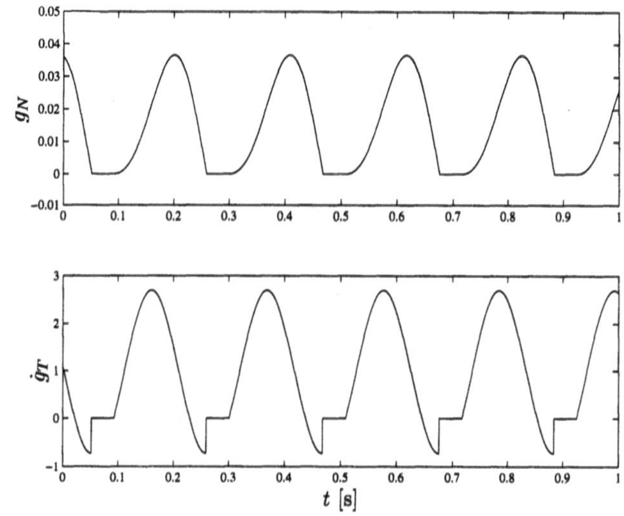

Figure 8. Time history of the Frictional Impact Oscillator, $\mu = 1$.

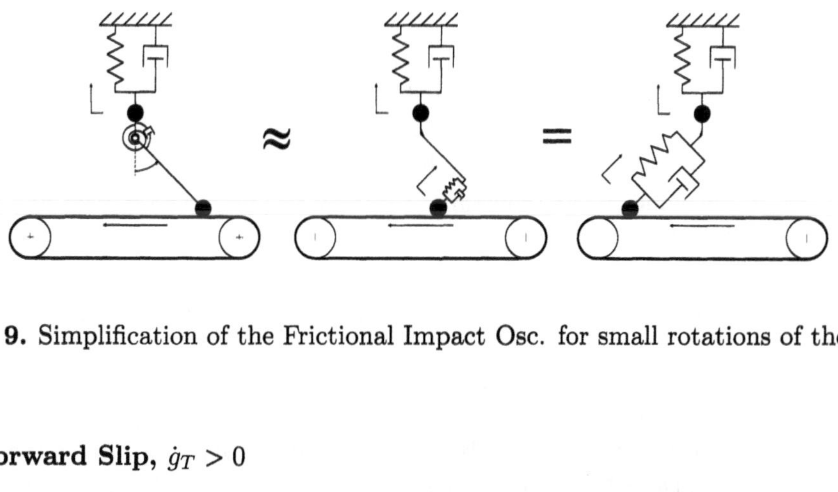

Figure 9. Simplification of the Frictional Impact Osc. for small rotations of the rod.

5.2 Forward Slip, $\dot{g}_T > 0$

The motion of the system in the forward slip mode is constraint with $\lambda_N \geq 0$ and $\lambda_T = -\mu\lambda_N$. The equations of motion become

$$\begin{aligned} m_1\ddot{u}_1 + m_1 \sin\gamma\, \ddot{u}_2 + c_1\dot{u}_1 + k_1 u_1 &= (\sin\gamma - \mu\cos\gamma)\lambda_N - m_1 g \sin\gamma \\ m_1 \sin\gamma\, \ddot{u}_1 + (m_1 + m_2)\ddot{u}_2 + c_2\dot{u}_2 + k_2 u_2 &= \lambda_N - (m_1 + m_2)g. \end{aligned} \tag{5.5}$$

Elimination of λ_N together with the conditions $g_N = \dot{g}_N = \ddot{g}_N = 0$ gives a second order differential equation

$$\mathcal{M}_\gamma \ddot{u}_1 + \mathcal{C}_\gamma \dot{u}_1 + \mathcal{K}_\gamma u_1 = \mathcal{F}_\gamma, \tag{5.6}$$

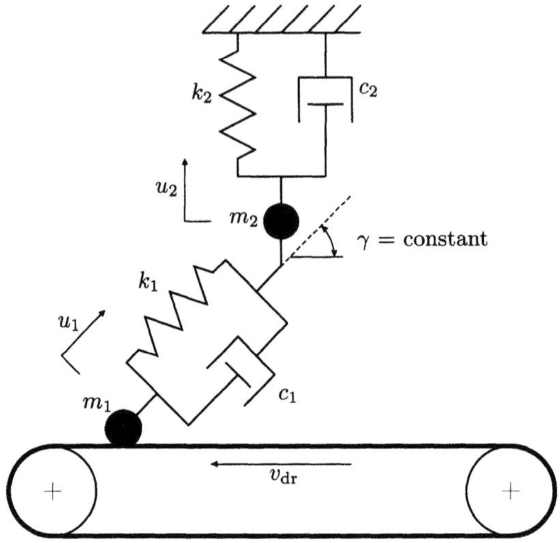

Figure 10. Model of the Simplified Frictional Impact Oscillator.

with

$$\begin{aligned} \mathcal{M}_\gamma &= m_1 \cos^2\gamma + m_2 \sin\gamma(\sin\gamma - \mu\cos\gamma) \\ \mathcal{C}_\gamma &= c_1 + c_2 \sin\gamma(\sin\gamma - \mu\cos\gamma) \\ \mathcal{K}_\gamma &= k_1 + k_2 \sin\gamma(\sin\gamma - \mu\cos\gamma) \\ \mathcal{F}_\gamma &= -\big((m_1+m_2)\mu\cos\gamma - m_2\sin\gamma\big)g, \end{aligned} \tag{5.7}$$

where the $\mathcal{M}_\gamma, \mathcal{C}_\gamma, \mathcal{K}_\gamma$ and $\mathcal{F}_\gamma$ are constants contrary to (4.7). The system has an equilibrium u_{1eq} in the forward sliding mode

$$u_{1eq} = \frac{\mathcal{F}_\gamma}{\mathcal{K}_\gamma} = \frac{\big(-(m_1+m_2)\mu\cos\gamma + m_2\sin\gamma\big)g}{\big(k_1 + k_2\sin\gamma(\sin\gamma - \mu\cos\gamma)\big)}, \tag{5.8}$$

when $\lambda_N = -k_2 u_{1eq}\sin\gamma + (m_1+m_2)g \geq 0$ and $\mathcal{K}_\gamma \neq 0$.

The equilibrium is stable if $\mathcal{M}_\gamma > 0$, $\mathcal{C}_\gamma > 0$ and $\mathcal{K}_\gamma > 0$ or $\mathcal{M}_\gamma < 0$, $\mathcal{C}_\gamma < 0$ and $\mathcal{K}_\gamma < 0$. The equilibrium undergoes a Hopf bifurcation for $\mathcal{M}_\gamma > 0$, $\mathcal{K}_\gamma > 0$ when the friction coefficient μ passes the critical value

$$\mu_{d\gamma} = \tan\gamma + \frac{c_1}{c_2}\frac{1}{\sin\gamma\cos\gamma}. \tag{5.9}$$

for which $\mathcal{C}_\gamma = 0$.

5.3 Stick Mode

In the stick mode holds $g_N = \dot{g}_N = \ddot{g}_N = \dot{g}_T = \ddot{g}_T = 0$ and $|\lambda_T| < \mu\lambda_N$. It follows that $u_2 = -u_1\sin\gamma$ and $\dot{u}_1 = -\frac{v_{\mathrm{dr}}}{\cos\gamma}$. The normal contact force therefore becomes

$$\lambda_N = c_2 v_{\mathrm{dr}}\tan\gamma - k_2 u_1 \sin\gamma + (m_1+m_2)g. \tag{5.10}$$

The condition that the contact force must be in the interior of the friction cone, $|\lambda_T| < \mu\lambda_N$, results in an interval $u_{\text{stick}}^- < u_1 < u_{\text{stick}}^+$ with

$$u_{\text{stick}}^{\pm} = \frac{c_1 \frac{v_{\text{dr}}}{\cos\gamma} + (\sin\gamma \pm \mu\cos\gamma)\big(c_2 v_{\text{dr}} \tan\gamma + (m_1+m_2)g\big) - m_1 g \sin\gamma}{k_1 + k_2 \sin\gamma(\sin\gamma \pm \mu\cos\gamma)}. \tag{5.11}$$

The solutions to the detachment LCP for the stick mode exist and are unique which follows from

$$\begin{aligned} \ddot{g}_N &= \ddot{u}_1 \sin\gamma + \ddot{u}_2 \\ &= \frac{1}{m_1+m_2}\big(\lambda_N + c_2\dot{u}_1\sin\gamma + k_2 u_1 \sin\gamma\big) - g. \end{aligned} \tag{5.12}$$

The stick mode is therefore consistent and determinate.

5.4 Painlevé paradox

Similar to the Frictional Impact Oscillator we analyze the Painlevé paradox of the Simplified Frictional Impact Oscillator in the forward and backward slip mode.

The normal acceleration $\ddot{g}_N = \ddot{u}_1 \sin\gamma + \ddot{u}_2$ in the forward and backward slip mode becomes

$$\ddot{g}_N = \begin{cases} A_\gamma^+ \lambda_N + b_\gamma, & \dot{g}_T > 0 \\ A_\gamma^- \lambda_N + b_\gamma, & \dot{g}_T < 0 \end{cases} \tag{5.13}$$

with

$$A_\gamma^+ = \boldsymbol{W}_N^{\mathrm{T}} \boldsymbol{M}^{-1}(\boldsymbol{W}_N - \mu \boldsymbol{W}_T) = \frac{\cos^2\gamma}{N_\gamma}\big(1 + \frac{m_2}{m_1}\tan\gamma(\tan\gamma - \mu)\big), \tag{5.14}$$

$$A_\gamma^- = \boldsymbol{W}_N^{\mathrm{T}} \boldsymbol{M}^{-1}(\boldsymbol{W}_N + \mu \boldsymbol{W}_T) = \frac{\cos^2\gamma}{N_\gamma}\big(1 + \frac{m_2}{m_1}\tan\gamma(\tan\gamma + \mu)\big), \tag{5.15}$$

$$b_\gamma(u_1, \dot{u}_1) = \boldsymbol{W}_N^{\mathrm{T}} \boldsymbol{M}^{-1}\boldsymbol{h} + \bar{\boldsymbol{w}}_N = \frac{\sin\gamma}{N_\gamma}\big(\cos^2\gamma(c_2\dot{u}_1 + k_2 u_1) - \frac{m_2}{m_1}(c_1\dot{u}_1 + k_1 u_1)\big) - g \tag{5.16}$$

and

$$N_\gamma = m_1 \cos^2\gamma + m_2. \tag{5.17}$$

Note that $A_\gamma^\pm$ is not dependent on u_1 and $\dot{u}_1$. The value A_γ^+, belonging to the forward slip mode, can become negative for sufficiently large values of the friction coefficient μ. The critical value occurs at $\mu = \mu_{c\gamma}^+$ for which $A_\gamma^+ = 0$,

$$\mu_{c\gamma}^+ = \tan\gamma + \frac{m_1}{m_2}\frac{1}{\tan\gamma}. \tag{5.18}$$

The critical value $\mu_{c\gamma}^+$ is minimal when $\gamma = \arctan\sqrt{\frac{m_1}{m_2}}$ giving

$$\mu_{\text{cmin}} = 2\sqrt{\frac{m_1}{m_2}}. \tag{5.19}$$

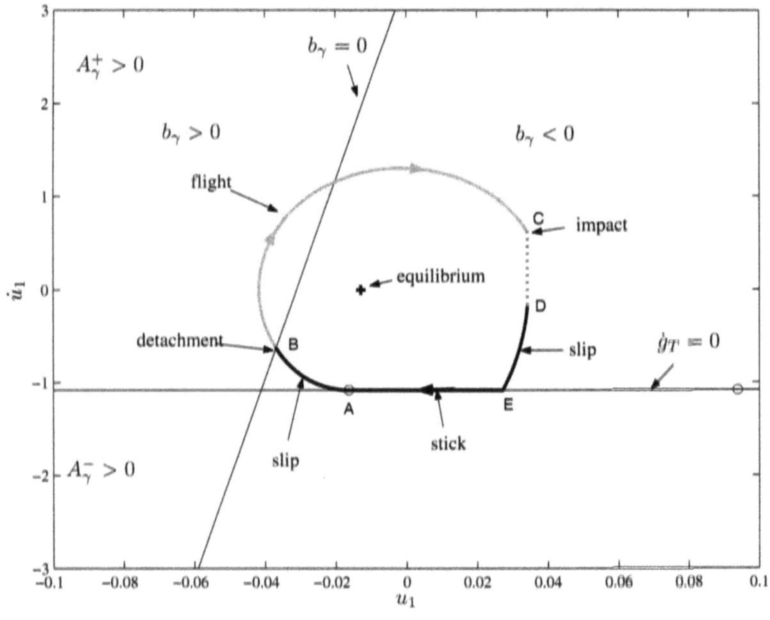

Figure 11. The $(\varphi, \dot{\varphi})$-plane of the Simp. Frictional Impact Osc., $\mu = 0.5$, $A_\gamma^+ = 0.5070$.

The Simplified Frictional Impact Oscillator has therefore the same critical friction coefficient for consistency and determinacy of the forward sliding mode as the Frictional Impact Oscillator. The piece-wise linear character of the Simplified Frictional Impact Oscillator yields:

Lemma 5.1. *If* $\mu = \mu_{c\gamma}^+$ *such that* $A_\gamma^+ = 0$, *then the point* $(u_1, \dot{u}_1) = (u_{\mathrm{stick}}^-, -\frac{v_{\mathrm{dr}}}{\cos\gamma})$ *is located on the line* $b_\gamma(u_1, \dot{u}_1) = 0$.

Lemma 5.2. *If* $\mu = \mu_{c\gamma}^+$ *such that* $A_\gamma^+ = 0$, *then the equilibrium point* $(u_1, \dot{u}_1) = (u_{1eq}, 0)$ *is located on the line* $b_\gamma(u_1, \dot{u}_1) = 0$.

The above lemmas show that the equilibrium as well as the left border of the friction cone pass to the left side of the line $b_\gamma = 0$ when μ passes the critical value for the Painlevé paradox. The equilibrium and the left border of the friction cone can therefore not be located in, or be bordered by, the inconsistent mode $A_\gamma^+ < 0, b_\gamma < 0$.

5.5 Analysis of Periodic Motion

The Simplified Frictional Impact Oscillator can exhibit periodic motion with slip, stick and flight phases similar to the Frictional Impact Oscillator. We will explore how the topology of the periodic solution depends on the regions defined by the value of $A_\gamma^\pm$ and the line $b_\gamma = 0$. We will consider two cases (similar to the two cases of the Frictional

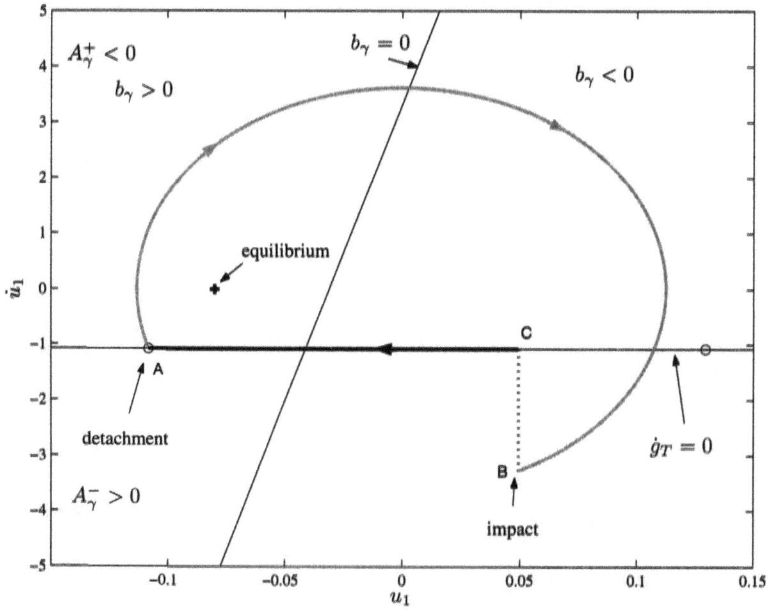

Figure 12. The $(\varphi, \dot{\varphi})$-plane of the Simp. Frictional Impact Osc., $\mu = 1$, $A_\gamma^+ = -1.1218$.

Impact Oscillator): case 1 with $\mu = 0.5 < \mu_{c\gamma}^+$ and case 2 with $\mu = 1 > \mu_{c\gamma}^+$, where $\mu_{c\gamma}^+ = 0.6556$. The other parameters are $m_1 = 0.1$ kg, $m_2 = 1$ kg, $l = 1$ m, $k_1 = 100$ N/m, $k_2 = 100$ N/m, $c_1 = 0$ N/(ms), $c_2 = 10$ N/(ms), $v_{\mathrm{dr}} = 1$ m/s, $g = 10$ Nm/s^2.

Case 1: $\mu = 0.5$. The values of $A_\gamma^\pm$ are not dependent on u_1 for the Simplified Frictional Impact Oscillator and are both positive for the parameter set of case 1. The forward and backward slip modes are therefore always consistent and determinate. The lines $b_\gamma = 0$ (5.16) and $\dot{g}_T = 0$ are drawn in the plane $(u_1, \dot{u}_1)$ of Figure 11. The borders of the friction cone on the line $\dot{g}_T$ in Figure 11 are given by (5.11) and are depicted by a small circle ($\circ$). The part of the line $\dot{g}_T$ between the two $\circ$ signs is therefore the stick mode. Note that the friction cone lies, for this parameter set, totally outside the detachment region $b_\gamma > 0$. A periodic solution is found numerically and consists of the following phases: stick–slip–flight–slip. The numerically obtained Floquet multipliers indicate that the periodic solution is stable, which agrees with its (locally) attracting nature. The stick to slip transition occurs at the border of the friction cone. The body remains some time in forward slipping contact with the belt in the space $b_\gamma < 0$ until the line $b_\gamma = 0$ is hit. The sign of b_γ changes at the line $b_\gamma = 0$, giving the LCP solution $\lambda_N = 0, \ddot{g}_N > 0$, which means that the body detaches from the belt. The body remains in flight (unconstrained free motion) until an impact occurs ($g_N = 0$). The impact causes the velocities $\dot{u}_1$ and $\dot{u}_2$ to jump. The impact results in slipping contact of the body to the belt. Other coexisting periodic solutions might exist theoretically but were not found numerically.

Case 2: $\mu = 1$. The value of A_γ^+ is negative for case 2. The friction cone is now partly bordered by the region $b_\gamma > 0$ (the indeterminate mode of the detachment LCP of the forward slip mode) and partly by the region $b_\gamma < 0$ (the inconsistent mode of the detachment LCP of the forward slip mode). A periodic solution was found which contains a stick phase. Detachment occurs at the end of the stick phase, when the solution is on the border of the friction cone. The problem is identical to case 2 of the Frictional Impact Oscillator. The detachment LCP has two solutions:

1. $\lambda_N = -\frac{b_\gamma}{A_\gamma^+} \wedge \ddot{g}_N = 0$ implying forward slip.
2. $\lambda_N = 0 \wedge \ddot{g}_N = b_\gamma$, detachment occurs followed by a flight phase (like the depicted periodic solution in Figure 12).

The undeterminacy of the detachment LCP at the end of the stick phase for Case 2 suggests that the solution $(\boldsymbol{q}(t), \dot{\boldsymbol{q}}(t))$ might be non-unique after this transition. We will have to check rigourously uniqueness at this point. The relationship between uniqueness of solutions of the LCP and uniqueness of solutions $(\boldsymbol{q}(t), \dot{\boldsymbol{q}}(t))$ is not simple in general, see (Brogliato, 1999) §2.2.3 and Remark 5.18 therein, but the following result can be obtained (Leine et al., 2002)

Lemma 5.3. *Let $\mu > \mu_{c\gamma}^+$, $\mathcal{K}_\gamma > 0$ and*

$$(\boldsymbol{q}(t_0), \dot{\boldsymbol{q}}(t_0)) = \begin{bmatrix} u_{\text{stick}}^- & -u_{\text{stick}}^- \sin\gamma & -\frac{v_{\text{dr}}}{\cos\gamma} & v_{\text{dr}} \tan\gamma \end{bmatrix}^{\mathrm{T}}.$$

The solution $\boldsymbol{q}(t)$ to the initial value problem of the Simplified Frictional Impact Oscillator with initial condition $(\boldsymbol{q}_0, \dot{\boldsymbol{q}}_0)$ at $t = t_0$ is locally unique and the solution proceeds in a flight phase $g_N > 0$.

Uniqueness of the stick-to-flight transition has been proven. Until now nothing has been said about the existence and uniqueness of the solution after the impact. For the following lemma we will introduce the notation:

$$A_\gamma(\alpha) = \boldsymbol{W}_N^{\mathrm{T}} \boldsymbol{M}^{-1}(\boldsymbol{W}_N - \alpha\mu \boldsymbol{W}_T) = \frac{\cos^2\gamma}{N_\gamma}(1 + \frac{m_2}{m_1}\tan\gamma(\tan\gamma - \alpha\mu)), \quad (5.20)$$

$$B_\gamma(\alpha) = \boldsymbol{W}_T^{\mathrm{T}} \boldsymbol{M}^{-1}(\boldsymbol{W}_N - \alpha\mu \boldsymbol{W}_T) = \frac{\cos^2\gamma}{N_\gamma}(\tan\gamma\frac{m_2}{m_1} - \alpha\mu(1 + \frac{m_2}{m1})), \quad (5.21)$$

with $-1 \leq \alpha \leq 1$. It follows that $A_\gamma^+ = A_\gamma(1)$, $A_\gamma^- = A_\gamma(-1)$ and we introduce the abbreviation $B_\gamma^+ = B_\gamma(1)$, $B_\gamma^- = B_\gamma(-1)$. Then, see (Leine et al., 2002):

Lemma 5.4. *Let $(\boldsymbol{q}(t_0), \dot{\boldsymbol{q}}(t_0^-))$ be an initial condition for which holds $g_N(t_0) = 0$ and $\dot{g}_N(t_0^-) < 0$ implying an impact at $t = t_0$. Let $\mu > \mu_{c\gamma}^+$. The post-impact solution $(\boldsymbol{q}(t_0), \dot{\boldsymbol{q}}(t_0^+))$ is locally unique and proceeds in either the backward slip mode, the stick mode or the flight mode.*

6 Conclusions

In the previous sections we analyzed the Frictional Impact Oscillator and in more detail the Simplified Frictional Impact Oscillator. It was shown that both models have the same critical value of the friction coefficient for the Painlevé paradox. A peculiarity of these models is that the critical friction coefficient depends only on a mass ratio and can therefore be made arbitrary small, i.e.

$$\mu_{\mathrm{cmin}} = 2\sqrt{\frac{m_1}{m_2}}.$$

The Painlevé Paradox can therefore occur at physically realistic values of the friction coefficient. This fact was not apparent from the classical Painlevé example (Figure 2).

As stated in the introduction, many systems exhibit a kind of frictional hopping motion (e.g. a finger or piece of chalk pushed over a table, robot in contact with an obstacle, brake systems). In fact, the Frictional Impact Oscillator (and its simplified version) can be seen as a sort of archetype of these systems, carrying the basic mechanism for the frictional hopping/bouncing phenomenon. What can we conclude from the analysis of the (Simplified) Frictional Impact Oscillator with respect to the existence of hopping/bouncing motion in these type of systems?

- If the mass of the end-effector of the structure, m_1, is small with respect to the mass of the supporting structure m_2, then the Painlevé paradox can occur at physically realistic values of the friction coefficient. Take for instance $m_1/m_2 = 1/36$, which is certainly realistic in robotic applications, then the critical friction coefficient becomes $\mu_{\mathrm{cmin}} = 1/3$.
- Hopping motion can occur when the friction coefficient μ is large enough such that either
  1. the linear damping terms vanish/become negative ($\mu > \mu_{d\gamma}$ for the Simplified Frictional Impact Oscillator),
  2. the Painlevé paradox occurs ($\mu > \mu_{c\gamma}$ for the Simplified Frictional Impact Oscillator).
- For many practical applications it might very well be that $\mu_{c\gamma} < \mu_{d\gamma}$ and that the Painlevé paradox is the actual cause of (undesirable) periodic motion. To avoid the hopping phenomenon in these type of systems, one should therefore increase the mass m_1 of the end-effector.
- The support stiffness k and damping c (or k_2 and c_2 of the Simplified Frictional Impact Oscillator) can be looked upon as the PD-action of a position controller for the link with mass m_2. When the friction is high enough such that $\mu > \tan\varphi$, then the D-action will diminish the dissipation in the system. In fact, the friction causes negative feedback, which can lead to instability of the forward sliding equilibrium.

The $(\varphi, \dot{\varphi})$-planes with lines $A = 0$, $b = 0$ and $\dot{g}_T = 0$ (like Figures 7, 5 and 12, 11) proved to be valuable tools when analyzing periodic solutions induced by the Painlevé paradox. The lines $A = 0$, $b = 0$ and $\dot{g}_T = 0$ determine the subspaces where detachment, indeterminacy or inconsistency occurs, which greatly affects the topology of periodic solutions and equilibria.

Bibliography

B. Brogliato. *Nonsmooth Mechanics.* Springer, London, 2 edition, 1999.

F. Génot. *Contributions à la modélisation et à la commande des systèmes avec contraintes unilatérales.* Ph.D. thesis, Institut National Polytechnique de Grenoble, France, 1998.

F. Génot and B. Brogliato. New results on Painlevé Paradoxes. *Eur. J. Mech.A/Solids,* 18:653–677, 1999.

Ch. Glocker. *Dynamik von Starrkörpersystemen mit Reibung und Stößen,* volume 18, no. 182 of *Fortschr.-Ber. VDI.* VDI Verlag, Düsseldorf, 1995.

Ch. Glocker. Scalar force potentials in rigid multibody systems. In F. Pfeiffer and Ch. Glocker, editors, *Multibody Dynamics with Unilateral Contacts,* volume 421 of *CISM Courses and Lectures,* pages 69–146. Springer, Wien, 2000.

Ch. Glocker. *Set-Valued Force Laws, Dynamics of Non-Smooth Systems,* volume 1 of *Lecture Notes in Applied Mechanics.* Springer-Verlag, Berlin, 2001.

R. A. Ibrahim. Friction-induced vibration, chatter, squeal and, chaos; Part II: Dynamics and modeling. *ASME Applied Mechanics Reviews,* 47(7):227–253, 1994.

J. H. Jellet. *Treatise on the Theory of Friction.* Hodges, Foster and Co, 1872.

L. Lecornu. Sur le frottement de glissement. *Comptes Rendu des Séances de l'Academie des Sciences,* 140(6):635–637,847–848, March 1905.

R. I. Leine, B. Brogiato, and H. Nijmeijer. Periodic motion and bifurcations induced by the Painlevé paradox. *European Journal of Mechanics - A/Solids,* 21(5):869–896, 2002.

R. I. Leine, Ch. Glocker, and D. H. Van Campen. Nonlinear dynamics and modeling of various wooden toys with impact and friction. *Journal of Vibration and Control,* 9 (1):25–78, 2003.

P. Lötstedt. Coulomb friction in two-dimensional rigid-body systems. *Z. Angew. Math. Mech.,* 61:605–615, 1981.

M. T. Mason and Y. Wang. On the inconsistency of rigid-body frictional planar machanics. *Proc. IEEE Int. Conf. Robotics and Automation,* pages 524–528, 1988.

J. J. Moreau. Dynamique de systèmes à liaisons unilatérales avec frottement secéventuel; essais numériques. Technical Report 85-1, LMGC, Montpellier, France, 1986.

J. J. Moreau. Unilateral contact and dry friction in finite freedom dynamics. In J. J. Moreau and P. D. Panagiotopoulos, editors, *Non-Smooth Mechanics and Applications,* volume 302 of *CISM Courses and Lectures,* pages 1–82. Springer, Wien, 1988.

P. Painlevé. *Leçon sur le frottement.* Hermann, Paris, 1895a.

P. Painlevé. Sur les lois du frottement de glissement. *Comptes Rendu des Séances de l'Academie des Sciences,* 121:112–115, 1895b.

P. Painlevé. Sur les lois du frottement de glissement. *Comptes Rendu des Séances de l'Academie des Sciences,* 141:401–405 and 546–552, August 1905.

F. Pfeiffer and Ch. Glocker. *Multibody dynamics with unilateral contacts.* Wiley, New York, 1996.

E. V. Wilms and H. Cohen. The occurence of Painlevé's paradox in the motion of a rotating shaft. *Journal of Applied Mechanics,* 64:1008–1010, 1997.

Coordination of Rigid and Flexible Joint Robot Manipulators

A. Rodriguez-Angeles[†][*], H. Nijmeijer[‡], and H. A. van Essen [#]

[†] Research Program in Applied Mathematics and Computation, Mexican Petroleum Institute, Mexico

[‡] Department of Mechanical Engineering, Eindhoven University of Technology The Netherlands

[#] Industrial Design, Eindhoven University of Technology, The Netherlands

Abstract. In this paper we address the problem of position coordination of robot manipulators. Two coordinating controllers are presented, one for rigid joint robots, and one for flexible joint robots. In both cases the coordination system is based on a master-slave scheme and only position measurements are required.

The proposed controllers consist of a feedback control law and two nonlinear observers. The controller for rigid joint robots yields semi-global ultimate uniformly boundedness of the closed loop errors and a relation between the bound of the errors and the gains on the controller is established. For flexible joint robots only local convergence is established. Experimental results show the predicted convergence performance.

1 Introduction

In manufacturing processes, automotive applications and tele-operated systems there is a high requirement on flexibility, manoeuvrability and high performance of the involved systems. In most of these processes the use of robot manipulators is widely spread, and their variety in uses is practically endless, e.g. ensembling, transporting, painting, welding, grasping. All the mentioned tasks require large manoeuvrability and manipulability from the robots, such that some of the tasks can not be carried out by a single robot. In those cases the use of multi robot systems, working under cooperative or coordinated schemes, has been considered as an option.

The coordinated and cooperative schemes give flexibility and manoeuvrability that can not be achieved by an individual system, e.g. multi finger robot-hands, multi robot systems, multi-actuated platforms, (Brunt, 1998, and Liu et al., 1999), vibro-machinery (Blekhman, 1988), tele-operated master-slave systems (Dubey and Everett, 1997, Lee and Chung, 1998, Hills and Jensen, 1998, and Guthart and Salisbury, 2000). In robot

*The first author acknowledges support from the CONACYT (National Council for Science and Technology), Mexico, Scholarship No. 72368.

coordination the basic problem is to ascertain synchronous motion of two (or more), either identical or different, robotic systems. This is obviously a control problem that implies the design of suitable controllers to achieve the required coordinated (synchronized) motion.

Robot coordination can be seen as tracking between two (or more) robots. Although it seems to be a straightforward extension of classical tracking controllers, the synchronization problem implies challenges that are not considered in tracking controllers. These challenges arise by the interaction (interconnection) between the robots. This interaction can not be neglected, since it is precisely what generates the flow of information necessary to guarantee, and at the same time to determine, the coordinated behavior.

This work addresses the problem of position coordination of robotic systems under master-slave schemes. Since the pioneering work of Goertz (1954), most of the master-slave robot manipulator systems – if not all – are based on full knowledge of the dynamic model and joint variables (position, velocity and acceleration) of the master and slave robots (Komatsu and Akabane, 1998, and Kosuge et al., 1990). However, in practice, robot manipulators are equipped with high precision position sensors, such as encoders, but very often the velocity measurements are obtained by means of tachometers, which are contaminated by noise. Velocity sensoring equipment is even omitted due to the savings in cost, volume and weight that can be obtained. New technologies have been designed for measuring velocities and accelerations, e.g. brushless AC motors with digital servo-drivers, accelerometers, etc. (Laopoulos and Papageorgiou, 1996, Lygouras et al., 1998, and Ovaska and Vliviita, 1998). However such techniques are not very common in applications yet.

A further complication on coordination of robot manipulators is due to the kind of joints in the robots, i.e. rigid or flexible joints. In case of a rigid joint the position of the link connected to the joint and the motor rotor, that actuates the joint, are the same; meanwhile, for a flexible joint they are different. Joint flexibility (also called joint elasticity) is caused by transmission elements such as harmonic drives, belts or long shafts. Joint flexibility can be modelled by considering the position and velocity of the motor rotor and the position and velocity of the link (Book, 1984, and Spong, 1987). Therefore the dynamic model of a joint flexible robot has twice the dimension of an equivalent rigid robot, consequently the controllers for flexible joint robots are more complex that those for rigid joint robots. It has been shown that joint flexibility considerably affects the performance of robot manipulators being a major source of oscillatory behavior (Good et al., 1985). Therefore, to improve the performance of robot manipulators, joint flexibility has to be taken into account in the modelling and control of such systems.

Joint flexibility is not the only source of oscillatory behavior in robot manipulators. In high performance robotic systems, friction can severely deteriorate the performance. Bad compensation of the friction phenomenon generates oscillatory behavior like limit cycles or stick-slip oscillations, introduces tracking errors, and in some cases can generate instability of the system (Armstrong-Hélouvry, 1993, and Olsson and Åström, 2001).

In this paper we consider rigid and flexible joint robots working in a cooperative system, under a master-slave scheme. Coordinating controllers that only require position measurements for both type of joints are presented. The proposed coordinating controller for rigid joint robots yields semi-global ultimate uniformly boundedness of the closed loop errors and a relation between the bound of the errors and the gains on the controller is established. Meanwhile, for flexible joint robots local convergence of the closed loop errors is obtained. The stability analysis of the proposed coordinating controller for rigid joint robots is discussed in detail in (Rodriguez-Angeles and Nijmeijer, 2001), and practical issues for implementation of the controller are addressed in (Rodriguez-Angeles et al., 2002).

The proposed coordinating controllers consist of a feedback control law and two nonlinear model-based observers. In case of rigid joint robots, the observers reconstruct the master and slave joint velocities and accelerations which are used in the feedback controller. For flexible joint robots the link velocities and accelerations and the motor rotor positions and velocities are reconstructed. Of course, presently other ways of estimating velocity and accelerations signals, like numerical differentiation or low pass filters (Ovaska and Vliviita, 1998), are available, and in principle such alternatives could be used in the here developed coordinating controllers. These alternative techniques have the advantage of simplicity in implementation. Nevertheless they present a reduced bandwidth and in general there is not (or it is too difficult to determine) an analytical method to guarantee stability of the closed loop systems.

Two different setups are considered. The first setup considers two rigid joint robots, such that the motion of one of the robots is independent of the other one. This robot is the dominant one and is referred to as the master robot. The master robot is driven by a control $\tau_m(\cdot)$, that in the ideal case, ensures convergence of the master robot angular positions and velocities $q_m, \dot{q}_m$ to a given desired trajectory $q_d, \dot{q}_d$. Then, the goal is to design interconnections and a feedback controller for the non-dominant robot, hereafter referred to as slave, such that its position and velocity $q_s, \dot{q}_s$ coordinate (synchronize) to those of the master robot $q_m, \dot{q}_m$. However, the input torque τ_m, the dynamic model and parameters of the master robot, as well as the velocity and acceleration variables $\dot{q}_m, \ddot{q}_m$, are not available for the design of the slave control law $\tau_s(\cdot)$. Therefore for the design of the slave interconnections and controller only master and slave angular positions q_m, q_s are available by means of measurements.

The second setup considers flexible joint robots, such that the master robot is driven by a control $\tau_m(\cdot)$, that in the ideal case, ensures convergence of the master robot link angular positions and velocities $q_m, \dot{q}_m$ to a given desired trajectory $q_d, \dot{q}_d$. The goal is to design interconnections and a feedback controller for the slave robot, such that its link position and velocity $q_s, \dot{q}_s$ coordinate (synchronize) to the link position and velocity of the master robot $q_m, \dot{q}_m$. For design of the controller only link position measurements are available, and thus link velocities and accelerations, as well as motor rotor position and velocities must be reconstructed.

Notice that in both cases the goal is to ensure coordination between the slave robot trajectories $q_s, \dot{q}_s$ and the master robot trajectories $q_m, \dot{q}_m$, and not to the master desired trajectories $q_d, \dot{q}_d$ which may not be realized due to model uncertainties or disturbances in the system, e.g. noise, unknown loads, friction, etc..

Most of the master-slave robot systems are designed to interact with their environment, and thus force-position controllers are required (Kosuge et al., 1990, and Komatsu and Akabane, 1998). This paper only focuses on the position coordination problem. Nevertheless in case of a master-slave system interacting with its environment, passive compliance or end effector compliance models can be used in order to control the interacting forces between the slave robot and the environment.

Throughout this paper standard notation is used, in particular, vector norms are Euclidean, and for matrices the induced norm $\|A\| = \sqrt{\lambda_{\max}(A^T A)}$ is employed, with $\lambda_{\max}(\cdot)$ the maximum eigenvalue. Moreover, for any positive definite matrix A we denote by A_m and A_M its minimum and maximum eigenvalue respectively.

The paper is organized as follows. Section 2 presents the dynamic model of the master and slave robots. The coordinating controller for rigid joint robots is presented in Section 3. It includes a gain tuning procedure that ensures the convergence properties of the coordinating system. Section 4 introduces the dynamic model of flexible joint robots. The coordinating controller for flexible joint robots is presented in Section 5.

In Section 6 experimental results for the case of rigid joint robots are presented and discussed. Some remarks about design and implementation of the proposed controllers are discussed in Section 7. Section 8 presents general conclusions and some further extensions of the proposed controllers.

2 Dynamic model of the rigid joint robot

Without loss of generality and considering that the friction phenomena can be compensated separately, it is assumed that the robots are frictionless. Consider a pair of fully actuated rigid robots, each with the same number of rotational joints, i.e. $q_i \in \mathbb{R}^n$, where $i = m, s$ identifies the master (m) and slave (s) robot. This does not mean, however, that they are identical in their parameters (masses, inertias, etc.).

For each of the robots, the kinetic energy is given by $T_i(q_i, \dot{q}_i) = \frac{1}{2}\dot{q}_i^T M_i(q_i)\dot{q}_i$, $i = m, s$, with $M_i(q_i) \in \mathbb{R}^{n \times n}$ the symmetric, positive-definite inertia matrix, and the potential energy is denoted by $U_i(q_i)$. Hence, applying the Euler-Lagrange formalism (Spong and M., 1989) the dynamic model of the robot is given by

$$M_i(q_i)\ddot{q}_i + C_i(q_i, \dot{q}_i)\dot{q}_i + g_i(q_i) = \tau_i \qquad i = m, s \tag{2.1}$$

where $g_i(q_i) = \frac{\partial}{\partial q_i} U_i(q_i) \in \mathbb{R}^n$ denotes the gravity forces, $C_i(q_i, \dot{q}_i)\dot{q}_i \in \mathbb{R}^n$ represents the Coriolis and centrifugal forces, and τ_i is the $[n \times 1]$ vector of input torques.

3 Coordinating controller for rigid robots

As mentioned, it is assumed that only angular joint positions q_m, q_s are measured. Therefore, the slave control τ_s can only depend on direct position measurements q_m, q_s. And thus estimated values for the velocities $\dot{q}_m, \dot{q}_s$ and accelerations $\ddot{q}_m, \ddot{q}_s$ are required to implement controllers based on velocity and acceleration feedback.

3.1 Feedback control law

Under the assumption that the required estimated velocities and accelerations are available, and that the nonlinearities and parameters of the slave robot are known, we propose the controller τ_s for the slave robot as

$$\tau_s = M_s(q_s)\widehat{\ddot{q}}_m + C_s(q_s, \widehat{\dot{q}}_s)\widehat{\dot{q}}_m + g_s(q_s) - K_d\widehat{\dot{e}} - K_p e \tag{3.1}$$

were $\widehat{\dot{q}}_s, \widehat{\dot{e}}, \widehat{\dot{q}}_m, \widehat{\ddot{q}}_m \in \mathbb{R}^n$ represent the estimates of $\dot{q}_s, \dot{e}, \dot{q}_m$ and $\ddot{q}_m$ respectively.

The coordination errors $e, \dot{e} \in \mathbb{R}^n$ are defined by

$$e := q_s - q_m, \qquad \dot{e} := \dot{q}_s - \dot{q}_m, \tag{3.2}$$

$M_s(q_s)$, $C_s(q_s, \widehat{\dot{q}}_s)$, $g_s(q_s)$ are defined as in equation (2.1), and $K_p, K_d \in \mathbb{R}^{n \times n}$ are positive definite gain matrices.

3.2 An observer for the coordination errors $(e, \dot{e})$

Estimated values for the coordination errors $e, \dot{e}$ (3.2) are denoted by $\hat{e}, \widehat{\dot{e}}$. These estimated values are obtained by the full state nonlinear Luenberger observer

$$\begin{aligned} \frac{d}{dt}\hat{e} &= \widehat{\dot{e}} + \Lambda_1 \tilde{e} \\ \frac{d}{dt}\widehat{\dot{e}} &= -M_s(q_s)^{-1}\left[C_s(q_s, \widehat{\dot{q}}_s)\widehat{\dot{e}} + K_d \widehat{\dot{e}} + K_p \hat{e}\right] + \Lambda_2 \tilde{e} \end{aligned} \tag{3.3}$$

where the estimation position and velocity coordination errors $\tilde{e}, \widetilde{\dot{e}}$ are defined by

$$\tilde{e} := e - \hat{e}, \quad \widetilde{\dot{e}} := \dot{e} - \widehat{\dot{e}}, \tag{3.4}$$

and $\Lambda_1, \Lambda_2 \in \mathbb{R}^{n \times n}$ are positive definite gain matrices.

3.3 An observer for the slave joint state $(q_s, \dot{q}_s)$

Lets $\hat{q}_s, \widehat{\dot{q}}_s$ denote estimated values for $q_s, \dot{q}_s$, to compute these estimated values, we define the full state nonlinear observer

$$\begin{aligned} \frac{d}{dt}\hat{q}_s &= \widehat{\dot{q}}_s + L_{p1}\tilde{e}_q \\ \frac{d}{dt}\widehat{\dot{q}}_s &= -M_s(q_s)^{-1}\left[C_s(q_s, \widehat{\dot{q}}_s)\widehat{\dot{e}} + K_d \widehat{\dot{e}} + K_p e\right] + L_{p2}\tilde{e}_q \end{aligned} \tag{3.5}$$

where the estimation position and velocity errors $\tilde{e}_q$ and $\widetilde{\dot{e}}_q$ are defined by

$$\tilde{e}_q := q_s - \hat{q}_s \qquad \widetilde{\dot{e}}_q := \dot{q}_s - \widehat{\dot{q}}_s, \tag{3.6}$$

and $L_{p1}, L_{p2} \in \mathbb{R}^{n \times n}$ are positive definite gain matrices.

3.4 Estimated values for $\dot{q}_m, \ddot{q}_m$

As stated, the master robot variables $\dot{q}_m, \ddot{q}_m$ are not available, therefore estimated values for $\dot{q}_m, \ddot{q}_m$ are used in τ_s (3.1). From (3.2) and the definition of the estimated variables $\hat{e}, \widehat{\dot{e}}$, $\hat{q}_s, \widehat{\dot{q}}_s$, we can consider that estimated values for $q_m, \dot{q}_m, \ddot{q}_m$ are given by

$$\begin{aligned} \hat{q}_m &= \hat{q}_s - \hat{e} \\ \widehat{\dot{q}}_m &= \widehat{\dot{q}}_s - \widehat{\dot{e}} \\ \widehat{\ddot{q}}_m &= \frac{d}{dt}\left(\widehat{\dot{q}}_s - \widehat{\dot{e}}\right) \end{aligned} \tag{3.7}$$

from the definition of the observers (3.3), (3.5) it follows that

$$\widehat{\ddot{q}}_m = -(M_s(q_s)^{-1}K_p + L_{p2})\tilde{e} + L_{p2}\tilde{e}_q$$

which gives a clear insight of how $\widehat{\ddot{q}}_m$ is reconstructed and why by increasing some appropriate gains, specifically K_p, L_{p2}, the closed loop errors decrease in magnitude.

Remark 3.1. Note that, in (3.3) and (3.4) the estimate for $\dot{e}$ is given by $\widehat{\dot{e}}$, not by $\dot{\hat{e}}$. This definition introduces an extra correcting term in $\dot{\tilde{e}}$, as it follows from (3.3), (3.4) that

$$\dot{\tilde{e}} = \dot{e} - \dot{\hat{e}} = \widetilde{\dot{e}} - \Lambda_1\tilde{e},$$

The term $\Lambda_1\tilde{e}$ gives faster estimation performance, especially during transients, but it has some negative effects on noise sensitivity, since it amplifies noise measurements on $\tilde{e}$.

The same can be said for observer (3.5) and the estimation position and velocity errors (3.6).

Figure 1 shows a schematic representation of the proposed coordinating controller for rigid joint robots.

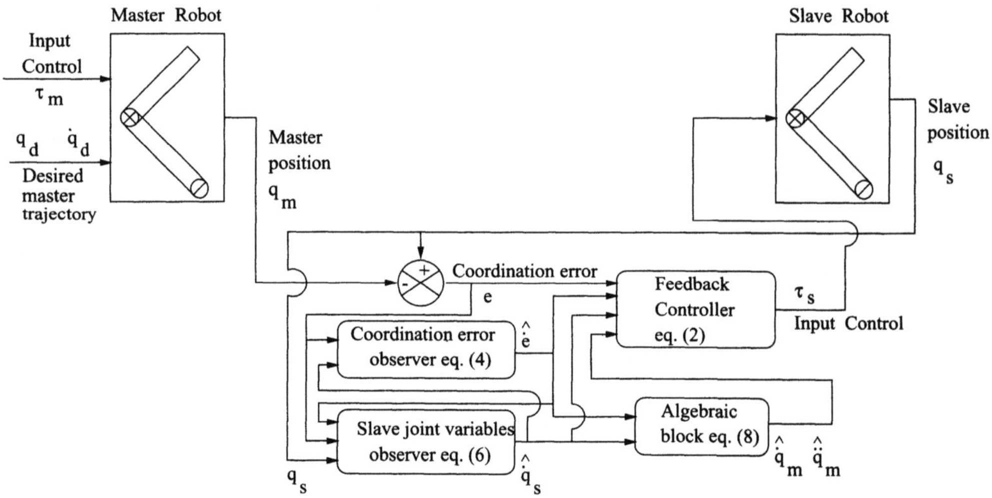

Figure 1. The coordinating controller for rigid joint robots.

3.5 Ultimate boundedness of the closed loop system

For the sake of simplicity in the stability analysis and without loss of generality the following assumptions are considered.

Assumption 3.2. The gain matrices Λ_1, Λ_2 and L_{p1}, L_{p2} satisfy

$$\Lambda_1 = L_{p1}, \qquad \Lambda_2 = L_{p2}. \tag{3.8}$$

Assumption 3.3. The gains K_p, K_d, L_{p1}, L_{p2} are symmetric positive definite matrices.

On the other hand the following assumption is required to prove boundedness of the coordination closed loop system.

Assumption 3.4. The signals $\dot{q}_m(t)$ and $\ddot{q}_m(t)$ are bounded, i.e. there exist finite positive scalars V_M and A_M such that

$$V_M = \sup_t \|\dot{q}_m(t)\|, \tag{3.9}$$

$$A_M = \sup_t \|\ddot{q}_m(t)\|. \tag{3.10}$$

In practice, it is often not difficult to obtain on the basis of the desired motion $q_d(t), \dot{q}_d(t)$ and $\ddot{q}_d(t)$ of the master robot bounds on $\dot{q}_m(t)$ and $\ddot{q}_m(t)$. However due to friction effects, tracking errors, etc., the actual motion of the master robot may differ from its desired motion, but still could induce suitable values for V_M and A_M.

Theorem 3.5. *Consider the master and slave rigid joint robots that are described by (2.1), and the slave robot in closed loop with the control law (3.1) and both observers (3.3), (3.5). If the assumptions 3.2, 3.3, and 3.4 are satisfied, then there exist conditions on the gain matrices K_d, K_p, L_{p1}, L_{p2}, specifically in their minimum eigenvalues, such that the errors $\dot{e}$, e, $\dot{\tilde{e}}$, $\tilde{e}$, $\dot{\tilde{e}}_q$, $\tilde{e}_q$ in the closed loop system are semi-globally uniformly ultimately bounded. Moreover, the bound of the errors can be made small by a proper choice of the minimum eigenvalues of the gains K_p, L_{p2}.*

Proof: The proof can be found in (Rodriguez-Angeles and Nijmeijer, 2001).

In (Rodriguez-Angeles and Nijmeijer, 2001) a tuning gain procedure to ensure the semi-globally uniformly ultimately bounded result is given. A sketch for this procedure is as follows.

1. Determine the bounds of the master trajectories $\dot{q}_m, \ddot{q}_m$, the physical parameters $M_s(q_s)$, $C_s(q_s, \dot{q}_s)$ and their partial derivatives with respect to q_s.
2. Choose the weighting factors $\lambda_0 > 0$, $\varepsilon_0 > 0$, $\mu_0 > 0$ and $\gamma_0 > 0$ and a bound for the maximum eigenvalue of the gains K_p, K_d, L_{p1} and L_{p2}.
3. Consider the matrix $Q_N = Q_N^T$

$$Q_N = \begin{bmatrix} Q_{11_N} & Q_{12_N} & Q_{13_N} \\ Q_{12_N}^T & Q_{22_N} & Q_{23_N} \\ Q_{13_N}^T & Q_{23_N}^T & Q_{33_N} \end{bmatrix} \tag{3.11}$$

with the block matrices

$$Q_{11_N} = \varepsilon_o \begin{bmatrix} K_{d,m} - \lambda_o M_{s,M} & \frac{1}{2}\lambda_o V_M \left(C_{s,M} - M_{s,pM}\right) \\ \frac{1}{2}\lambda_o V_M \left(C_{s,M} - M_{s,pM}\right) & \lambda_o K_{p,m} \end{bmatrix}$$

$$Q_{12_N} = \frac{\varepsilon_o}{2} \begin{bmatrix} -M_{s,M} L_{p1,M} & K_{p,M} - K_{d,M} L_{p1,M} - C_{s,M} L_{p1,M} V_M \\ -\lambda_o M_{s,M} L_{p1,M} & \lambda_o \left(K_{p,M} - K_{d,M} L_{p1,M} - C_{s,M} L_{p1,M} V_M\right) \end{bmatrix}$$

$$Q_{13_N} = \frac{\varepsilon_o}{2}\begin{bmatrix} -K_{d,M} & -K_{d,M}L_{p1,M} \\ -\lambda_o K_{d,M} & -\lambda_o K_{d,M}L_{p1,M} \end{bmatrix}$$

$$Q_{22_N} = \begin{bmatrix} L_{p1,m} - 2\mu_o & \frac{1}{2}\left(M_{s,m}^{-1}K_{p,M} + \mu_o L_{p1,M}\right) \\ \frac{1}{2}\left(M_{s,m}^{-1}K_{p,M} + \mu_o L_{p1,M}\right) & \mu_o\left(M_{s,m}^{-1}K_{p,m} + L_{p2,m}\right) \end{bmatrix}$$

$$Q_{23_N} = \begin{bmatrix} 0 & \frac{1}{2}M_{s,m}^{-1}K_M \\ \frac{1}{2}\left(M_{s,m}^{-1}K_{p,M} + L_{p2,M}\right) & \frac{1}{2}\left((\mu_o + \gamma_o)M_{s,m}^{-1}K_{p,M} + \gamma_o L_{p2,M}\right) \end{bmatrix}$$

$$Q_{33_N} = \begin{bmatrix} L_{p1,m} - 2\gamma_o + 2M_{s,m}^{-1}C_{s,M}V_M & q_{56} \\ q_{56} & q_{66} \end{bmatrix}$$

$$q_{56} = \frac{1}{2}\left(M_{s,m}^{-1}K_{p,M} + \gamma_o L_{p1,M}\right) + M_{s,m}^{-1}C_{s,M}V_M\left(L_{p1,M} + \gamma_o\right)$$
$$q_{66} = \gamma_o\left(M_{s,m}^{-1}K_{p,m} + L_{p2,m} + 2M_{s,m}^{-1}C_{s,M}V_M L_{p1,m}\right)$$

4. By considering the matrix Q_N given by (3.11) determine the minimum eigenvalue of the gain matrices K_p, K_d, L_{p1} and L_{p2}, such that Q_N is a positive definite matrix. For this a set of nonlinear algebraic equations must be solved.

Let ΔQ_i denotes the determinant of the i-th leading minor of Q_N. Then sufficient conditions for positive definiteness of Q_N are given by

$$\begin{aligned} L_{p2,m} &> \max\left\{\mu_o^2,\ \gamma_o^2, L_{p2q4}, L_{p2q5}, L_{p2q6}\right\}, \\ L_{p1,m} &> \max\left\{2\mu_o, L_{p1q3}, L_{p1q5,a}, L_{p1q5,b}\right\}, \\ K_{p,m} &> \max\left\{K_{pq2}, K_{pq6}\right\}, \\ K_{d,m} &> \max\left\{K_{dq1}\right\}, \end{aligned} \tag{3.12}$$

with $L_{p2q4}, L_{p2q5}, L_{p2q6}, L_{p1q3}, L_{p1q5,a}, L_{p1q5,b}, K_{pq2}, K_{pq6}, K_{dq1}$ given by

- $K_{dq1} = \lambda_o M_{s,M}$,
- $K_{pq2} = \frac{\lambda_o V_M (M_{s,pM} - C_{s,M})^2}{4(K_{d,m} - \lambda_o M_{s,M})}$,
- L_{p1q3}: denotes the solution of the equation $\Delta Q_3 = a_1 L_{p1q3} + a_2 = 0$,with a_1, a_2 the resultant coefficients in the factorization of $L_{p1,m}$ in ΔQ_3, and $L_{p1,m}$ substituted by L_{p1q3}.
- L_{p2q4}: denotes the solution of the equation $\Delta Q_4 = b_1 L_{p2q4} + b_2 = 0$, with b_1, b_2 the resultant coefficients in the factorization of $L_{p2,m}$ in ΔQ_4, and $L_{p2,m}$ substituted by L_{p2q4}.
- $L_{p1q5,a} = 2(\gamma_o - C_{s,M}V_M M_{s,M}^{-1})$,
- $L_{p1q5,b}$: denotes the largest solution of the equation $\Delta Q_5 = c_0 + c_1 L_{p1q5} + c_2 L_{p1q5}^2 = 0$, with c_0, c_1, c_2 the resultant coefficients in the factorization of $L_{p1,m}$ in ΔQ_5, and $L_{p1,m}$ substituted by L_{p1q5}.
- L_{p2q5}: denotes the solution of the equation $c_2 = r_1 L_{p2q5} + r_2 = 0$, with c_2 as in $L_{p1q5,b}$, r_1, r_2 the resultant coefficients in the factorization of $L_{p2,m}$ in c_2, and $L_{p2,m}$ substituted by L_{p2q5}.

- L_{p2q6}: denotes the largest solution of the equation $\Delta Q_6 = t_o + t_1 L_{p2q6} + t_2 L_{p2q6}^2 = 0$, with t_0, t_1, t_2 the resultant coefficients in the factorization of $L_{p2,m}$ in ΔQ_6, and $L_{p2,m}$ substituted by L_{p2q6}.
- K_{pq6}: denotes the solution of the equation $t_2 = s_1 K_{pq6} + s_2 = 0$, with t_2 as in L_{p2q6}, s_1, s_2 the resultant coefficients in the factorization of $K_{p,m}$ in t_2, and $K_{p,m}$ substituted by K_{pq6}.

Remark 3.6. The above steps readily provide a tuning gain procedure for master-slave coordination of rigid joint robots.

Remark 3.7. Note that the conditions implied by Theorem 3.5 and the Assumptions 3.2 - 3.4 are only sufficient, but not necessary, to ensure stability and boundedness of the coordinating system. Hence for different values of λ_0, ε_0, μ_0 and γ_0, different minimum eigenvalues of the gains K_p, K_d, L_{p1} and L_{p2} would be obtained.

4 Reduced dynamic model of the flexible joint robot

From a modelling point of view, two dynamic models for the flexible joint robot have been considered. In (Book, 1984 and Tomei, 1991) an extended model for flexible joint robots is presented. This extended model includes the full nonlinear dynamic interactions among joint flexibilities and inertial properties of links and actuators. If it is assumed that the kinetic energy of the electrical actuators is due only to their own rotor spinning, then a reduced model is obtained (Spong, 1987). This reduced model satisfies the conditions for full state linearization and decoupling via static state feedback. Meanwhile, it has been proved (see De Luca and Lanari, 1995) that the extended model is fully linearizable and decouplable via dynamic state feedback and a general algorithm has been proposed by De Luca and Lucibello (1998).

The dynamic model of a flexible joint robot can be obtained by extending the procedures already used for rigid robots (Book, 1984). Consider a flexible joint robot, with n rigid links, all joints being flexible, rotational, and actuated by electrical drives. Let $q_i \in \mathbb{R}^n$ be the link positions and $\theta_i \in \mathbb{R}^n$ be the rotor positions, as reflected through the gear ratios, the subindex $i = m, s$ identifies the master (m) and slave (s) robot respectively. The difference $q_{ij} - \theta_{ij}$ is the j-th joint deformation, in view of small deformations, elasticity is modelled as a linear spring. The rotors of the motors are modelled as balanced uniform bodies having their centre of mass on the rotation axis, so that the inertia matrix and the gravity term in the dynamic model are independent from the motor position θ_i.

Assuming that the motion of the rotors can be considered as pure rotations with respect to an inertial frame, then the kinetic energy of each rotor is due to its own spinning. Therefore, the inertial coupling between links and rotors can be neglected and a reduced dynamic model is obtained (see Spong, 1987). Following (Spong, 1987) and the Lagrangian approach (De Luca and Tomei, 1996), we obtain the reduced dynamic model given by

$$M_i(q_i)\ddot{q}_i + N(q_i, \dot{q}_i) + K_i(q_i - \theta_i) = 0 \qquad i = m, s \tag{4.1}$$

$$J_i\ddot{\theta}_i + K_i(\theta_i - q_i) + B_{id}\dot{\theta}_i = \tau_i. \tag{4.2}$$

$$N(q_i, \dot{q}_i) = C_i(q_i, \dot{q}_i)\dot{q}_i + g_i(q_i);$$

where the symmetric positive definite inertia matrix $M_i(q_i) \in \mathbb{R}^{n\times n}$, the Coriolis and centrifugal term $C_i(q_i, \dot{q}_i)\dot{q}_i \in \mathbb{R}^n$, and the gravity term $g_i(q_i) \in \mathbb{R}^n$ are all related to the rigid links, $J_i \in \mathbb{R}^{n\times n}$ is the constant diagonal inertia matrix of the motors, $K_i \in \mathbb{R}^{n\times n}$ is the constant diagonal matrix of the joint stiffness, $B_{id} \in \mathbb{R}^{n\times n}$ is the diagonal positive definite viscous friction coefficient matrix, and $\tau_i(\cdot)$ is the $n-$vector of torques supplied by the motors.

The dynamic model (4.1), (4.2) can be transformed into a linear one via static state feedback (Spong, 1987). The relation from the new input to the output q_i is given by n independent chains of 4 integrators.

5 Coordinating controller for flexible joint robots

For clarity, first a coordinating controller based on full access to all position and velocities is presented, then it is modified to consider estimated variables.

5.1 Nominal coordinating controller

Consider that (4.1), (4.2) define the dynamics of two robots which are interconnected in a coordinating master slave scheme. Based on inverse dynamics computation and according to De Luca and Lucibello (1998), the slave robot with dynamic model (4.1), (4.2) can be fully linearized and decoupled via the static feedback control law

$$\tau_s = K_s(\theta_s - q_s) + B_{sd}\dot{\theta}_s + J_s K_s^{-1}\left[M_s(q_s)v(t) + \alpha(q_s, \dot{q}_s, \ddot{q}_s, q_s^{(3)})\right] \tag{5.1}$$

where

$$\alpha(q_s, \dot{q}_s, \ddot{q}_s, q_s^{(3)}) = 2\dot{M}_s(q_s, \dot{q}_s)q_s^{(3)} + (\ddot{M}_s(q_s, \dot{q}_s, \ddot{q}_s) + K_s)\ddot{q}_s + \ddot{N}(q_s, \dot{q}_s, \ddot{q}_s, q_s^{(3)}). \tag{5.2}$$

On the other hand (4.1) establishes a relation between $\ddot{q}_s, q_s^{(3)}$ and lower order variables $q_s, \dot{q}_s, \dot{\theta}_s$. From (4.1) it follows that

$$\ddot{q}_s = -M_s^{-1}(q_s)\left(N(q_s, \dot{q}_s) + K_s(q_s - \theta_s)\right) \tag{5.3}$$

$$q_s^{(3)} = -M_s^{-1}(q_s)\left(\dot{M}_s(q_s, \dot{q}_s)\ddot{q}_s + \dot{N}(q_s, \dot{q}_s, \ddot{q}_s) + K_s(\dot{q}_s - \dot{\theta}_s)\right) \tag{5.4}$$

Therefore (5.1) can be written as function of $q_s, \dot{q}_s, \theta_s, \dot{\theta}_s$, i.e.

$$\tau_s = K_s(\theta_s - q_s) + \beta(q_s, \dot{q}_s, \theta_s, \dot{\theta}_s) + \varphi(q_s)v(t) \tag{5.5}$$

with

$$\begin{aligned} \beta(q_s, \dot{q}_s, \theta_s, \dot{\theta}_s) &= B_{sd}\dot{\theta}_s + J_s K_s^{-1}\alpha(q_s, \dot{q}_s, \ddot{q}_s, q_s^{(3)}) \\ \varphi(q_s) &= J_s K_s^{-1} M_s(q_s) \end{aligned}$$

Applying the control law (5.1) to the system (4.1), (4.2) yields the linear decoupled closed loop system

$$q_s^{(4)} = v(t) \tag{5.6}$$

To ensure position coordination between the slave and the master robot an additional feedback for $v(t)$ is proposed as

$$v(t) = q_m^{(4)} - K_3 e^{(3)} - K_2 \ddot{e} - K_1 \dot{e} - K_0 e \tag{5.7}$$

with the coordination errors defined by

$$e = q_s - q_m \qquad \dot{e} = \dot{q}_s - \dot{q}_m \tag{5.8}$$

where q_m is the master robot link position, that has to be at least four times differentiable, i.e. $q_m \in C^4$, and $K_i \in \mathbb{R}^n$ are positive definite gain matrices.

Clearly, there exist general choices for the gain matrices K_i such that the closed loop system is stable, but for simplicity and without lost of generality, we assume that $K_i = k_i I_n$, $i = 0, 1, 2, 3$, with k_i positive scalars. Then, it is straightforward to conclude that the coordination error e is exponentially stable, if the scalars k_i, $i = 0, 1, 2, 3$ are chosen such that the polynomial $s^4 + k_3 s^3 + k_2 s^2 + k_1 s + k_0$ is Hurwitz.

5.2 Coordinating controller based on estimated variables

As stated in the introduction, it is assumed that only the link positions q_m, q_s are measured, therefore (5.1) can not be implemented. The feedback control $v(t)$ (5.7) can neither be implemented since it depends on higher order derivatives of q_s and q_m.

Let $\widehat{q}_s, \widehat{\dot{q}}_s, \widehat{\theta}_s, \widehat{\dot{\theta}}_s$ denote estimated values for the link and rotor variables $q_s, \dot{q}_s, \theta_s, \dot{\theta}_s$, and $v_e(t)$ denotes the implementation of $v(t)$ based on estimated coordination errors. Assuming that $\widehat{q}_s, \widehat{\dot{q}}_s, \widehat{\theta}_s, \widehat{\dot{\theta}}_s$ and $v_e(t)$ are available, (5.5) can be modified as

$$\tau_e = K_s(\widehat{\theta}_s - \widehat{q}_s) + \beta(q_s, \widehat{\dot{q}}_s, \widehat{\theta}_s, \widehat{\dot{\theta}}_s) + \varphi(q_s) v_e(t) \tag{5.9}$$

with $\beta(q_s, \widehat{\dot{q}}_s, \widehat{\theta}_s, \widehat{\dot{\theta}}_s)$ given by

$$\beta(q_s, \widehat{\dot{q}}_s, \widehat{\theta}_s, \widehat{\dot{\theta}}_s) = B_{sd} \widehat{\dot{\theta}}_s + J_s K_s^{-1} \alpha(q_s, \widehat{\dot{q}}_s, \widehat{\ddot{q}}_s, \widehat{q_s^{(3)}})$$

$$\alpha(q_s, \widehat{\dot{q}}_s, \widehat{\ddot{q}}_s, \widehat{q_s^{(3)}}) = 2\dot{M}_s(q_s, \widehat{\dot{q}}_s)\widehat{q_s^{(3)}} + (\ddot{M}_s(q_s, \widehat{\dot{q}}_s, \widehat{\ddot{q}}_s) + K_s)\widehat{\ddot{q}}_s + \ddot{N}(q_s, \widehat{\dot{q}}_s, \widehat{\ddot{q}}_s, \widehat{q_s^{(3)}})$$

where according to (5.3), (5.4), estimates for $\ddot{q}_s, q_s^{(3)}$ are given by

$$\widehat{\ddot{q}}_s = -M_s^{-1}(q_s)\left(N(q_s, \widehat{\dot{q}}_s) + K_s(\widehat{q}_s - \widehat{\theta}_s)\right) \tag{5.10}$$

$$\widehat{q_s^{(3)}} = -M_s^{-1}(q_s)\left(\dot{M}_s(q_s, \widehat{\dot{q}}_s)\widehat{\ddot{q}}_s + \dot{N}(q_s, \widehat{\dot{q}}_s, \widehat{\ddot{q}}_s) + K_s(\widehat{\dot{q}}_s - \widehat{\dot{\theta}}_s)\right). \tag{5.11}$$

5.3 An observer for the coordination errors

Based on the work of Berghuis and Nijmeijer (1994), we propose the modified controller $v(t)$ given by

$$v_e(t) = -K_3 \dot{w}_2 - K_2 \dot{w}_1 - K_1 \dot{\hat{e}} - K_0 \hat{e} \tag{5.12}$$

where $\hat{e}, \dot{\hat{e}}, \dot{w}_1, \dot{w}_2$ represent estimates for $e, \dot{e}, \ddot{e}, e^{(3)}$ respectively. They are obtained by the observer

$$\begin{aligned} \dot{\hat{e}} &= w_1 + \gamma_1 \tilde{e} \\ \dot{w}_1 &= w_2 + \gamma_2 \tilde{e} \\ \dot{w}_2 &= w_3 + \gamma_3 \tilde{e} \\ \dot{w}_3 &= \gamma_4 \tilde{e} \end{aligned} \tag{5.13}$$

with the estimation coordination errors defined by

$$\tilde{e} = e - \hat{e} \qquad \dot{\tilde{e}} = \dot{e} - \dot{\hat{e}} \tag{5.14}$$

and $\gamma_i \in \mathbb{R}^{n\times n}$, $i = 1, 2, 3, 4$ diagonal positive definite gain matrices.

5.4 An observer for the slave variables $(q_s, \dot{q}_s, \theta_s, \dot{\theta}_s)$

Based on the dynamic model (4.1), (4.2), we propose the full state nonlinear Luenberger observer

$$\begin{aligned} \dot{z}_1 &= z_2 + \mu_1 \tilde{q} \\ \dot{z}_2 &= -M_s^{-1}(q_s)\left(N(q_s, z_2) + K_s(z_1 - r_1)\right) + \mu_2 \tilde{q} \end{aligned} \tag{5.15}$$

$$\begin{aligned} \dot{r}_1 &= r_2 + \mu_3 \tilde{q} \\ \dot{r}_2 &= J_s^{-1}\left(\tau_e(q_s, z_1, z_2, r_1, r_2) - K_s(r_1 - z_1) - B_{sd} r_2\right) + \mu_4 \tilde{q} \end{aligned} \tag{5.16}$$

where z_1, z_2, r_1, r_2 correspond to the estimates of q_s, $\dot{q}_s$, θ_s, $\dot{\theta}_s$, i.e. $\widehat{q}_s, \widehat{\dot{q}}_s, \widehat{\theta}_s, \widehat{\dot{\theta}}_s$, and $\mu_1, \mu_2, \mu_3, \mu_4 \in \mathbb{R}^{n\times n}$ are diagonal positive definite gain matrices. The estimation joint position and velocity errors are defined by

$$\begin{aligned} \tilde{q} &= q_s - z_1 \\ \widetilde{\dot{q}} &= \dot{q}_s - z_2 \\ \widetilde{\theta} &= \theta_s - r_1 \\ \widetilde{\dot{\theta}} &= \dot{\theta}_s - r_2. \end{aligned} \tag{5.17}$$

Remark 5.1. Note that in a similar way to the observers given in Sections 3.2 and 3.3 (see Remark 3.1), in (5.15) and (5.16) the estimates for $\dot{q}_s, \dot{\theta}_s$ are given by z_2, r_2, not by $\dot{z}_2, \dot{r}_2$. These definitions introduce extra correcting terms in $\dot{\tilde{q}}, \dot{\widetilde{\theta}}$, as it follows from (5.15) and (5.16) that

$$\dot{\tilde{q}} = \widetilde{\dot{q}} - \mu_1 \tilde{q}, \qquad \dot{\widetilde{\theta}} = \widetilde{\dot{\theta}} - \mu_3 \tilde{q}$$

5.5 Stability analysis

The stability analysis uses a Lyapunov function, whose derivative can be bounded in terms of the closed loop errors. It can be proven that the bound is negative in an annulus around the origin. The stability analysis is greatly simplified by the fact that the coordination closed loop error dynamics resembles a cascaded system formed by a perturbed linear system and a nonlinear system, that is related to the estimation joint position and velocity error dynamics.

Closed loop error dynamics Consider the master and slave flexible joint robots, which are described by (4.1) and (4.2), the slave robot in closed loop with the coordinating controller τ_e (5.9) and the observers (5.13), (5.15), and (5.16). Define states $x_1, \ldots, x_8 \in \mathbb{R}^n$ as $x_1 = e$, $x_2 = \dot{e}$, $x_3 = \ddot{e}$, $x_4 = e^{(3)}$, $x_5 = \tilde{e}$, $x_6 = \dot{\tilde{e}}$, $x_7 = \ddot{\tilde{e}}$, $x_8 = \tilde{e}^{(3)}$, and $y \in \mathbb{R}^{4n}$ as $y_1 = \tilde{q}$, $y_2 = \dot{\tilde{q}}$, $y_3 = \tilde{\theta}$, $y_4 = \dot{\tilde{\theta}}$. Such that the state vectors x and y are given by $x = [\ x_1^T\ \ x_2^T\ \ x_3^T\ \ x_4^T\ \ x_5^T\ \ x_6^T\ \ x_7^T\ \ x_8^T\ ]^T$ and $y = [\ y_1^T\ \ y_2^T\ \ y_3^T\ \ y_4^T\ ]^T$. Then the closed loop error dynamics can be written as

$$\begin{aligned} \dot{x} &= Ax - BM_s^{-1}(q_s)\Phi(q_s, \dot{q}_m, \ddot{q}_m, \overset{(3)}{q_m}, x_2, x_3, x_4, y_1, y_2, y_3, y_4) + B\overset{(4)}{q_m} \\ \dot{y} &= f(q_s, \dot{q}_m, x, y) \end{aligned} \tag{5.18}$$

where the matrices $A \in \mathbb{R}^{8n \times 8n}$, $B \in \mathbb{R}^{8n \times n}$ are given by

$$A = \begin{bmatrix} 0 & I_n & 0 & 0 & 0 & 0 & 0 & 0 \\ 0 & 0 & I_n & 0 & 0 & 0 & 0 & 0 \\ 0 & 0 & 0 & I_n & 0 & 0 & 0 & 0 \\ -K_0 & -K_1 & -K_2 & -K_3 & K_0 & \pi_1 & \pi_2 & K_3 \\ 0 & 0 & 0 & 0 & 0 & I_n & 0 & 0 \\ 0 & 0 & 0 & 0 & 0 & 0 & I_n & 0 \\ 0 & 0 & 0 & 0 & 0 & 0 & 0 & I_n \\ -K_0 & -K_1 & -K_2 & -K_3 & K_0 - \gamma_4 & \pi_1 - \gamma_3 & \pi_2 - \gamma_2 & K_3 - \gamma_1 \end{bmatrix}, \quad B = \begin{bmatrix} 0 \\ 0 \\ 0 \\ I_n \\ 0 \\ 0 \\ 0 \\ I_n \end{bmatrix} \tag{5.19}$$

$$\pi_1 = K_3\gamma_2 + K_2\gamma_1 + K_1, \qquad \pi_2 = K_3\gamma_1 + K_2$$

with $0, I_n \in \mathbb{R}^{n \times n}$ the zero and identity matrices.

The nonlinear vector function $f(q_s, \dot{q}_m, x, y)$ is given by

$$f(q_s, \dot{q}_m, x, y) = \begin{bmatrix} y_2 \\ -M_s^{-1}(q_s)\left(2C_s(q_s, x_2 + \dot{q}_m) - C_s(q_s, y_2 + \mu_1 y_1)\right)(y_2 + \mu_1 y_1) + \\ -M_s^{-1}(q_s)K_s(y_1 - y_3) - \mu_1 y_2 - \mu_2 y_1 \\ y_4 \\ -J_s^{-1}K_s(y_3 - y_1) - J_s^{-1}B_{sd}(y_4 + \mu_3 y_1) - \mu_3 y_2 - \mu_4 y_1 \end{bmatrix} \tag{5.20}$$

Remark 5.2. The term $\Phi(q_s, \dot{q}_m, \ddot{q}_m, \overset{(3)}{q_m}, x_2, x_3, x_4, y_1, y_2, y_3, y_4)$ represents the deviation from the nominal linear system due to considering estimates for positions and velocities. The deviation term satisfies $\Phi(q_s, \dot{q}_m, \ddot{q}_m, \overset{(3)}{q_m}, 0, 0, 0, 0, 0, 0, 0) = 0$, so, it can be considered as a vanishing disturbance that affects (5.18). Also notice that (5.18) presents a linear structure in the first set of equations, which greatly simplifies the stability analysis.

In order to derive bounds on the derivative of the Lyapunov function the following assumption is required.

Assumption 5.3. The signals $\dot{q}_m, \ddot{q}_m, q_m^{(3)}, q_m^{(4)}$ are bounded for all $t \in [t_0, \infty)$, i.e.

$$V_M = \sup_t \|\dot{q}_m\|, \qquad A_M = \sup_t \|\ddot{q}_m\|,$$

$$D_M = \sup_t \left\|q_m^{(3)}\right\|, \qquad E_M = \sup_t \left\|q_m^{(4)}\right\|. \tag{5.21}$$

For simplicity in the stability analysis and without loss of generality, it is assumed that.

Assumption 5.4. All gains in the feedback controller (5.9), (5.12), and the observer (5.13) are of the form $K = kI$, where k is a positive scalar and I denotes the identity matrix.

Using the Assumptions 5.3 and 5.4 and by considering the closed loop error dynamics (5.18) the following theorem can be formulated.

Theorem 5.5. *Consider that the Assumptions 5.3 and 5.4 are satisfied. If the gain matrices K_i and γ_j, $i = 0, 1, 2, 3$, $j = 1, 2, 3, 4$ are chosen such that A, given by (5.19), is Hurwitz, and additionally the minimum eigenvalues of the gains μ_l, $l = 1, 2, 3, 4$ satisfy certain conditions determined from the stability analysis. Then the estimation joint position and velocity errors $\tilde{q}, \dot{\tilde{q}}, \widetilde{\theta}, \dot{\widetilde{\theta}}$, the coordination errors $e, \dot{e}, \ddot{e}, e^{(3)}$ and the estimation coordination errors $\tilde{e}, \dot{\tilde{e}}, \ddot{\tilde{e}}, \tilde{e}^{(3)}$,, in the closed loop error dynamics (5.18), are locally uniformly ultimately bounded.*

Proof: The proof is long and requires large algebraic computations. Nevertheless, the Lyapunov function and the boundedness of its derivative along the error dynamics are quite simplified because of the structure of the closed loop error dynamics (5.18). As a sketch of the proof the Lyapunov function and its time derivative are presented.

Lyapunov function A Lyapunov function $V(q_s, x, y)$, for the system defined by (5.18) is given by

$$V(q_s, x, y) = x^T P_x x + \frac{1}{2} y^T P_y(q_s, y) y \tag{5.22}$$

where the positive definite symmetric matrix P_x is the solution of the Lyapunov equation $P_x A + A^T P_x = -Q_x$, for any given symmetric positive definite matrix Q_x, and $P_y(q_s, y)$ is given by

$$P_y(q_s, y) = \begin{bmatrix} \begin{bmatrix} K_s + 2\lambda_0 \mu_1 + \beta_1 I_n & 2\lambda_0 I_n \\ 2\lambda_0 I_n & M_s(q_s) \end{bmatrix} & 0 \\ 0 & \begin{bmatrix} K_s + \mu_4 + \beta_2 I_n & 2\eta(y_3) I_n \\ 2\eta(y_3) I_n & J_s \end{bmatrix} \end{bmatrix}.$$

with β_1, β_2 and $\eta(y_3)$ defined by

$$\eta(y_3) := \frac{\eta_0}{1 + \|y_3\|}$$

$$\begin{aligned}\beta_1 &= 2C_M V_M (2M_{sm}^{-1}\lambda_0 + \mu_{1M}) - M_{sM}\mu_{2M} \\ \beta_2 &= 2\eta_0 J_{sm}^{-1} B_{sdM} - \mu_{4M}\end{aligned}$$

and the scalars $\lambda_0, \eta_0 > 0$ are constants to be determined.

Existence of P_x is guaranteed if A, which is given by (5.19), is Hurwitz. By Assumption 5.4 a sufficient condition for A being Hurwitz is that the scalar polynomial $\lambda(s)$, given by

$$\begin{aligned}\lambda(s) &= s^8 + \gamma_1 s^7 + \gamma_2 s^6 + \gamma_3 s^5 + (\gamma_4 + k_1\gamma_1 + k_2\gamma_2 + k_3\gamma_3)s^4 + \\ &+ (k_0\gamma_1 + k_1\gamma_2 + k_2\gamma_3 + k_3\gamma_4)s^3 + (k_0\gamma_2 + k_1\gamma_3 + k_2\gamma_4)s^2 + (k_0\gamma_3 + k_1\gamma_4)s + k_0\gamma_4,\end{aligned} \tag{5.23}$$

is Hurwitz. Therefore by choosing the gains K_i and γ_j, $i = 0, 1, 2, 3$, $j = 1, 2, 3, 4$ it can be ensured that the matrix A is Hurwitz. And thus, there exists a positive symmetric matrix P_x, which satisfies $P_x A + A^T P_x = -Q_x$, for any given symmetric positive definite matrix Q_x. Moreover P_x is unique, see (Khalil, 1996).

There exist conditions on the observer gains μ_j, $j = 1, 2, 3, 4$ such that $P_y(q_s, y)$ is positive definite. If A is Hurwitz and $P_y(q_s, y)$ is positive definite, then the Lyapunov function $V(q_s, x, y)$ is positive, moreover it satisfies that

$$\frac{1}{2}P_m \left\|\xi\right\|^2 \leq V(q_s, x, y) \leq \frac{1}{2}P_M \left\|\xi\right\|^2$$

with P_m, P_M positive constants and the vector ξ defined as $\xi^T = \left[\begin{array}{cc} x^T & y^T \end{array}\right]$.

After a long algebraic computation and considering the Assumptions 5.3 and 5.4, the derivative of $V(q_s, x, y)$ along the coordination error dynamics (5.18) can be upper-bounded as

$$\dot{V} \leq \left\|\xi_N\right\| \left(r_0 - r_1 \left\|\xi_N\right\| + r_2 \left\|\xi_N\right\|^2 + r_3 \left\|\xi_N\right\|^3 + r_4 \left\|\xi_N\right\|^4\right)$$

where $\zeta_N = \left[\begin{array}{cccccc} \|x_1\| & \cdots & \|x_8\| & \|y_1\| & \cdots & \|y_4\| \end{array}\right]^T$. The coefficients r_0, r_2, r_3, r_4 are positive and r_1 depends on the controller and observer gains, such that there exist conditions to ensure that $\dot{V}$ is negative in an annulus around the origin. Then the convergence properties of the coordinating controller for flexible joint robots follows in a similar way to the rigid robot case presented in (Rodriguez-Angeles and Nijmeijer, 2001), see also (Berghuis and Nijmeijer, 1994).

6 Experimental results

The proposed coordinating controller for rigid joint robots has been implemented on a one degree of freedom (DOF) master and slave system. The master and slave system correspond to the medium link of two identical transposer robots fabricated by the Centre for Manufacturing Technology (CFT) Phillips Laboratory. The robots are equipped with DC brushless servomotors and encoders attached to the shaft of the motors with a resolution of 2000 PPR. Although the shaft of the motors and the corresponding links are connected by means of belts, the pair servomotor-link prove to be stiff enough to be considered as a rigid joint.

The transposer robots have 4 DOF, but for the sake of brevity along this experiment only the middle link of the upper arm was considered. The other 3 DOF of the robots were locked such that only the middle links were able to move. Figure 2 shows the medium link of the considered transposer robots, this link can rotate around its axis mounted in the based of the transposer robot.

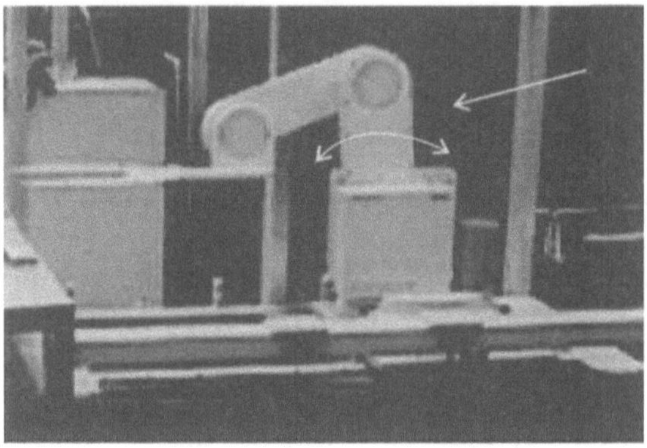

Figure 2. The transposer robot and its middle link.

In the position in which the 3 DOF of the transposer robot were locked, the model of the middle link can be described by

$$J\ddot{q}_s - gm_g\cos(-q_s - \alpha) + \tau_f(\dot{q}_s) = \tau \tag{6.1}$$

with $g = 9.81\ m/s^2$ the acceleration due to the gravity. The torque $\tau_f(\dot{q}_s)$ models the friction phenomena and according to (Hensen et al., 2000), it can be described by

$$\tau_f(\dot{q}_s) = B_v\dot{q}_s + B_{f1}\left(1 - \frac{2}{1+e^{2w_1\dot{q}_s}}\right) + B_{f2}\left(1 - \frac{2}{1+e^{2w_2\dot{q}_s}}\right) \tag{6.2}$$

where B_v represents the viscous friction coefficient and the other terms model the Coulomb and Stribeck friction effects.

The physical parameters of the dynamic model (6.1, 6.2) have been estimated using an extended Kalman Filter in a similar way to the work presented in (Kostic et al., 2001). The estimated physical parameters are

$J\ [Kg \cdot m^2]$	$m_g\ [Kg \cdot m]$	$B_v\ [Kg \cdot m^2/s]$	$B_{f1}\ [Kg \cdot m^2/s^2]$
1.7575	3.2934	3.5384	6.9590
$B_{f2}\ [Kg \cdot m^2/s^2]$	w_1	w_2	$\alpha\ [rad]$
6.6233	46.1224	9.9428	2.5606

As mentioned the friction phenomena is precompensated, therefore τ is given by

$$\tau = \tau_f(\widehat{\dot{q}}_s) + \tau_c \tag{6.3}$$

with τ_c the coordinating controller given by (3.1). The friction compensating term $\tau_f(\widehat{\dot{q}}_s)$ is function of the estimated velocity $\widehat{\dot{q}}_s$, that is obtained by the observer (3.5).

It is supposed that the master robot perfectly follows the desired trajectory

$$q_m(t) = 1.9805 + 0.15\sin(2\pi\varpi t) + 0.05\sin(4\pi\varpi t) - 0.03\sin(6\pi\varpi t) + 0.02(8\pi\varpi t)$$

with the frequency $\varpi = 0.4$.

From the tuning gain procedure given on Subsection 3.5, it follows that for the weighting factors $\lambda_0 = 5\times10^{-7}$, $\varepsilon_0 = 0.0003$, $\mu_0 = 0.0001$, $\gamma_0 = 0.02$, the coordinating system is stable and the errors are bounded if the scalar gains satisfy $K_p > 0, K_d > 0, L_{p1} > 28.91$, and $L_{p2} > 0$. After a series of experiments in order to decrease the coordination position error e, the gains on the controller were set as $K_p = 1000, K_d = 80, L_{p1} = 100$ and $L_{p2} = 1900$.

For the sake of comparison a PD controller was also implemented on the slave system. The gains of the PD controller were set equal to the proportional and derivative gains in the coordinating controller, i.e. $K_p = 1000, K_d = 80$. The velocity in the friction compensation term $\tau_f(\widehat{\dot{q}}_s)$ was obtained by using the numerical differentiator block supplied in the standard Simulink distribution for Matlab version 6.0 Release 12.

The controllers were implemented using the real time workshop platform supplied in Simulink for Matlab. The real time controllers were implemented in a Laptop with a processor Pentium II and 192 MB in RAM. The sampling frequency was 500 Hz. The initial position of the link was $q_s(0) = 1.98$ $[rad]$, the initial conditions in the observers (3.3), (3.5) were chosen as $\hat{e} = 0.1, \widehat{\dot{e}} = 0$, $\hat{q}_s = 1.98, \widehat{\dot{q}}_s = 0$.

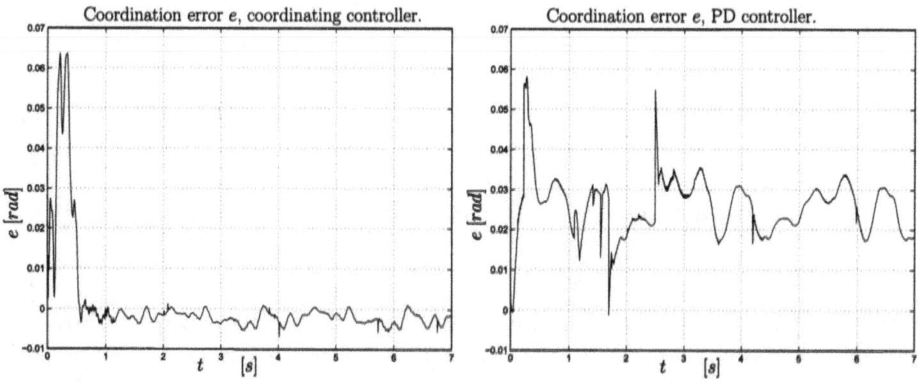

Figure 3. Coordination position error e for the coordinating controller (3.1) and PD.

Figure 3 shows the coordination errors for both the proposed coordinating controller (3.1) and the PD controller. The master and slave joint positions are shown in Figure 4. From Figures 3 and 4 it can be concluded that the proposed controller not only reduces the amplitude of the coordinating error e, but also eliminates the offset in the coordination error obtained with the PD controller. Figure 5 presents the control torques

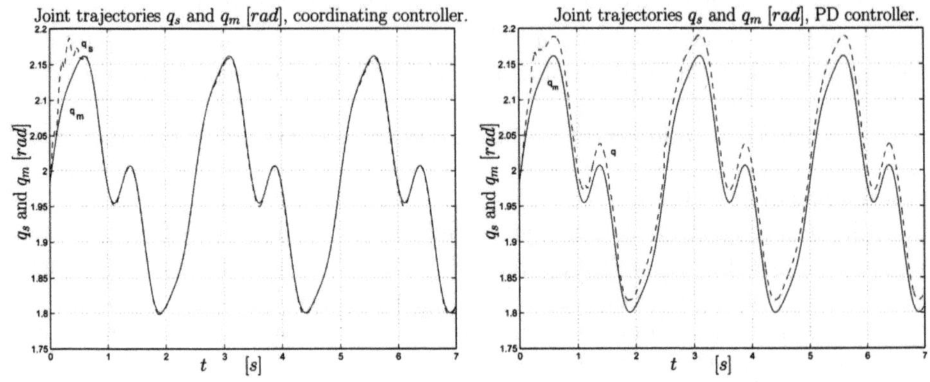

Figure 4. Joint positions q_s and q_m for the coordinating controller (3.1) and PD.

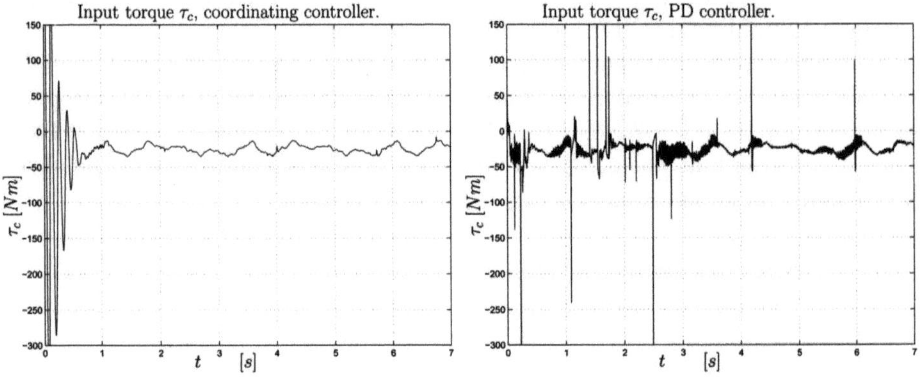

Figure 5. Input torques τ_c for the coordinating controller (3.1) and PD.

generated by the proposed coordinating controller and the PD controller. From Figure 5 it can be concluded that the proposed controller is more robust against noise in the position measurements and therefore it results in a less noisy torque in the servomotors than the PD controller. Figure 6 presents the estimation coordination errors and slave position estimation errors obtained with the observers (3.3), (3.5)

7 Remarks and discussion

Remark 7.1. The conditions mentioned in Theorems 3.5 and 5.5 are very conservative. However, even without knowledge of the required physical bounds, the closed loop systems can be made uniformly ultimately bounded. This may be achieved by selecting the control and observer gains large enough. However, such high gain implementations

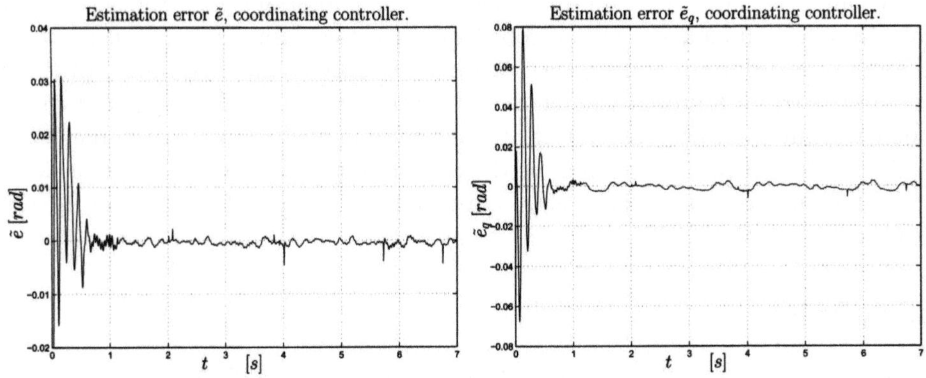

Figure 6. Estimation errors $\tilde{e}$ and $\tilde{e}_q$, observers (3.3) and (3.5).

are not always desirable in practical circumstances, since these may amplify unavoidable noise.

Remark 7.2. The conditions on Theorems 3.5 and 5.5 imply a minimum value of the eigenvalues of certain gain matrices (mainly in the observers), which resembles -with out being- high gain observers. As mentioned the proposed observers are model-based and allow a physical interpretation of the observer structure and the observer gains.

Remark 7.3. The coordinating controller for rigid joint robots (3.1) is similar to computer torque controllers, but it is based on measurements of both the master and the slave robots. This definition of the controller does not cancel the dynamics of the slave robot, but dominates it and forces a dynamics similar to that of the master robot (what is called a desired compensation controller). In this sense the slave robot coordinates to the master one in a more natural way, since the dynamics of the slave robot is not cancelled but emulates the master robot dynamics.

On the other hand the coordinating controller for flexible joint robots (5.9) is designed to cancel the slave dynamics giving rise to a linear systems. Because of this there is not enough information to reconstruct the fourth order derivative of the master robot link position q_m. Therefore the implementation of the feedback controller $v(t)$ (5.7) reduces to a function of estimates of the coordinating errors (see $v_e(t)$ given by 5.12). The lack of an estimate for the fourth order derivative of q_m is the main limitation to reduced the ultimate bound of the coordination errors e and $\dot{e}$.

8 Conclusions and further extensions

Position coordinating controllers for multi-robot systems with rigid and flexible joints, working in master-slave schemes have been presented. The proposed controllers are independent of the master robot dynamics and its physical parameters, and only require position measurements. The coordinating controller for rigid joint robots yields semi-global ultimate boundedness of the coordination closed loop errors, in case of flexible joint robots the resultant convergence prove to be only local. The bound of the errors can be decreased by a proper tuning on the controllers gains. The proposed controllers are robust again parametric uncertainties and noise in the position measurements.

A further extension of the proposed coordinating controllers is to consider other schemes different to master-slave, e.g. cooperative schemes. This extension is reported in (Rodriguez-Angeles, 2002).

Bibliography

B. Armstrong-Hélouvry. Stick-slip and control in low-speed motion. *IEEE Transactions on Automatic Control*, 38:1483–1496, 1993.

H. Berghuis and H. Nijmeijer. Robust control of robots via linear estimated state feedback. *IEEE Transactions on Automatic Control*, 39:2159–2162, 1994.

I.I. Blekhman. *Synchronization in science and technology.* ASME Press Translations, New York, 1988.

W.J. Book. Recursive Lagrangian dynamics of flexible manipulator arms. *International Journal of Robotics Research*, 3:87–101, 1984.

M. Brunt. *Coordination of redundant systems.* PhD thesis, Technical University Delft, The Netherlands, 1998.

A. De Luca and L. Lanari. Robots with elastic joints are linearizable via dynamic feedback. In *Proceedings of the 34th IEEE Conference on Decision and Control*, pages 3895–3897, 1995.

A. De Luca and P. Lucibello. A general algorithm for dynamic feedback linearization of robots with elastic joints. In *Proceedings of the IEEE Conference on Robotics and Automation*, pages 504–510, 1998.

A. De Luca and P. Tomei. *Elastic joints.* Theory of Robot Control. Springer, Berlin, 1996.

Chan T.F. Dubey, R.V. and S.E. Everett. Variable damping impedance control of a bilateral telerobotics system. *IEEE Control and Systems*, 17:37–45, 1997.

R.C. Goertz. Manipulator systems development at ANL. In *Proceedings of the 12th Conference on Remote Systems Technology*, pages 117–136. American Nuclear Society, 1954.

M.C. Good, L.M. Sweet, and K.L. Strobel. Dynamic models for control systems design of integrated robot and drive systems. *ASME Journal of Dynamic Systems, Measurement, and Control*, 107:53–59, 1985.

G.S. Guthart and J.K. Jr. Salisbury. The intuitive^{TM} telesurgery system: overview and application. In *Proceedings of the IEEE International Conference on Robotics and Automation*, pages 618–621, 2000.

R.H.A. Hensen, G.Z. Angelis, M.J.G. v.d. Molengraft, A.G. de Jager, and J.J. Kok. Grey-box modeling of friction: An experimental case-study. *European Journal of Control*, 6:258–267, 2000.

J.W. Hills and J.F. Jensen. Telepresence technology in medicine: principles and applications. In *Proceedings of the IEEE*, volume 86, pages 569–580, 1998.

H.K. Khalil. *Nonlinear Systems.* Prentice Hall, Upper Saddle River, 1996.

T. Komatsu and T. Akabane. Control of a space flexible master-slave manipulator based on parallel compliance models. In *Proceedings of the IEEE International Conference on Robotics and Automation*, pages 1932–1937, 1998.

D. Kostic, R. Hensen, B. de Jager, and M. Steinbuch. Modeling and identification of an RRR-robot. In *Proceedings of the IEEE Conference on Decision and Control*, pages 1144–1149, 2001.

K. Kosuge, J. Ishikawa, K. Furuta, and M. Sakai. Control of single-master slave manipulator systems using VIM. In *Proceedings of the IEEE International Conference on Robotics and Automation*, volume 2, pages 1172–1177, 1990.

T. Laopoulos and C. Papageorgiou. Microcontroller-based measurements of angular position, velocity and acceleration. In *Proceedings of the IEEE Instrumentation and Measurement Technology Conference*, pages 73–77, 1996.

H.K. Lee and M.J. Chung. Adaptive controller of a master-slave system for transparent teleoperation. *Journal of Robotic Systems*, 15:465–475, 1998.

Y.H. Liu, Y. Xu, and M. Bergerman. Cooperation control of multiple manipulators with passive joints. *IEEE Transactions on Robotics and Automation*, 15:258–267, 1999.

J.N. Lygouras, K.A. Lalakos, and P.G. Tsalides. High-performance position detection and velocity adaptive measurements for closed-loop position control. *IEEE Transactions on Instrumentation and Measurement*, 47:978–985, 1998.

H. Olsson and K.J. Åström. Friction generated limit cycles. *IEEE Transactions on Control Systems Technology*, 9:629–636, 2001.

S.J. Ovaska and S. Vliviita. Angular acceleration measurement: A review. *IEEE Transactions on Instrumentation and Measurement*, 47:1211–1217, 1998.

A. Rodriguez-Angeles. *Synchronization of mechanical systems.* PhD thesis, Eindhoven University of Technology, The Netherlands, 2002.

A. Rodriguez-Angeles and H. Nijmeijer. Coordination of two robot manipulators based on position measurements only. *International Journal of Control*, 74:1311–1323, 2001.

A. Rodriguez-Angeles, H. Nijmeijer, and H.A. van Essen. Coordination of robot manipulators. In *Proceedings of the 8th Mechatronics Forum International Conference (Mechatronics 2002)*, 2002.

M.W. Spong. Modelling and control of elastic joint robots. *ASME Journal of Dynamic Systems, Measurement, and Control*, 109:310–319, 1987.

M.W. Spong and Vidyasagar M. *Robot Dynamics and Control.* Wiley, New York, 1989.

P. Tomei. A simple PD controller for robots with elastic joints. *IEEE Transactions on Automatic Control*, 36:1208–1213, 1991.

Structural Control for Dynamic Hazard Mitigation

Fabio Casciati

Department of Structural Mechanics, University of Pavia, Italy
fabio@dipmec.unipv.it

Abstract. After a review of the structural control implementations expected in the area of Civil and Architectural Engineering, these notes approach some associated research fields. Special care is devoted to the interdisciplinary character of the achievable progress and to the role played by laboratory testing in the scientific and technical development of structural control and structural control applications.

1 Introduction

In civil engineering, active systems for structural control are achieving an increasing importance in the fields of vibration attenuation and wind and earthquake hazard mitigation. But one deals with rather massive structures which are driven by mechanical devices with: i) the application of large forces; ii) fast reaction times and iii) relatively low accuracy.

As a consequence, structural control applications in civil engineering are rather different from the implementations of control theory which have been developed in electrical engineering and applied mechanics. Also the required tools cannot be compared with the ones adopted for the control of space structures.

The developments achieved in the areas listed above provide a help but do not solve the problems of structural control in civil engineering (Housner & Masri, 1994; Soong, 1991; Housner te al.,1997; Song & Dargush, 1997).

This growing interest, and the need for a co-ordinated national and international research activity, led to the establishment of the US and Japan Panels on structural control research, under the auspices of the US National Science Foundation and of the Japan Science Council, respectively. In Europe there are not super-national responsible institutions. This suggested the European Association for the Control of Structures (*E*ACS) to pursue a sort of research co-ordination which led to two main actions: CONVIB (2001), within the European Science Foundation and SAMCO (2001) within the European Union program GROWTH.

*E*ACS was founded in July 1993 and is linked to the International Association for Structural Control (IASC) which organises every four years the World Conference on Structural Control: Pasadena, USA, (Housner et al., 1994), Kyoto, Japan, (Kobori et al., 1998) and Como, Italy, (Casciati, 2002).

*E*ACS itself organised the European Conferences on Structural Control in Barcelona, Spain (Baratta & Rodellar, 1996) and Paris, France (Bourquin et al., 2000).The next meeting will be in Vienna, Austria, in July 2004.

Reference should also be done to the Journal of Structural Control, the *E*ACS bulletin. The aim of this journal is to provide a forum in which survey articles and original research in the field of structural control can be communicated rapidly, towards a cross-fertilisation of information and ideas. An important aim is to achieve a balance between academic material and practical applications. The scope of the journal encompasses all aspects of structural control theory and technology. Special attention is devoted to civil and infrastructure engineering applications. One meets the control of linear and non-linear systems, the control of uncertain systems, actuator design, dynamic stability, interaction of primary and secondary systems, artificial intelligence tools, measurement techniques, monitoring, intelligent systems, and smart materials.

2 Structural Control Implementations

Innovative architectural solutions or increased standards of availability and comfort push toward actively controlled systems. Nevertheless, the associated relatively high costs prevent from a fast spreading of applications, which only address the construction stage or the mitigation of wind induced effects on flexible structures. Examples of bridges and buildings are provided to show the innovation in Europe and in Asia, throughout the nineties.

However an actual promotion of innovative design solutions requires that semi-active, hybrid and active structural control be preliminary conceived and tested in a laboratory environment, as discussed in Section 3.

2.1 Background

A block diagram of active/semi-active structural control is sketched in Figure 1. In the control literature special attention is focused on the coupling of control-structure interaction (CSI) with actuator/sensor dynamics.

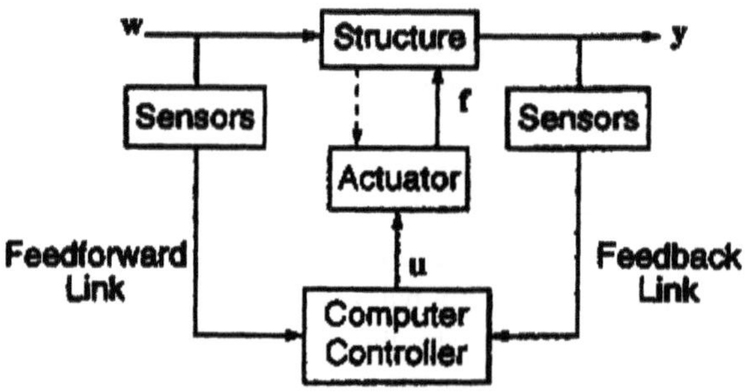

Figure 1 – Block diagram of active/semi-active structural control.

In a passive system, sensors and controller are absent and the actuator is actually a device coupled with the structural system in order to isolate it or to help it in dissipating the incoming energy.

2.2 Bridges

Despite the design of many bridges belong to the class of the never ended tales, as for the bridge across the Messina Strait, in Italy, the last decade has seen many outstanding bridges studied, designed and built or still in construction. The latter group consists of three main links which are supposed to double the existing capability by completely new parallel connections. All of them are located in Southern Europe countries and cover the areas of Lisbon in Portugal and Istanbul in Turkey, while a third design concerns a bridge in Greece, the Rion-Antirion bridge. All these bridges are located in seismic regions and for all of them the designers planned the adoption of passive devices. Since they are still in construction, it is likely that they will host, at least in the construction stage, some semi-active or active control solutions (Casciati & Faravelli, 1999).

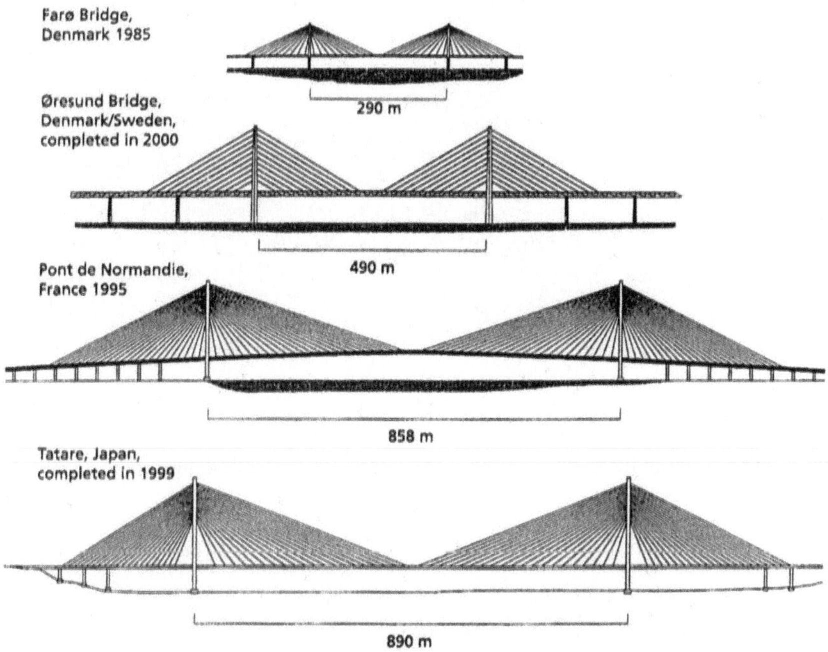

Figure 2 – A comparison of cable-stayed bridges showing the span of the Normandie bridge and the Oresund bridge.

Passive control solutions against wind loading were adopted during the construction of the Normandie bridge, in France (Figures 2 and 3), which was recently completed. Only passive appendages (flaps) were designed for The Great Belt Bridge in Denmark, finished in June 1998 (Figure 4). It is the second-longest suspension bridge in the world (the longest suspension bridge is the Akashy Kaikyo bridge in Japan). No active vibration control was used, but researches are in progress towards the adoption of actively controlled flaps, i.e. towards a semi-active control solution (Hansen, 1998; Hansen and Thoft-Christensen, 1998). The Oresund bridge is a further component of the link between Denmark and Sweden and is briefly illustrated in Figures 2: it also carries the railway.

Similarly to the Normandie bridge, during the construction of the tower (height = 119m) of the Rainbow suspension bridge in Tokyo (Figure 5), an actively controlled mass was coupled with a tuned mass damper, i.e. a hybrid solution was adopted (Spencer & Sain, 1997). Figure 6 is devoted to the Oklahoma State bridge whose retrofitting led to one of the first exploitation of the concept of semi-active control.

a) b)

Figure 3 – The Normandie bridge in France (a), passively controlled during the construction (b).

Figure 4 – The Great Belt suspended bridge, in Denmark, passively controlled by flaps.

Figure 5 – The tower of the Rainbow bridge

Figure 6 – The Oklahoma State bridge

Figure 7– The large scale test on a suspended bridge in progress at the Joint Research Centre in Ispra.

Actively controlled solutions are pursued in large scale tests at the Joint Research Centre in Isptra as illustrated in Figure 7 (Magonette et al., 1999).

2.3 Buildings

Moving from bridges to buildings, in addition to several base isolated constructions, (Martelli & Forni, 1996), two are the towers which are worth being mentioned in Europe. The tower in Frankfurt, Germany (Figure 8), was simply equipped by an active mass driver (AMD) on the top to mitigate the wind induced vibration. The same purpose is pursued in London, UK (Figure 9), where the top equipment of a skyscraper under design is just a commercial AMD produced in Japan.

It is worth noting the architectural character of the building in Figure 8: its skeleton is made of three towers. In the presence of similar architectural solution, in Japan there are many implementations of towers of different heights (i.e. different natural frequencies) linked by hydraulic dampers (Figure 10). But Japan is also the country which worked as pioneer in the area of actively controlled structures (Figure 11) and has the larger number of implementations as shown in Table 1.

Figure 8 - Two different views of the Frankfurt tower actively controlled by an active mass driver against wind loading. (Commerzbank: 260 m tall; 62 stores)

Figure 9– A sketch of the London tower to be equipped by an active mass driver. (London Millennium Tower)

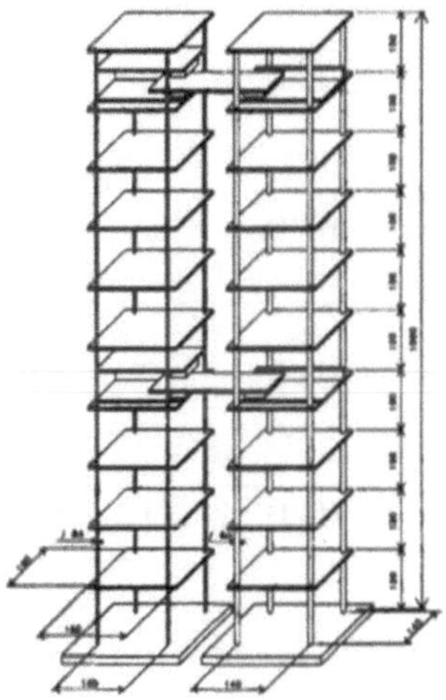

Figure 10 – Towers of different frequencies with hydraulic damper links.

Figure 11 – The building which represents the pioneer of the actively controlled structural systems: Kyobashi Saiwa.

Table 1. Some actively controlled buildings/towers (re-elaborated from Spencer & Sain, 1997)

Full Scale Structure	**Location**	**Year Completed**	**Scale of Building**	**Control System Employed**	**AMD/HMD**		**Actuation mechanism**
					No.	**Mass (Tons)**	
Kyobashi Seiwa	Tokyo, Japan	1989	33m, 400 ton, 11 stories	AMD	2	5	Hydraulic
Kajima Research. Institute Katri N° 21 Building	Tokyo, Japan	1990	12m, 400 ton, 3 stories	Active Variable Stiffness System(6 devices)			Hydraulic
Sendagaya INTES	Tokyo, Japan	1992	58m, 3280ton, 11 stories	AMD	2	72	Hydraulic
Applause Tower	Osaka, Japan	1992	161m, 13943 ton, 34 stories	HMD	1	480	Hydraulic
Kansai Int. Airport Control Tower	Osaka, Japan	1992	86m, 2570 ton, 50 stories	HMD	2	10	Servo motor
Osaka Resort City 2000	Osaka, Japan	1992	200m, 56980 ton, 50 stories	HMD	2	200	Servo motor
Yokohama Land Mark Tower	Yokohama, Kanagawa, Japan	1993	296m, 260610 ton, 70 stories	HMD	2	340	Servo motor
Long Term Credit Bank	Tokyo, Japan	1993	129m, 40000 ton, 21 stories	HMD	1	195	Hydraulic
Ando Nishikicho	Tokyo, Japan	1993	54m, 2600 ton, 14 stories	HMD (DUOX)	1	22	Servo motor
Hotel Nikko Kanazawa	Kanazawa, Ishikawa, Japan	1994	131m, 27000 ton, 29 stories	HMD	2	100	Hydraulic
Hiroshima Riehga Royal Hotel	Hiroshima, Japan	1994	150m, 3000 ton, 35 stories	HMD	1	80	Servo motor

Table 1 – continued

Full Scale Structure	Location	Year Com-pleted	Scale of Building	Control System Employed	AMD/HMD		Actuation mechanism
					No.	Mass (Tons)	
Shinjuku Park Tower	Tokyo, Japan	1994	227m, 130000 ton, 52 stories	HMD	3	330	Servo motor
MHI Yokohama Building	Yokohama, Kanagawa, Japan	1994	152m, 61800 ton, 34 stories	HMD	1	60	Servo motor
Hamamatsu ACT Tower	Hamamatsu, Shizuoka, Japan	1994	212m, 107500 ton, 46 stories	HMD	2	180	Servo motor
Riverside Sumida	Tokyo, Japan	1994	134m, 52000 ton, 33 stories	AMD	2	30	Servo motor
Hikarigaoka J-City	Tokyo, Japan	1994	110m, 29300 ton, 26 stories	HMD	2	44	Servomotor
Miyazaki Phoenix Hotel Ocean 45	Miyazaki, Japan	1994	154m, 83650 ton, 43 stories	HMD	2	240	Servo motor
Osaka WTC Building	Osaka, Japan	1994	252m, 80000 ton, 52 stories	HMD	2	100	Servo motor
Dowa Kasai Phoenix Tower	Osaka, Japan	1995	145m, 26000 ton, 28 stories	HMD (DUOX)	2	84	Servo motor
Rinku Gate Tower North Building	Osaka, Japan	1995	255m, 75000 ton, 56 stories	HMD	2	160	Servo motor
Hirobe Miyake Building	Tokyo, Japan	1995	31m, 273 ton, 9 stories	HMD	1	2.1	Servo motor
Plaza Ichihara	Chiba, Japan	1995	61 m, 6760 ton, 12 stories	HMD	2	14	Servo motor
TC tower	Kao Hsung, Taiwan	1996	85 stories	HMD	2	350	Servo motor
Nanjing Tower	Najing, China	1997/98	310m	AMD	1	60	Hydraulic

AMD= Active Mass Damper; HMD= Hybrid Mass Damper

2.4 Other classes of structures

Significant realisations of active structural control were pursued in the field of television towers and large construction crane. The more famous implementation in the former area was realised in Nanjing China (Figure 12). It exploits a hydraulic active mass damper. On the other side, Figure 13 shows the use of jet actuators in the active control of the crane used during the Rainbow bridge construction in Japan. Investments, but no realisation till now, are characterising two special classes of structural systems: offshore platforms (Nielsen et al. 1999) and monumental buildings (Anton and Casciati, 1998; Casciati et al. 1998). The advantages they offer are mainly two: (i) the intrinsic value of the structural system is able to justify the high costs the innovation requires; (ii) they are ruled by specific codes and the resulting design may disagree with the present Eurocodes.

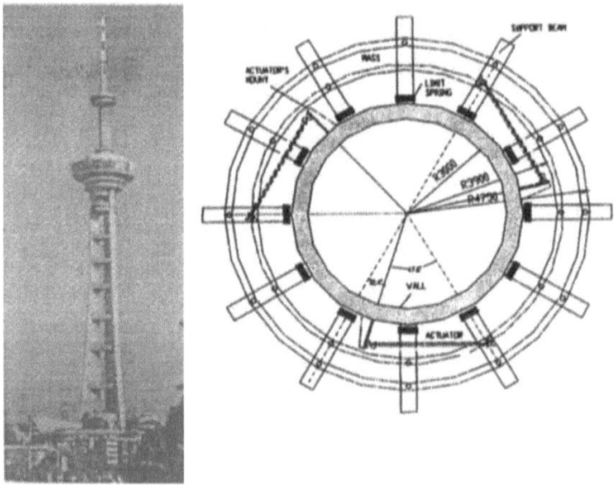

Figure 12 – The Nanjing television tower (from Lecture Notes (Soong, 1998)).

Figure 13 – Actively controlled crane during the Rainbow bridge construction (from Lecture Notes (Spencer, 1997)).

3 Testing the Innovation

While the most fascinating implementations are all located in Asia, the deepest experimental investigations in laboratory were planned toward the exploitation of active control in seismic engineering: they were conducted in the United States. The early experimental studies are:

- SDOF active tendon controlled structure (Chung, Reinhorn & Soong, 1988) with governing equations and reference system in Figure 14:

$$m\ddot{z} + c\dot{z} + kz = -(4k_c \cos\theta)\, u - m\ddot{z}_g$$
$$u = -k_1 z - k_2 \dot{z}$$

  with an evident meaning of the symbols: m = mass; c = damping; k = stiffness; z = cinematic variable; u = control force, function of the state variables; $\ddot{z}_g$ = ground acceleration; θ is related to the tendon geometry and k_c is the associated stiffness. In the following, capital letters will denote the same quantities where a matrix notation is required.

- 3DOF active tendon controlled structure (Chung, Lin, Soong & Reinhorn, 1989)

$$\mathbf{M}\ddot{\mathbf{z}} + \mathbf{C}\dot{\mathbf{z}} + \mathbf{K}\mathbf{z} = -\mathbf{\Gamma}_1 u - \mathbf{M}\mathbf{\Gamma}_2 \ddot{z}_g$$
$$u = -\mathbf{K}_1 \mathbf{z} - \mathbf{K}_2 \dot{\mathbf{z}}$$

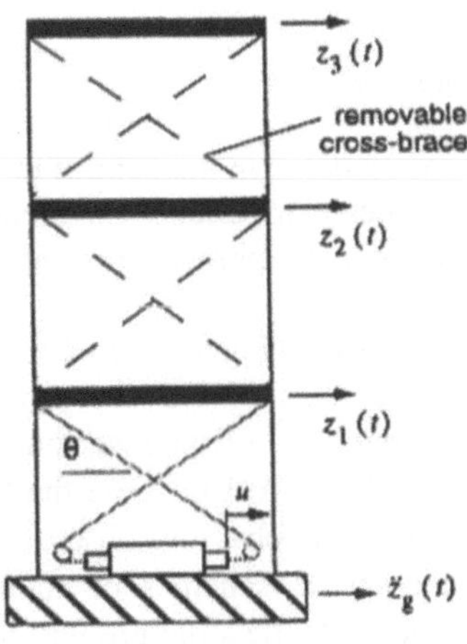

Figure 14 – Actively controlled system facility at Buffalo, New York.

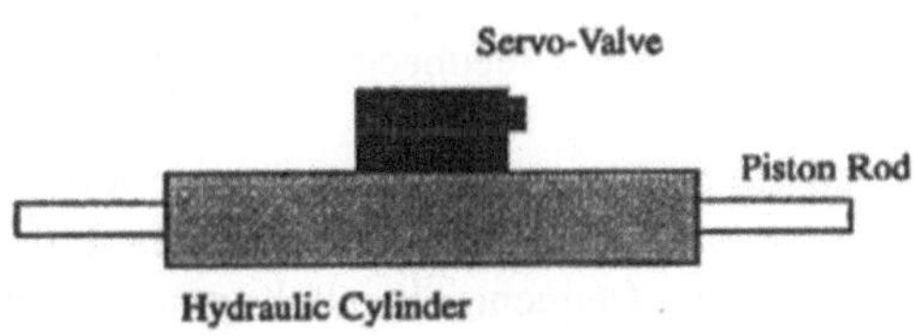

Figure 15 – Hydraulic actuator scheme.

Initial tests resulted in the closed-loop structural system being unstable. There was an apparent *delay between* when the control signal was commanded by the computer and when the actuator displacement was realized.

Hydraulic Actuator Modeling

- Flow equations linearized about the origin

servo-valve: $q = k_q c - k_c \frac{f}{A}$

hydraulic actuator: $q = A\dot{z} + \frac{V}{2\beta A}\dot{f}$

$$\dot{f} = \frac{2\beta}{V}\left(Ak_q c - k_c f - A^2 \dot{z}\right)$$

- Block diagram of open-loop servo-valve/actuator model

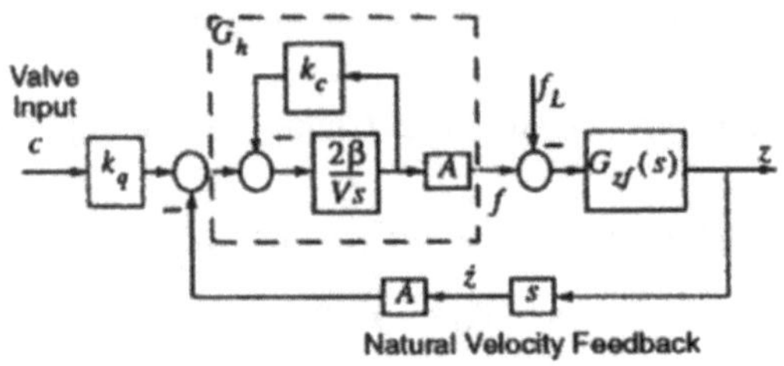

Figure 16 – Hydraulic actuator model.

A number of methods were tried to accommodate this apparent delay in the control loop. In the experimental studies conducted at NCEER (National Center Earthquake Engineering Research) using state feedback controllers, the phase compensation approach proved the most useful.

Indeed the problem was due to the adoption of hydraulic actuators (Figure 15) which mount a servo-valve. Typically, servo-valves have bandwidths greater than 50 Hz. One might argue that since this frequency is much greater than structural frequencies of interest, the dynamics of the actuators can be neglected. But with the information of Figure 16, where the hydraulic actuator is modelled, one enters Figure 17 and understand the real behaviour observed in laboratory. Figure 18 summarises this conclusion for three common schemes of actuation. In words, when the feedback interaction path is present, the poles of the structure will appear as zeros of the transfer function from the command input to the force applied to the structure. For actuators attached to lightly damped structures in which the feedback interaction path is present, the ability of the actuator to apply forces at the structure's natural frequencies is greatly limited. In particular, hydraulic actuators have an implicit feedback interaction path that occurs due to the natural velocity feedback of the actuator response. However, simple models can be employed to represent the dynamics of the hydraulic actuator and the associated control-structure interaction (from Lecture Notes (Spencer, 1997)).

General Actuator-Structure Configuration

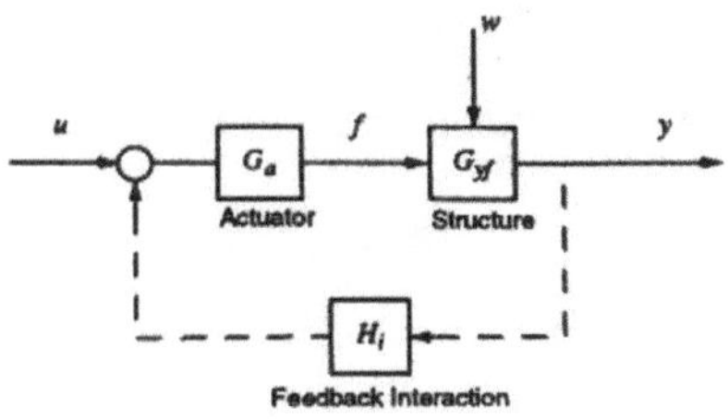

- Using block diagram algebra

$$G_{yu} = \frac{G_{yf}G_a}{1 + G_{yf}G_aH_i}$$

Representing the transfer functions as ratios of numerator and denominator polynomials in s

$$G_{yu} = \frac{n_{yu}}{d_{yu}} = \frac{n_{yf}\,n_a d_i}{d_{yf}\,d_a d_i + n_{yf}\,n_a n_i}$$

- The actuator and structure dynamics are tightly coupled!

- Similarly, the actuator transfer function can be determined

$$G_{fu} = \frac{G_a}{1 + G_{yf}\,G_aH_i}$$

In terms of the numerator and denominator polynomials

$$G_{fu} = \frac{n_{fu}}{d_{fu}} = \frac{d_{yf}\,n_a d_i}{d_{yf}\,d_a d_i + n_{yf}\,n_a n_i}$$

- The transfer functions G_{yu} and G_{fu} have the same poles, which are different than those of the uncontrolled structure, G_{yf}.

- The poles of the uncontrolled structure appear as zeros in the transfer function for the actuator.

- This is precisely the behavior observed in the laboratory!

Figure 17 – Structural system/Hydraulic actuator interaction.

Neglecting the dynamics of the actuator, as well as the control structure interaction effect, is equivalent to assuming that the transfer function G_{fu} is constant and magnitude with zero phase. In general, neither the phase nor the magnitude of G_{fu} will be constant. Ignoring phase differences between the command input u and the resulting force f, will result in a time lag associated with generation of the control forces. Appropriate modeling of the actuator dynamics and control structure interaction accommodates this time lag. In a system with active bracing, neglecting the actuator dynamics results in larger accelerations than in the case where the control-structure interaction is considered. Also neglecting actuator dynamics results in less achievable performance of the controller since the system easily becomes closed loop unstable.

Active Tendon

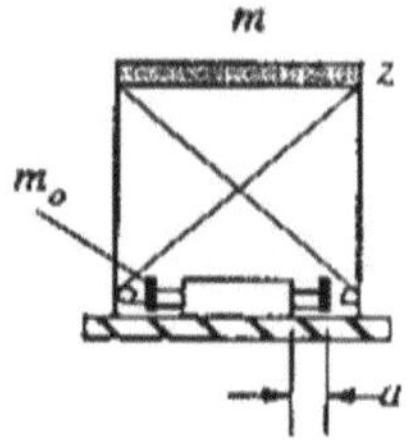

- Equations of motion

$$m\ddot{z} + (c+c_o)\dot{z} + (k+k_o)z + c_o(\cos\theta)\dot{a} + k_o(\cos\theta)a = -m\ddot{z}_g$$

$$m_o\ddot{a} + c_o\dot{a} + k_o a + \frac{c_o}{\cos(\theta)}\dot{z} + \frac{k_o}{\cos(\theta)}x = -m_o\ddot{z}_g + f$$

Incorporating unity gain displacement feedback

$$\dot{f} = \frac{2\beta}{V}\left(Ak_q\gamma(u-a) - k_c f - A^2\dot{a}\right)$$

- The state vector is defined as $\mathbf{x} = [a \;\; \dot{a} \;\; z \;\; \dot{z} \;\; f]'$

Active Mass Driver

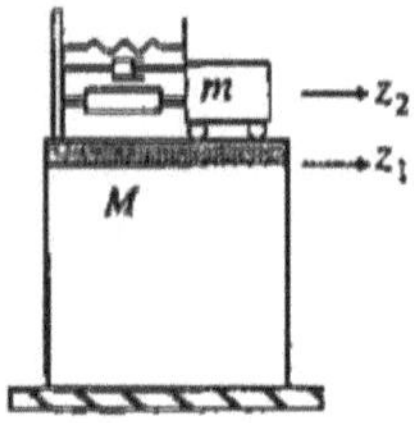

- Equation of motion

$$M\ddot{z}_1 + (c_1+c_2)\dot{z}_1 - c_2\dot{z}_2 + (k_1+k_2)z_1 - k_2z_2 = -f - M\ddot{z}_g$$

$$m\ddot{z}_2 + c_2(\dot{z}_2-\dot{z}_1) + k_2(z_2-z_1) = f - m\ddot{z}_g$$

Incorporating unity gain displacement feedback

$$\dot{f} = \frac{2\beta}{V}\left(Ak_q\gamma\{u-(z_2-z_1)\} - k_c f - A^2(\dot{z}_2-\dot{z}_1)\right)$$

- Defining the state vector of the system as $\mathbf{x} = [z_1 \;\; z_2 \;\; \dot{z}_1 \;\; \dot{z}_2 \;\; f]'$

Active Bracing

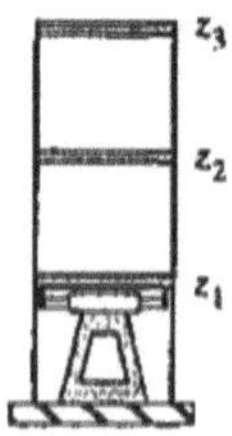

- Equations of motion

$$\mathbf{M}_s\ddot{\mathbf{z}} + \mathbf{C}_s\dot{\mathbf{z}} + \mathbf{K}_s\mathbf{z} = \mathbf{B}_s f - \mathbf{M}_s\mathbf{\Gamma}_s\ddot{z}_g$$

Incorporating unity gain displacement feedback

$$\dot{f} = \frac{2\beta}{V}\left(Ak_q\gamma(u-z_1) - k_c f - A^2\dot{z}_1\right)$$

- The state vector is defined as

$$\mathbf{x} = [\mathbf{z}' \;\; \dot{\mathbf{z}}' \;\; f]'$$

Figure 18 – Structural system/Hydraulic actuator interactions

Better results are obtained if the compressibility of the hydraulic fluid is taken into account rather than if the fluid is treated as incompressible, as shown in the AMD example. Also in the latter case, the achievable performance level of the controller is significantly reduced.
In general, modeling errors resulting from neglecting actuator dynamics and control-structure interaction can be expected to decrease both the stability and performance robustness of the controlled structure.

Such approaches are first generation studies (Housner, Soong & Masri, 1994). The next generation of control strategies had to focus on *implementability!* Because displacements and velocities are not absolute, but dependent upon the inertial reference frame in which they are taken, their direct measurement at arbitrary locations on large-scale structures is difficult to achieve. Effective implementation of full-state feedback control strategies is impracticable.

Accelerometers can provide inexpensive and reliable measurements of the accelerations at strategic points on the structure. However, early efforts to develop and implement acceleration feedback control strategies proved unsuccessful. But *output feedback* control algorithms that make use of acceleration feedback have been systematically developed (Spencer, et al., 1991; Spencer et al., 1994; Suhardjo et al., 1992).
The results can be summarised as follows:

- Measurement of the absolute accelerations can effectively be used to control the structure.
- Structural models were developed which fully incorporate actuator dynamics and control-structure interaction.
- H_2/LQG control strategies have been shown to be effective for these control designs.
- The results indicate that acceleration feedback strategies are practically implementable, and can achieve high performance

All the previous results were obtained in laboratories equipped with large to medium scale shaking tables (Casciati, 1998; Casciati et al., 1998). It is worth mentioning, however, the different strategy experienced at the European Joint Research Centre in Ispra.
The European Laboratory for Structural Assessment (ELSA) of the Joint Research Centre was and is currently engaged in research related to active control of large scale structures, organised in the frame of international consortia.

One of the activities covered the Active Control in Civil Engineering (ACE program). The ACE program was mainly oriented to the study of active tendon control of cable-stayed bridges (as in Figure 7). Improvements in materials led to the construction of progressively longer, structurally more efficient and slender bridges. But consequently, structures are more and more flexible. Deck and cable vibrations have become a major issue in cable-stayed bridge design.
The technical objectives of this project were, therefore, (a) to improve the understanding of the induced vibrations of cable-supported structures, (b) to develop an appropriate software package capable of analysing the behaviour of cable-supported structures, (c) to develop an active system to control induced vibrations of cable-supported structures, (d) to develop the appropriate actuators and (e) to validate the active control system with high scale mock-ups and measurements of existing structures (Magonette et al., 2001)

Another activity at ELSA was related to the development of the Continuous Pseudo-dynamic (PsD) Method.
From the beginning of the development of PsD testing (Donea et al., 1996), it has been suggested that a part of the structure where damage is expected (column, beam-to-column joint, damping devices, etc) is fabricated and tested in laboratory, and the rest of the structure, assumed to be linear elastic, is modelled and analysed by computer. This technique was designated as the substructure PsD testing method.
The JRC is currently developing a structural dynamics testing system capable of performing high speed substructuring tests, which considerably reduces the test duration and opens the way to accurate testing of the earthquake response of structures with velocity-dependant devices.

4 Interdisciplinary character of the required research

In the meantime passive structural control (Martelli & Forni, 1997) was implemented in a number of buildings, structural active control is reserved to solve transient construction problems, large facility special aspects and undesirable interaction problems (Anton & Casciati, 1998).
Indeed European designers are presently sceptic about the possibility of a direct application of active control devices to counteract wind and earthquake excitations. The low seismicity which characterises large areas of Southern European countries, in fact, prevents the constructor from concentrating investments in one area rather than another. Also to maintain operative the control device along the wide periods of inactivity would put the global cost of the design out of the market.
In a seismic context, two fields are mainly investigated: the implementation of hybrid control systems and the one of semi-active devices. The problem is still the identification of a suitable class of buildings to be equipped . Their intrinsic value must be very high to justify the additional cost of active control devices. The idea is to focus attention on the wide European architectural heritage (Casciati et al., 1999). Hybrid and semi-active devices seem to be appropriate for this purpose, provided the necessary tests be conducted.

Active structural control is therefore confined in laboratory implementations where specific problems are detected, discussed and solved, e.g.:

1. monitoring;
2. controller implementation:
3. actuator design;
4. reliability aspects (Battaini et al., 1999).

This is essentially made at a medium size level, i.e. by exploiting a shaking table of the size of 1 by 1 meter, of either hydraulic or electromagnetic type. The facility available at the University of Pavia is if this type; it mounts a hydraulic actuator.
To provide an example of the interdisciplinary character of such a research, Figures from 19 to 24 summarise the core of some experiments carried out.

4.1 Tuned liquid dampers

A special class of passive devices, which offers affordable costs and good efficiency, is the one of the so-called tuned liquid columns (TLC). Column is the appropriate word because most of the existing applications were realised by cylinders (Figure 19). Their inconvenience is the narrow band of frequencies of effectiveness. A first idea was to move from cylindrical tanks to conic tanks (Figure 20), with two main advantages: a lower mass of liquid can be used because of the wider area where the water can spread around and, the possibility of moving from one frequency to another by simply adding a thin layer of water (Tagliente, 1999). As Figures 20 shows, the conic tank develops its action over a broader range of frequencies, thus reducing the effects of inadequate tuning.
Nevertheless, the efficiency is largely increased only if semiactive devices are added. They are made of bodies which are introduced in the water on command so that the disorder of fluid motion increases the amount of energy dissipated. Once again the whole device requires sensors and a control law built in a standalone board.

Figure 19 – A set of cylinders to be used as tuned liquid dampers.

Figure 20 - A set of cylinders to be used as tuned liquid dampers

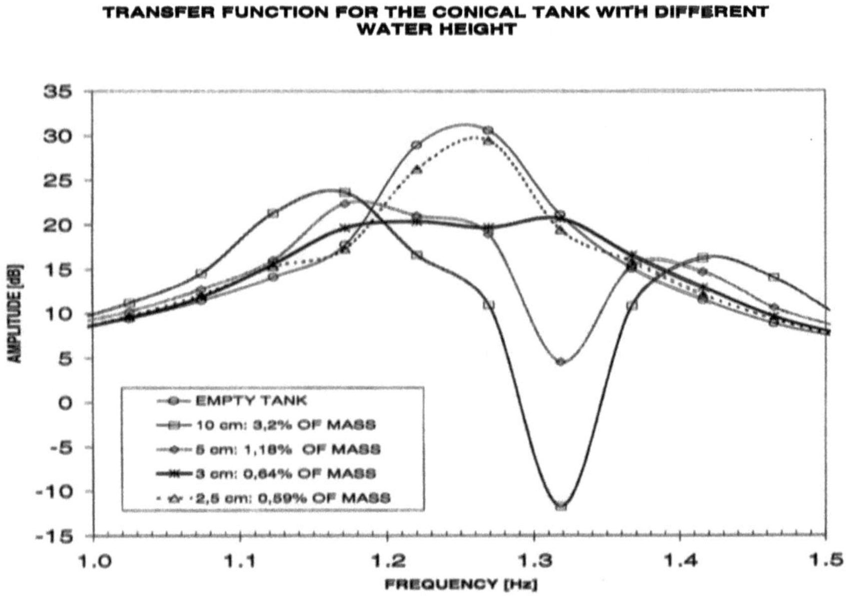

Figure 21 - Comparison between the transfer functions of the system in Figure 20 whit different heights of water in the cylindrical tanks.

4.2 Active air-jet control system

Following an idea coming from aerospace engineering and already investigated in a structural context by Caughey and Masri, (Beckley et al., 1982), the adoption of air-jet actuators was first investigated in the laboratory of the University of Pavia with reference to a cantilever beam (Figure 22) (Brambilla et al., 1998; Brambilla, 1999).

The main advantage of this active system is that the energy that will be required to activate the actuators during the external excitation is stored in the form of compressed air. This means that one has not to wonder of the actual availability of an external power source during a seismic event.

The frame of Figure 23 was built and the actuator consists of three electrovalves, each with its own nozzle, on the two directions.

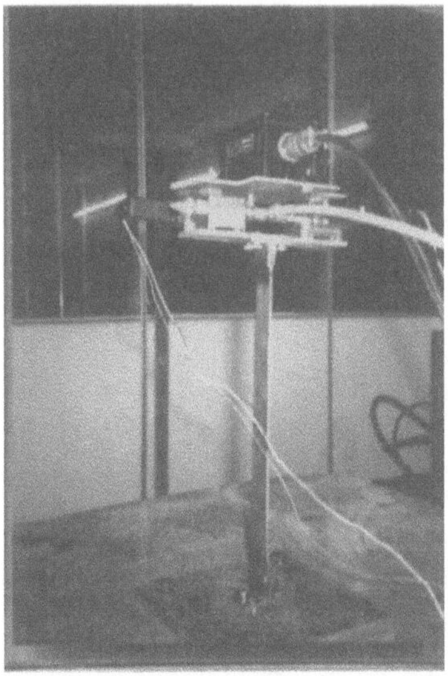

Figure 22 - Cantilever beam used in the laboratory tests. On the top of the beam an accelerometer acquires the absolute acceleration.

Figure 23 - Frame used in the laboratory tests. The controller device composed by the three electrovalves applies its force in correspondence to the barycentre of the structure (Fumagalli, 1999).

4.3 Controller

It is worth noting that all the schemes briefly described in the previous sections present a strong nonlinear behaviour, in the sense that the control device is expected to be active only after some extreme conditions are achieved. The central dead band must exist to avoid instabilities. One possible solution for such a situation is the adoption of fuzzy control, which also allows one to pursue a form of distributed control (Casciati, Faravelli & Torelli, 1999 ; Faravelli & Rossi, 2002).

Indeed, fuzzy control theory does not provide a single controller block that processes a set of feedback variables and produces a control output but, partitioning the controller block into simpler parts can give the designer a deeper physical insight into the whole system operation. For this reason, when controlling a multi-degree-of-freedom structure, it may prove convenient to split the controller into multiple controller blocks, each driving a single actuator (Figure 24). Each controller block should be aware of how all the others are working, as failure to do so may results in a worse performance or even instability.

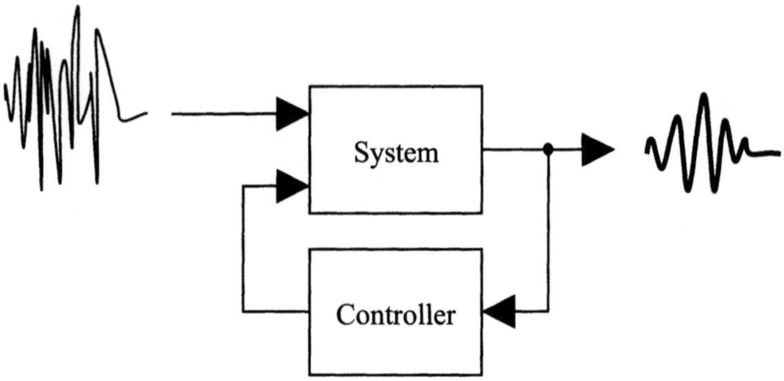

(a)

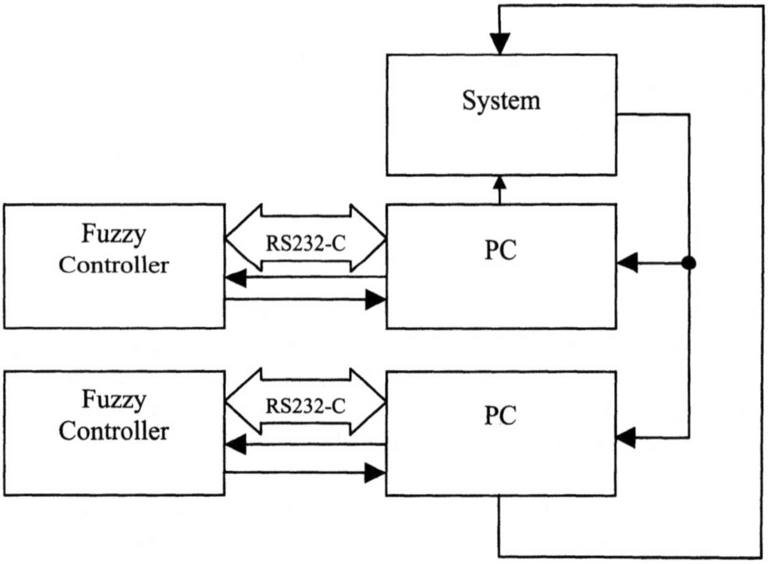

(b)

Figure 24 - Input-output model for a feedback controlled system (a) and alternative configuration of a controlled system (b): two controllers are deployed, each one driving a single actuator.

Communication between the controller blocks can be implemented either in direct (i.e. by physically interconnecting them) or indirect form (i.e. by providing them with the same amount of information about the system motion).

5. Semi-Active and Hybrid Structural Control: Opportunities and Challenges

Although active control systems have been designed and installed in full-scale structures, the engineering community has yet to fully embrace this technology.
This lack of acceptance stems, in part, from questions of

- cost effectiveness
- stability
- reliability
- power requirements

On the other side, passive control is unable to match situations different from the ones it was designed. As a consequence there is increasing technical interest in intermediate solutions as Figure 25 illustrates.

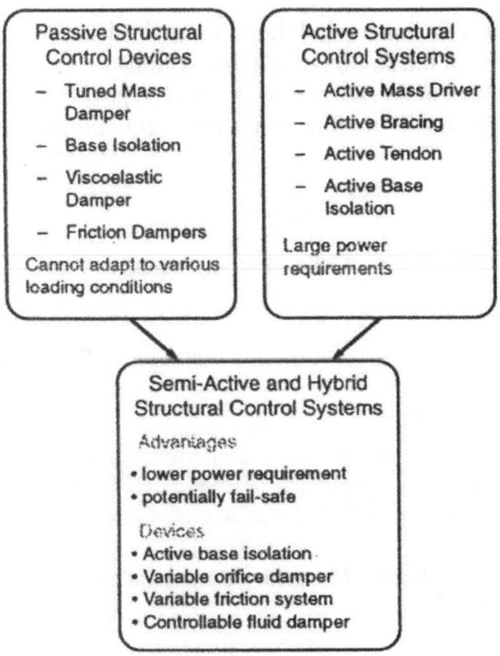

Figure 25 – Semiactive control vs. passive and active control.

5.1 Hybrid Devices

Hybrid control strategies combine active and passive control mechanism to reduce the energy required by fully active systems. A Hybrid Mass Damper (*e.g.*, the active pendulum damper in the Takenada building) system is linear for small motions.

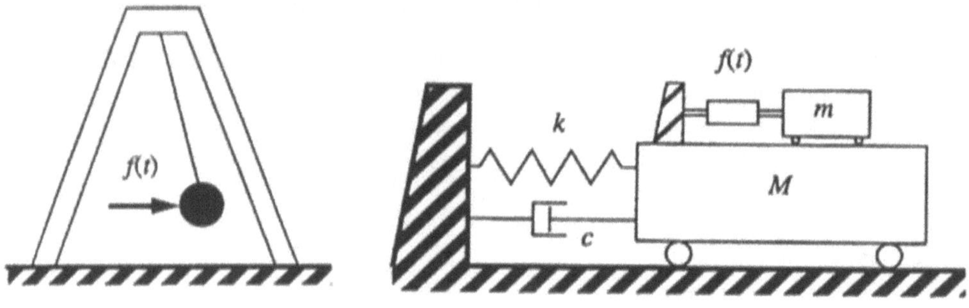

Figure 26 – Examples of HMS (Hybrid Mass Dampers)

The Shinjuku-Park Tower also has a pendulum type hybrid mass damper, while the DUOX system is the HMS installed in the Ando Nishikicho building.
A different concept is on the back of the Active Base Isolation proposed by (Reinhorn et al., 1993).
Note that the system is intrinsically nonlinear: idealized equations of motion are inadequate to model the system.

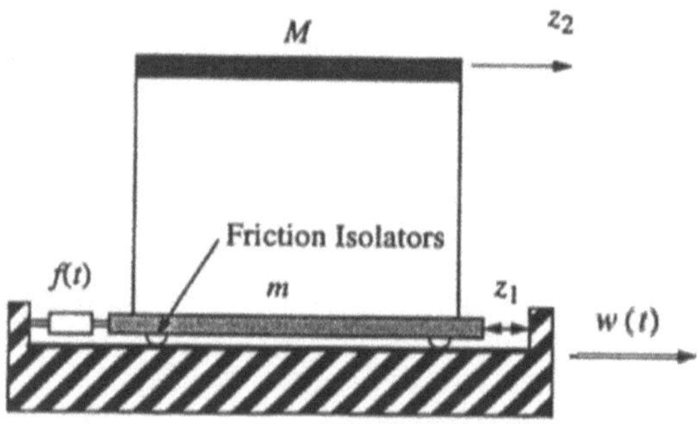

Figure 27 – Active Base Isolation

5.2 Semi-Active Devices

Semi active devices are characterized by low power requirement. A more restrictive definition also imposes that they do not counteract by opposite forces, but just by introducing a system modification (adaptability). They have:

- Potentially low cost
- Fail-safe operation (Cannot add energy to the structure)

Despite preliminary studies indicate that semi-active dampers can potentially achieve the majority of the performance of fully active systems, they are intrinsically nonlinear.
The Variable Orifice Damper sees the initial work done by Karnopp, Crosby and Harwood (1974). More recent efforts are described in Karnopp (1990) and Patten, Sack and He (1994). Figure 28 presents its basic features.

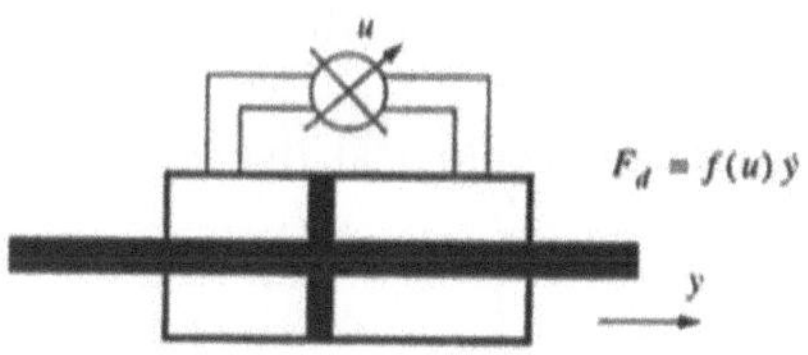

- Equation of Motion

$$m\ddot{z} + c_0\dot{z} + kz + f(u)\dot{y}\cos\theta = -mw$$

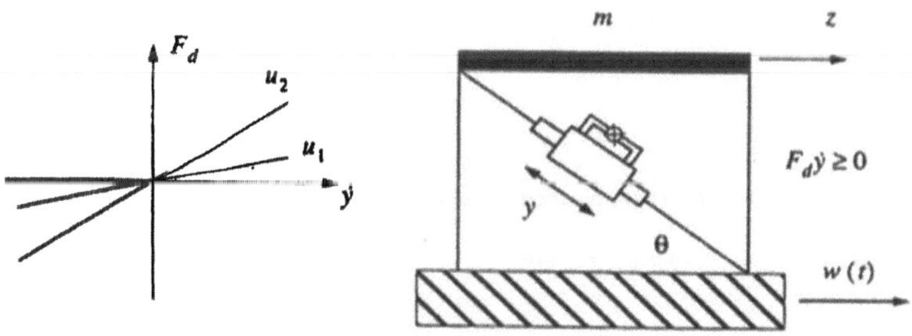

Figure 28 – Controllable Fluid Damper

The Variable Friction Damper sees the initial work done by by Akbay and Aktan (1991). More recent efforts are described in Dowdell and Cherry (1994).
Figure 29 presents its basic features.

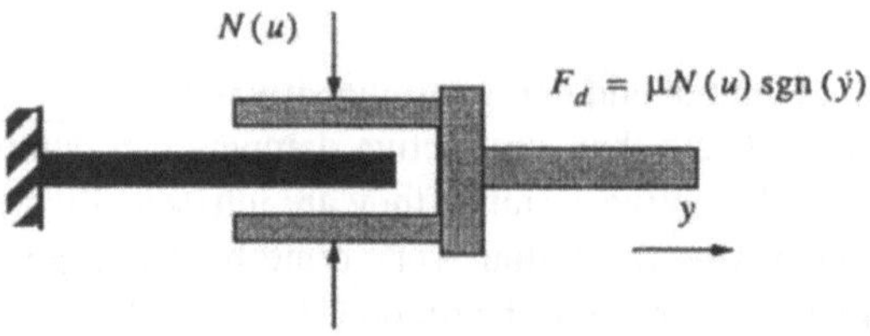

Figure 29 - Variable Friction Damper

The Controllable Fluid Damper collects both the Electro-Rheological and the Magneto-Rheological dampers.
The Conceptual sketch of a magneto-Rheological fluid damper is in Figure 30 (from Lecture Notes (Spencer, 1997)).

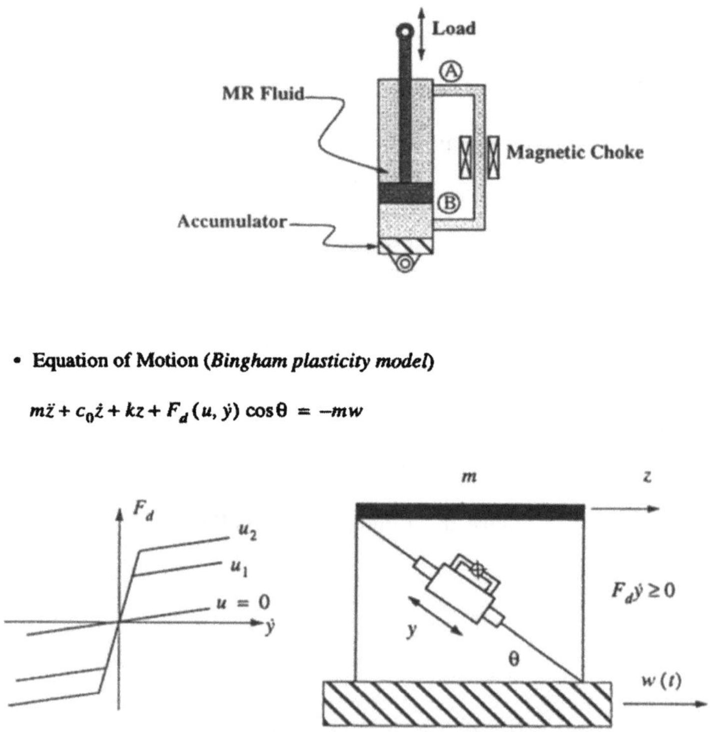

Figure 30 – Controllable Fluid Damper

6. A Case Study of Energy Dissipation in Bridges

Energy dissipation devices are used in bridge design against ultimate limit states, i.e. against the strongest design response spectrum the code introduces. They are conceived to serve as high stiffness components below a force threshold and to undergo large hysteretic cycles when the threshold is crossed.

This section discusses the selection of this threshold as the result of a cost-benefit analysis: lower the threshold is, higher the bridge reliability becomes. On the other side, lower the threshold is, higher the probability that the bridge supports undergo permanent deformation under the serviceability excitation becomes (Casciati & Faravelli, SPIE 2001)

When designing an energy dissipation design for a bridge in a seismic region, one meets two basic problems:

1. the structural codes imposes material partial safety factors which alters the actual dynamic behaviour of the structural system;
2. one must design the device accounting for both serviceability and ultimate limit states.

Such a situation suggests to investigate the feasibility of semi-active devices which can switch from one behaviour to another according to the severity of the ground motion. The goal is mainly to emphasise the need for the availability, in a damper, of different force-displacement curves, each of them suitable for different levels of seismic intensity.

6.1 The seismic excitation

The use of dissipation devices introduces nonlinearities in the structural system and requires a time step dynamic analysis assigning the seismic excitation as an acceleration time history.
In order to evaluate the structural system seismic response, the dynamic analyses should be conducted by several accelerograms compatible with the elastic spectrum of Eurocode 8.
The response spectrum of a single acceleration have been calculated for damping values of 2% and 5% and shown in Figure 31 together with the Eurocode 8 spectrum for ultimate state load condition (ULS).
For the stress checks at the elastic limit, the accelerogram will then be scaled by a factor of 3.5.

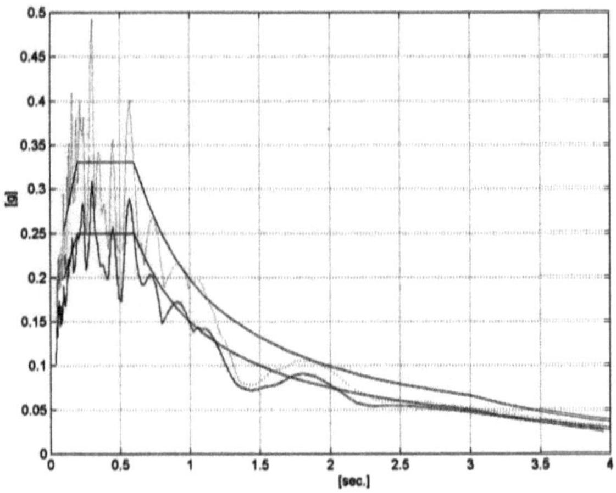

Figure 31 – Accelerogram response spectra for damping 2% and 5% compared to the Eurocode 8 design spectrum for the ultimate limit state check

The dynamic analyses can be performed by a finite element code. The assigned base accelerogram should be filtered in order to obtain a zero mean displacement signal by filtering the frequencies between 0.1 and 0.15 Hertz and between 25 and 27 Hertz. The filtered displacement input signal is shown in Figure 32.

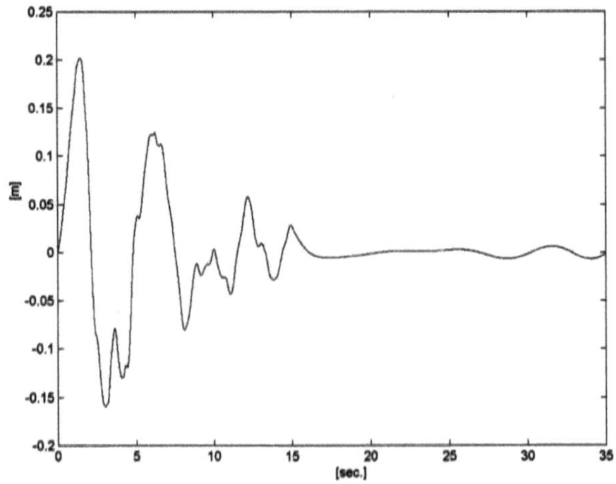

Figure 32 – Base displacement time history

6.2 The structural system

The simplified structural system used in the dynamic analyses is shown in Figure 33. It represents a simply supported bridge deck including the pier model. The dissipation device is located between the pier top and the deck while the other deck support is simply a vertical support representing a sliding bearing free to move along the longitudinal direction.

This simplified model represents the static scheme of an highway viaduct in Turkey recently subject to an exceptional earthquake with a peak ground acceleration more that twice the expected one. The viaduct was seismically isolated by elasto-plastic energy dissipation devices manufactured by ALGA S.p.A.. The seismic isolation project showed a good behaviour during the earthquake and although very high excitation level the structure did not collapse and the main damages where located in the energy dissipation devices as expected.
The study presented in this paper uses the same isolation concept of this highway viaduct.

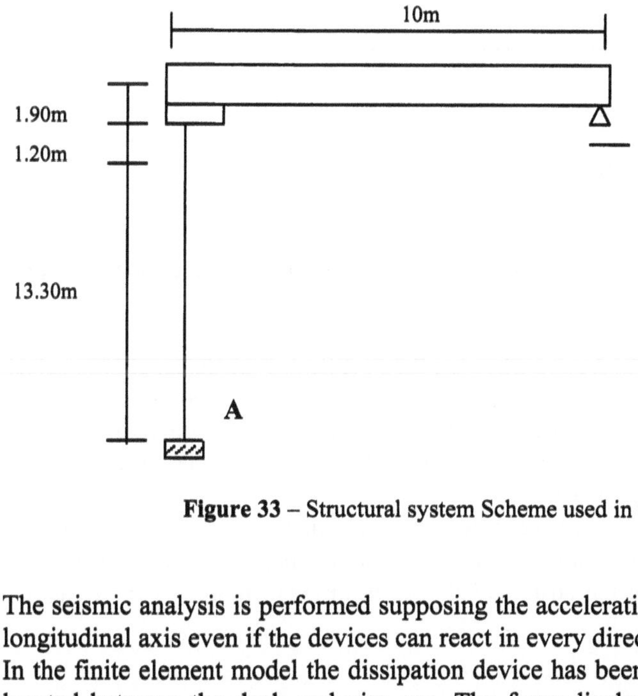

Figure 33 – Structural system Scheme used in the dynamic analyses

The seismic analysis is performed supposing the acceleration direction parallel to the structure longitudinal axis even if the devices can react in every direction of the horizontal plane.
In the finite element model the dissipation device has been modelled as an elasto-plastic truss located between the deck and pier cap. The force-displacement response law of the elasto-plastic device is shown in Figure 34. Three levels of yielding force have been considered; the mean one and the other two with 75 % and 125% yielding force as shown in Figure 34.

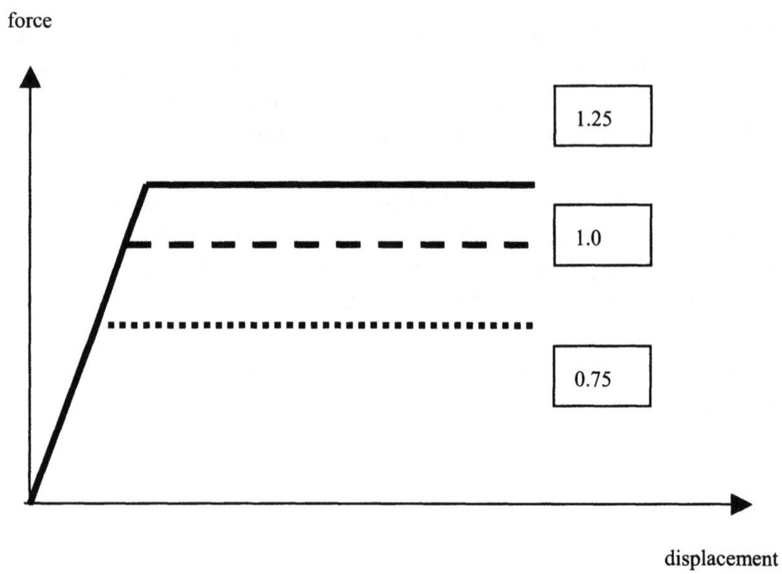

Figure 34 – Force – Displacement diagram for different yielding F_y

For each of the three devices two dynamic analyses were performed considering two levels of external excitation corresponding to service load condition (SLS) and ultimate load condition (ULS).
For each yielding level the device reacts as an elastic spring for SLS excitation while it dissipates energy for ULS excitation. At ULS, in particular, the problem of the designer is to select a suitable force-displacement response curve: increasing the yielding it is possible to reduce the device displacement (i.e. the relative displacement between deck and piles) but a higher shear force is sent to substructures. Decreasing the yielding force it is possible to reduced the shear forces, but the displacement is increased.
The optimal design must research the best compromise between the two previous aspects. These considerations show that the availability of devices able to adapt their properties to the structure requirement could optimise the structural design in order to minimise the dynamic response and costs.

6.3 Final remarks

In conclusion, the performed analyses showed a significant conflict between the goal and the codes requirements. The goals are simply that:

- the energy dissipation device must be inactive for the SLS load condition
- the energy dissipation device must reduce the structural response ULS load condition

But the code requires to introduce suitable safety coefficients with the risk of reducing the efficiency of the device at ULS.

If one requires a safety coefficient of 1.5 for the yielding level at SLS it can happen that the yielding force is too big at ULS and the device is unable to dissipate even a small energy because it yields at too high level.
The previous consideration shows that a semi active system could solve the problem adapting the dissipation level to the external excitation acceleration.
The open problems of the semi-active solution are:

- the definition of the optimal yielding values
- the device response time in varying the device properties
- the characterization of the plastic response in term of a force-displacement diagram after the yielding point has been reached.

7. Conclusions

Active and semi-active control is becoming one of the solution the designer considers in the design process of a new building or in conceiving solutions for the rehabilitation of an existing infrastructure component. Applications in the construction stage or to mitigate the vibration under wind loading are becoming more and more frequent.
Earthquake engineering applications are still lacking in Europe, mainly due to the strict constraints the structural codes offer and the impossibility of implementing a standard linear control. Non-linear control is the case for special hybrid, semi-active and active situations. The efficiency of such alternative solutions was tested in a laboratory environment and some results discussed through these notes.
Nevertheless, in preparing them the author imposed himself to avoid that the result would appear as a compact textbook of Structural Control. He mainly pursued the goal of recording the main steps of the five lectures held during the course week:

1. to show, by means of examples, the special features Structural Control assumes in the area of Civil and Architectural Engineering;
2. to introduce the fields of potential research toward the development of Structural Control applications in that area;
3. to discuss some recent ideas which exploit interdisciplinary results achieved in smart material, computer science and electronics.

Nevertheless, a list of references is provided to the reader who want achieve a global understanding of the Structural Control problems.

8. Acknowledgements

The research activity summarised in these note, was supported by different agencies. It is worth mentioning here the role of the Italian Space Agency (ASI; contract number I/R/187/01 with Franco Bernelli-Zazzera acting as national co-ordinator), of the Ministry of Instruction, University and Research (MIUR; contract COFIN-01 with Fabrizio Davì acting as national co-ordinator) and of the University of Pavia (FAR grant to Lucia Faravelli).
Moreover, the writer thanks the authors of the lecture notes which were distributed during short courses held at the Department of Structural Mechanic of the University of Pavia by Prof. T.T.

Soong and Prof. J.B. Spencer. The autor wishes also to acknowledge the help of Danilo Miozzari in formatting the material.

References

Akbay, Z., & Aktan, H.M. (1991) Actively Regulated Friction Slip Devices, *Proc. 6th Canadian Conf. on Earthquake Eng.*, pp. 367-374.

Anton M. and Casciati F. (1998). Structural Control Against Failure Interaction. *Journal of Structural Control*, Vol. 5 n. 1, p.63-73.

Baratta A. and Rodellar J. (1996). *Proceedings of the 1st European Conference on Structural Control*, World Scientific, Singapore.

Battaini M., Breitung K., Casciati F. and Faravelli L. (1998). Active Control and Reliability of a Structure under Wind Excitation, *Journal of Wind Engineering and Industrial Aerodynamics*,74-76, 1047-1055.

Beckey G.A. Caughey T.K. Masri S.F. (1982). On Line Control of Nonlinear Flexible Structure, *ASME Journal of Applied Mechanics*, n. 49, p.837-844.

Bourquin F. and Collet M. (2002), Proceedings of the 2nd European Conference on Structural Control, Duned, Paris, France.

Brambilla M. (1999). Design and Implementation of a Jet Actuator for Active Structural Control. *Journal of Structural Control*, Vol. 5 n. 1, p.117-128.

Brambilla M. Casciati F., Masri S.F. (1998). Testing Air Jet Actuator. *Proc. Second World Conference on Structural Control*, Kyoto, Japan, Vol 1, p.655-664.

Casciati F (2002), *Proceedings of the Third World Conference on Structural Control*, J. Wiley & Sons, Chichester, UK.

Casciati F. & Faravelli L.and Torelli G. (1999) A Fuzzy Chip Controller for Nonlinear Vibrations, *Nonlinear Dynamics* **20**: 85.98.

Casciati F. (1998). Scale Laboratory Testing towards Actual Actively Controlled Structural Systems. *Proc. Second World Conference on Structural Control*, Kyoto, Japan, Vol. 1, p.41-50.

Casciati F. Maceri F., Singh M.P., Spanos P. (1999). *Civil Infrastructure System: Intelligent Renewal*, World Scientific, Singapore.

Casciati F., and Faravelli L. (1999). Experimental Promotion of semiactive, active and Hybrid Control. *International Post-SMiRT Conference Seminar on Seismic Isolation, Passive Energy Dissipation and Active Control of Vibrations of Structures*, Cheju, Korea.

Casciati F., Faravelli L. (2001) Standalone Controller for a Bridge Semiactive Damper, *Proceedings of SPIE*, Vol. 4330, pp.339-404.

Casciati F., Faravelli L., Petrini L. (1998). Input-Signal Characterization for Shaking -Table. *Journal of Structural Control*, Vol. 5 n. 2, p.33-56.

Chung, L.L, Reinhorn, A.M., & Soong, T.T. (1988) Experiments on Active Control of Seismic Structures, *J. Eng. Mech., ASCE*, 114(2), 241-256.

Chung, L.L., R.C. Lin, T.T. Soong & A.M. Reinhorn (1989) Experiments on Active Control for MDOF Seismic Structures, *J. of Eng. Mech., ASCE*, Vol. 115, No. 8, pp. 1609-1627.

CONVIB (2001), http://sm26db.jrc.it/laboratory/convib/

Donea J., Magonette G., Negro P., Pegon P.,Pinto A., Verzeletti G. (1996). Pseudodynamic Capabilities of the ELSA Laboratory for Earthquake Testing of Large Structures. *Special Earthquake Spectra*. Volume 12, No 1.

Dowdell, D.J. & Cherry, S. (1994) Semiactive Friction Dampers for Seismic Response Control of Structures, *Proc. 5th US Nat. Conf. on Earthquake Engrg.*, 1:819-828.

Faravelli L., and Rossi R. (2002) Adaptive Fuzzy Control: theory vs Implementation. *Journal of Structural Control*, Volume 9, n°1.

Fumagalli D. (1999). Controllo Attivo delle Vibrazioni Strutturali con Attuatori a Getto, (in Italian), Master Degree Thesis, Dep. of Structural Mechanics, Univ. of Pavia.

Hansen H.I. (1998). Active Vibration Control of Long Suspension Bridges, Ph.D. Thesis, Dep. of Building Technology and Structural Engineering, Aalborg University, Denmark..

Hansen H.I. and Thof-Christensen P., (1998) Active Control of Long Bridge using Flaps, *Proc. Second World Conference on Structural Control*, Kyoto, Japan, Vol. 2, p.851-858.

Housner G.W. (1994) *Proceedings of the First World Conference on Structural Control,*

Housner G.W., and Masri S.F. (1994). Research Needs in the Field of Structural Control. *Journal of Structural Control*, Vol. 1, 9-22.

Housner G.W., Bergman L.A., Caughey T.K., Chassiakos A.G., Claus R.O., Masri S.F., Skelton R.E., Soong T.T., Spencer B.F. Jr. and Yao J.T.P.(1997). Structural Control: Past, Present and Future. *Journal of Engineering Mechanics*, ASCE, 123 (9), 897-971.

Housner, G.W., Soong, T.T. & Masri, S. (1994b) Second Generation of Active Structural Control in Civil Engineering, *Proc. First World Conference on Structural Control, Pasadena*, California, August 3-5, 1994, Vol. 1, pp. Panel:3-18.

Karnopp D. (1990) Design Principles for Vibration Control System Using Semiactive Dampers, *ASME Journal of Dynamics, Measurement and Control,* Vol. 112, No. 9, PP. 448-455.

Karnopp D., Crosby M.J. and Harwood R.A. (1974) Vibration Control Using Semi-Active Force Generators, *ASME Journal of Engineering for Industry,* Vol. 96, No. 2, pp. 619-626.

Kobori T. (1999) *Proceedings of the Second World Conference on Structural Control,* J. Wiley & Sons, Chichester, UK.

Magonette et al. (2001), Active Control in Civil Engineering: From Conception to Full Scale Application, *Journal of Structural Control,* 8-2, 123-188.

Magonette G., Marazzi F., Bossens F., Dumoulin C., (1999). Active Control Experiment of a Large Scale Cable-Stayed Bridge Mock-Up. Proc. of *the Third International Symposium on Cable Dynamics*, Trondhein, Norway, August 16-18.

Martelli A. and Forni M. (1997), Seismic Isolation, Passive Energy Dissipation and Active Control of Seismic Vibrations of Structures, *Proc. Int. Post-SmiRT Conference*, Taormina.

Nielsen S.R.K., Kirkegard P.H., Thesbjerg L. (1999). Active Vibration Control of a Monopile Offshore Structure, *Journal of Structural Control*, Vol 6, n.2.

Patten W.N., Sack R.L.and He Q. (1994) Suppression of Vehicle-Induced Bridge Vibration via a Semiactive Structural Controller, *Proceedings of the First World Conference on Structural Control,* Pasadena, California.

Reinhorn et al. (1993) Full Scale Implementation of Active Control. II: Installation and Performance, *Journal of Structural Engineering, ASCE*; **119**(6):1935-1960.

SAMCO (2001), http://dipmec.unipv.it/research/resintro.html

Soong T.T. (1991). Active Structural Control: Theory and Practice. *Longman Scientific and Technical*, Essex, UK.

Soong T.T. (1998) Lecture Notes, University of Pavia.

Soong T.T., and Dargush G.F. (1997). *Passive Energy Dissipation Systems in Structural Engineering*, J. Wiley & Sons, Chichester, UK.

Spencer B.F. Jr. and Sain M.K.(1997), Controlling Buildings: a New Frontier in Feedback, *Control Systems*, IEEE, 17, 8, 18-35.

Spencer J.B. (1997) Lecture Notes, University of Pavia.

Spencer Jr., B.F., Suhardjo, J., & Sain, M.K. (1991) Frequency Domain Control Algorithms for Civil Engineering Application, *Proc., Int. Workshop on Technol for Hong Kong's Infrastructure Developement*, The Commercial Press, Ltd. Hong Kong, 169-178

Spencer Jr., B.F., Suhardjo, J., & Sain, M.K. (1994) Frequency Domain Optimal Control Strategies for Aseismic Protection, *J. Eng. Mech., ASCE*, 120(1), 135-159.

Suhardjo, J., Spencer, B.F. & Kareem, A. (1992) Frequency Domain Optimal Control of Wind Excited Buildings, *J. Eng. Me ch., ASCE*, 118(12), 2463-2481.

Tagliente L.S. (1999). Controllo di Strutture Mediante Tuned Liquid Damper", (in Italian), Master Degree Thesis, Polytechnic of Turin.

Some Applications of Differential Geometry in Control

Kurt Schlacher, Stefan Fuchshumer and Johann Holl
Institute of Automatic Control and Control Systems Technology,
Johannes Kepler University of Linz, Austria

Abstract Differential geometry has been introduced to control about 25 years ago. This contribution shows, how one can identify dynamic systems with geometric objects defined on certain manifolds such that one obtains a coordinate free description of the systems. Based on this approach basic properties like accessibility and observability will be introduced. After that, tests, whether a system shows these properties, are presented. To show the power of the geometric approach for the control loop design, the two methods input to output and input to state linearization have been selected. Finally, it is worth mentioning that all the presented methods can easily be implemented in any advanced computer algebra system.

1 Introduction

Nonlinear dynamic systems, described by sets of ordinary differential equations, constitute an important class of systems for applied control as well as for control theory. Since more than two decades there is a rich theory for control available, which is based on geometric, or more precisely, on differential geometric methods. The main idea of this approach is to consider dynamic systems as geometric objects defined on certain manifolds, or even more to identify dynamic systems with these objects. Whenever this is possible, one can fall back to the pure geometric properties to exploit them for the purpose of system analysis and control system design. The drawback of this approach is that the reader needs some basic knowledge of differential geometry, which is presented in the chapter *Basic Differential Geometry for Mechanics and Control* of this book. More detailed treatments of the presented topics can be found in the books (Isidori, 1995), (Nijmeijer and van der Schaft, 1990), (Sastry, 1999). But this contribution deals with a more general class of nonlinear systems than the class of time invariant systems, which are affine with respect to the input.

According to the considerations above this contribution is organized as follows. Section 2 gives an overview of dynamic systems, which are described by explicit differential equations. This section puts the fundamental geometric properties of the systems at the reader's disposal, which allows us to analyse control systems more efficiently. Section 3 deals with the local observability and Section 4 with the local accessibility of control systems. The last two Sections 5 and 6 are devoted to the control loop design, where two methods, exact input to output and input to state linearization, are presented. Both turned out to have many applications in engineering problems.

2 Dynamic Systems

We consider dynamic systems described by n_x ordinary differential equations of the local form

$$\begin{aligned} \tfrac{\mathrm{d}}{\mathrm{d}t}x^{\alpha}(t) &= f^{\alpha}(t, x(t), u(t)) , \quad \alpha = 1, \ldots, n_x \\ y^{\alpha_y}(t) &= c^{\alpha_y}(t, x(t), u(t)) , \quad \alpha_y = 1, \ldots, n_y , \end{aligned} \tag{2.1}$$

where (x^{α}) denotes the n_x-dimensional state, (u^{α_u}), $\alpha_u = 1, \ldots, n_u$ the n_u-dimensional input and y the n_y-dimensional output. The independent variable is the time t, and we assume that all the functions f^{α}, c^{α_y} are smooth to avoid mathematical subtleties. We will study the behavior of these systems with respect to diffeomorphic coordinate changes of the form

$$\bar{t} = \varphi^1(t) , \quad \bar{x}^{\alpha} = \varphi^{\alpha}(t, x) , \quad \bar{u}^{\alpha_u} = \varphi^{\alpha_u}(t, x, u) \tag{2.2}$$

such that it takes the form

$$\begin{aligned} \tfrac{\mathrm{d}}{\mathrm{d}\bar{t}}\bar{x}^{\alpha}(\bar{t}) &= \bar{f}^{\alpha}(\bar{t}, \bar{x}(\bar{t}), \bar{u}(\bar{t})) , \quad \alpha = 1, \ldots, n_x \\ y^{\alpha_y}(\bar{t}) &= \bar{c}^{\alpha_y}(\bar{t}, \bar{x}(\bar{t}), \bar{u}(\bar{t})) , \quad \alpha_y = 1, \ldots, n_y . \end{aligned} \tag{2.3}$$

Although it is straightforward to include, we will not consider transformations of the output y here. In the following subsections we characterize several dynamic systems by their geometric properties, which are derived from the behavior of the systems with respect to coordinate transformations. Roughly speaking, we will be able to identify dynamic systems with geometric objects, defined on certain manifolds. This approach explains, why differential geometry plays a major role in the theory of dynamic systems. Therefore, we have to use the concepts and will use the notation introduced in the chapter *Basic Differential Geometry for Mechanics and Control* of this book. The interested reader is kindly asked to study this chapter before one continues reading this contribution.

2.1 Dynamic Systems without Input

To start with, we choose the autonomous system described locally by n_x ordinary differential equations. Let $\mathcal{M}$ denote the state manifold with local coordinates $(x^1, \ldots, x^{n_x})$ and let $\mathcal{T}(\mathcal{M})$ denote the tangent bundle of $\mathcal{M}$ with coordinates $(x^{\alpha}, \dot{x}^{\alpha})$, then this special case of (2.1) can be written as

$$\dot{x}^{\alpha} = f^{\alpha}(x) , \quad \alpha = 1, \ldots, n_x , \tag{2.4}$$

where we assume $f^{\alpha} \in C^{\infty}(\mathcal{M})$. A short look at the change of the coordinates for $\mathcal{T}(\mathcal{M})$ given by

$$\bar{x}^{\alpha} = \varphi^{\alpha}(x) , \quad \dot{\bar{x}}^{\alpha} = \partial_{\beta}\varphi^{\alpha}(x)\,\dot{x}^{\beta} , \quad \beta = 1, \ldots, n_x$$

and at the transformed system

$$\dot{\bar{x}}^{\alpha} = \bar{f}^{\alpha}(\bar{x}) , \quad \bar{f}^{\alpha} = \left((\partial_{\beta}\varphi^{\alpha}) f^{\beta}\right) \circ \varphi^{-1}$$

shows, that the system (2.4) can be considered as a vector field $f \in \Gamma(\mathcal{T}(\mathcal{M}))$,

$$f = f^\alpha \partial_\alpha \in \Gamma(\mathcal{T}(\mathcal{M})) \ . \tag{2.5}$$

Obviously, f can also be seen as map $f : C^\infty(\mathcal{M}) \to C^\infty(\mathcal{M})$ given by the differential operator (2.5). Furthermore, f admits locally a very simple description. Let ϕ_t denote the flow of (2.4), then we choose new coordinates $(\bar{x}^\alpha)$ near a point x such that

$$0 = \bar{\varphi}(x) \ , \quad \frac{\partial}{\partial t}\bar{\phi}_t^1(\bar{x})\bigg|_{t=0,\bar{x}=0} = \bar{f}^1(0) \neq 0$$

is met with the flow $\bar{\phi}_t = \bar{\varphi} \circ \phi_t \circ \bar{\varphi}^{-1}$ expressed in the new coordinates. This is always possible, iff $f(x) \neq 0$ is met, i.e. x is not an equilibrium. Now, we choose the new coordinates $(\tilde{x}^\alpha)$,

$$\bar{x} = \psi(\tilde{x}) \ , \quad \psi(\tilde{x}) = \bar{\phi}_{\tilde{x}^1}\left(0, \tilde{x}^2, \ldots, \tilde{x}^{n_x}\right) \ ,$$

which is always possible in a neighborhood of $\tilde{x} = 0$ because of

$$\partial_1 \psi^\alpha(0) = f^\alpha \ , \quad \partial_\alpha \psi^\beta(0) = \delta_\alpha^\beta \ , \quad \beta = 2, \ldots, n_x$$

and the inverse function theorem. Let $\tilde{\phi}_t = \psi^{-1} \circ \bar{\phi}_t \circ \psi$ be the flow in these coordinates then from

$$\begin{aligned} \tilde{\phi}_t(\tilde{x}) &= \psi^{-1} \circ \bar{\phi}_t \circ \bar{\phi}_{\tilde{x}^1}\left(0, \tilde{x}^2, \ldots, \tilde{x}^{n_x}\right) \\ &= \psi^{-1} \circ \bar{\phi}_{t+\tilde{x}^1}\left(0, \tilde{x}^2, \ldots, \tilde{x}^{n_x}\right) \\ &= \left(t + \tilde{x}^1, \tilde{x}^2, \ldots, \tilde{x}^{n_x}\right) \end{aligned}$$

it follows that the field f has the simple representation

$$\tilde{f}^1 = 1 \ , \quad \tilde{f}^\beta = 0 \ , \quad \beta = 2, \ldots, n_x$$

in the coordinates $(\tilde{x}^\alpha)$.

Now we consider the time variant case and introduce the bundle $(\mathcal{E}, \pi, \mathcal{B})$, where $\mathcal{B}$ denotes the time manifold with local coordinate (t) and $\mathcal{E}$ is the total manifold with coordinates (t, x^α), $\alpha = 1, \ldots, n_x$. Let $J(\mathcal{E})$ be the first jet manifold of $\mathcal{E}$ with coordinates $(t, x^\alpha, x_1^\alpha)$, then this special case of (2.1) can be represented by

$$x_1^\alpha = f^\alpha(t, x) \ , \quad \alpha = 1, \ldots, n_x \ . \tag{2.6}$$

A short look at the change of the coordinates

$$\bar{t} = \varphi^1(t) \ , \quad \bar{x}^\alpha = \varphi^\alpha(t, x) \ , \quad \bar{x}_1^\alpha \partial_1 \varphi^1 = \partial_1 \varphi^\alpha + \partial_\beta \varphi^\alpha x_1^\beta \ ,$$

and the transformed system

$$\bar{x}_1^\alpha = \bar{f}^\alpha(\bar{t}, \bar{x}) \ , \quad \bar{f}^\alpha = \left(\left(\partial_1 \varphi^\alpha + \partial_\beta \varphi^\alpha f^\beta\right)\left(\partial_1 \varphi^1\right)^{-1}\right) \circ \varphi^{-1}$$

shows, that the system (2.6) is a section of $J(\mathcal{E})$. Let $(t, x, \dot{t}, \dot{x})$ be coordinates of $\mathcal{T}(\mathcal{E})$, then we get with $f \in \Gamma(\mathcal{T}(\mathcal{E}))$,

$$f = \partial_1 + f^\alpha \partial_\alpha \tag{2.7}$$

the map $f : C^\infty(\mathcal{E}) \to C^\infty(\mathcal{E})$. A short test shows that the system (2.6) does not transform like (2.7), since in general $\dot{\bar{t}} = \partial_1 \varphi^{\bar{1}} \neq 1$ is met. But sections of $J(\mathcal{E})$ can be identified with the tensor

$$\Lambda = \mathrm{d}t \otimes f \,, \tag{2.8}$$

since the transformation rules of (2.6) coincide with the rules of (2.8). Therefore, (2.6) defines the connection (2.8), which allows us to split $\mathcal{T}(\mathcal{E})$,

$$\mathcal{T}(\mathcal{E}) = \mathcal{H}(\mathcal{E}) \oplus_{\mathcal{E}} \mathcal{V}(\mathcal{E}) \,,$$

where $\mathcal{V}(\mathcal{E})$ denotes the vertical bundle $\mathcal{V}(\mathcal{E}) = \{v \in \mathcal{T}(\mathcal{E}) \,|\, \pi_*(v) = 0\}$ and $\mathcal{H}$ is the horizontal bundle $\mathcal{H}(\mathcal{E}) = \mathrm{span}\{f\}$.

The system (2.6) admits another representation as a Pfaffian system given by

$$\omega^\alpha(t,x) = \mathrm{d}x^\alpha - f^\alpha(t,x)\,\mathrm{d}t \tag{2.9}$$

with $\omega^\alpha \in \Gamma(\mathcal{T}^*(\mathcal{E}))$. This representation is connected with the previous one by

$$f \rfloor \omega^\alpha = 0 \,, \quad \partial_1 \omega^\alpha \neq 0 \,.$$

Therefore, (2.9) is often said to be dual to (2.7). The tensor complementary to (2.8) is given by

$$\Delta = \omega^\alpha \otimes \partial_\alpha \,.$$

Both tensors are related by the relations $\Lambda \rfloor \Delta = 0$, $\Delta \rfloor \Lambda = 0$.

With the coordinate transformation

$$t = \bar{\varphi}^1(\bar{t}) \,, \quad x^\alpha = \bar{\varphi}^\alpha(\bar{t},\bar{x}) \,, \quad x_1^\alpha \partial_{\bar{1}} \bar{\varphi}^1 = \partial_{\bar{1}} \bar{\varphi}^\alpha + \partial_\beta \bar{\varphi}^\alpha \bar{x}_{\bar{1}}^\beta \,,$$

the transformed Pfaffian system is given by

$$\bar{\omega}^\alpha = \partial_\beta \bar{\varphi}^\alpha \mathrm{d}\bar{x}^\beta + \left(\partial_{\bar{1}} \bar{\varphi}^\alpha - \partial_{\bar{1}} \bar{\varphi}^1 f^\alpha \circ \bar{\varphi}\right) \mathrm{d}\bar{t} \,.$$

It is worth mentioning that this transformation would be possible, even if $\bar{\varphi}^1$ is invertible but $\bar{\varphi}$ not. But in this case the transformed system cannot be converted to an explicit one, since the matrix $\left[\partial_\alpha \bar{\varphi}^\beta\right]$ is singular. Finally, with $\sigma \in \Gamma(\mathcal{E})$ we recover the differential equations by

$$\sigma^*(\omega^\alpha) = \mathrm{d}\sigma^\alpha(t) - f^\alpha \circ \sigma(t)\,\mathrm{d}t = \left(\partial_1 \sigma^\alpha(t) - f^\alpha \circ \sigma(t)\right)\mathrm{d}t = 0$$

in a straightforward manner.

2.2 Systems with Input

Now we turn towards systems with input and consider the equations

$$\dot{x}^\alpha = f^\alpha(x,u) \,, \quad \alpha = 1,\ldots,n_x \tag{2.10}$$

together with the field

$$f = f^\alpha \partial_\alpha \,. \tag{2.11}$$

Let $\mathcal{M}$, $\mathcal{T}(\mathcal{M})$ denote the state manifold and its tangent bundle with local coordinates $(x^1, \ldots, x^{n_x})$ and $(x^\alpha, \dot{x}^\alpha)$, respectively. Obviously, $f \in \Gamma(\mathcal{T}(\mathcal{M}))$ cannot be met, since the system (2.10) defines a submanifold of $\mathcal{T}(\mathcal{M})$ parameterized by u. To simplify the description, we introduce the new bundle $(\mathcal{Z}, \pi_{\mathcal{Z}}, \mathcal{M})$ with coordinates (x^α, u^{α_u}) for $\mathcal{Z}$. Obviously, a section of $\mathcal{Z}$ is nothing else then a static control law. Let us consider the pullback bundle $\pi_{\mathcal{Z}}^*(\mathcal{T}(\mathcal{M})) = \{((x,u), \dot{x}) \in \mathcal{Z} \times_{\mathcal{M}} \mathcal{T}(\mathcal{M}),\ \pi_{\mathcal{Z}}(x,u) = x = \tau_{\mathcal{M}}(x, \dot{x})\}$. The rules concerning the coordinate changes for $\mathcal{Z}$ are given by

$$\bar{x}^\alpha = \varphi^\alpha(x) \ , \quad \bar{u}^{\alpha_u} = \varphi^{\alpha_u}(x,u) \ ,$$

and they read for $\mathcal{T}(\mathcal{M})$ as

$$\bar{x}^\alpha = \varphi^\alpha(x) \ , \quad \dot{\bar{x}}^\alpha = \partial_\beta \varphi^\alpha(x) \dot{x}^\beta \ .$$

Both together are the transformation laws for $\pi_{\mathcal{Z}}^*(\mathcal{T}(\mathcal{M}))$. Therefore, the system (2.10) is nothing else than a section of $\pi_{\mathcal{Z}}^*(\mathcal{T}(\mathcal{M}))$. Although f can be considered as a map $f : C^\infty(\mathcal{Z}) \to C^\infty(\mathcal{Z})$, this map does not behave well with respect to the substitution because differentiation along f and substitution of $h \in \Gamma(\mathcal{Z})$ do not commute, or we get

$$(f^\alpha \circ h) \partial_\alpha (g \circ h) \neq (f^\alpha \partial_\alpha (g)) \circ h \ , \quad h \in \Gamma(\mathcal{Z}) \ , \quad g \in C^\infty(\mathcal{Z})$$

in general. One possibility to overcome this problem is to replace f of (2.11) by an infinite dimensional field, which will be discussed shortly for the time variant case only.

The last case is the time variant one with input. Let us consider the equations

$$x_1^\alpha = f^\alpha(t, x, u) \ , \quad \alpha = 1, \ldots, n \tag{2.12}$$

and let $\mathcal{E}$, $\mathcal{T}(\mathcal{E})$ be the bundle $(\mathcal{E}, \pi, \mathcal{B})$ and its tangent bundle with local coordinates (t, x), $(t, x, \dot{t}, \dot{x})$. Now, we introduce the new bundle $(\mathcal{Z}, \rho, \mathcal{E})$ with coordinates (t, x, u) for $\mathcal{Z}$. Again, a section of $\mathcal{Z}$ is a static time dependent control law. Analogously to above, one can show that the system (2.12) is nothing else than a section f,

$$f = \partial_1 + f^\alpha \partial_\alpha \tag{2.13}$$

of $\pi_{\mathcal{Z}}^*(\mathcal{T}(\mathcal{E}))$ with the pull back bundle $\pi_{\mathcal{Z}}^*(\mathcal{T}(\mathcal{E})) = \{(t, x, u, \dot{t}, \dot{x}) \in \mathcal{Z} \times_{\mathcal{E}} \mathcal{T}(\mathcal{E}),\ \pi_{\mathcal{Z}}(t, x, u) = (t, x) = \tau_{\mathcal{E}}(t, x, \dot{t}, \dot{x})\}$. The interested reader is asked to check this statement by investigation of the transformation rules. Analogously to above, f can be considered as a map $f : C^\infty(\mathcal{Z}) \to C^\infty(\mathcal{Z})$, but the same problem shows up, since in general differentiation along f and substitution of $h \in \Gamma(\mathcal{Z})$ do not commute. One possibility to overcome this problem is to replace f of (2.13) by the infinite dimensional field f_e,

$$f_e = \partial_1 + f^\alpha \partial_\alpha + u_{I+1}^{\alpha_u} \partial_{\alpha_u}^I \ , \quad \partial_{\alpha_u}^I = \frac{\partial}{\partial u_I^{\alpha_u}} \ , \quad I = 0, \ldots, \infty \ . \tag{2.14}$$

The variables $u_I^{\alpha_u}$ are jet coordinates of the bundle $\mathcal{U} = (\mathcal{U}, \pi_{\mathcal{U}}, \mathcal{B})$, with local coordinates (t, u^{α_u}), or equivalently, the input u is considered as a smooth function of the time t. Similar considerations refer to (2.11).

The system (2.12) can also be described by the Pfaffian system

$$\omega^{\alpha} = \mathrm{d}x^{\alpha} - f^{\alpha}\mathrm{d}t \tag{2.15}$$

with $\omega^{\alpha} \in \Gamma\left(\pi_{\mathcal{Z}}^{*}\left(\mathcal{T}^{*}\left(\mathcal{E}\right)\right)\right)$ together with the contact forms

$$\theta_{I}^{\alpha_u} = \mathrm{d}u_{I}^{\alpha_u} - u_{I+1}^{\alpha_u}\mathrm{d}t\,, \quad I = 1,\ldots,\infty\,, \quad \theta_{I}^{\alpha_u} \in \Gamma\left(\mathcal{T}^{*}\left(J^{\infty}\left(\mathcal{U}\right)\right)\right)\,. \tag{2.16}$$

Again, both descriptions, (2.14) and (2.15), (2.16), are related by

$$f_e \rfloor \omega^{\alpha} = 0\,, \quad f_e \rfloor \theta_{I}^{\alpha_u} = 0\,.$$

Unfortunately, fields like f_e have a different behavior from finite dimensional ones. The reader should be careful, since several results of the chapter *Basic Differential Geometry for Mechanics and Control*, like the Theorem of Frobenius, are restricted to the finite dimensional case. E.g. in (Saunders, 1989) the reader will find a more detailed introduction to this topic.

3 Observability

Roughly speaking, the system (2.12) with the output

$$y^{\alpha_y} = c^{\alpha_y}\left(t,x,u\right)\,, \quad \alpha_y = 1,\ldots,n_y \tag{3.1}$$

is observable if one can reconstruct the state x from measurements of the input u and the output y. More precise, given sections $h \in \Gamma\left(\mathcal{Z}\right)$, $\sigma \in \Gamma\left(\mathcal{E}\right)$ such that

$$\left(x_{1}^{\alpha} - f^{\alpha} \circ h\right) \circ j\left(\sigma\right) = 0$$

is met on the interval $T = [t_1, t_2] \ni t$, $t_1, t_2 \in \mathbb{R}$, can one determine $x^{\alpha}\left(t_1\right) = \sigma^{\alpha}\left(t_1\right)$ from the measurement of $y^{\alpha_y}\left(t\right) = c^{\alpha_y} \circ h \circ \sigma\left(t\right)$, $u^{\alpha_u}\left(t\right) = h \circ \sigma\left(t\right)$? If this is possible, then the system is said to be observable along $\sigma\left(T\right)$, $h \circ \sigma\left(T\right)$. We call the system locally observable at (t,x), if this property holds for all trajectories sufficiently close to (t,x) with $\sigma^{\alpha}\left(t\right) = x^{\alpha}$ and a suitable choice of h. It is worth mentioning, that here local observability depends on the choice of h. Now, one can show that the system (2.12) with the output (3.1) is locally not observable, iff there exists a map φ,

$$t = \varphi^{1}\left(t\right)\,, \quad \bar{x}^{\bar{\alpha}} = \varphi^{\bar{\alpha}}\left(t,x\right)\,, \quad \bar{\alpha} = 1,\ldots,n_{\bar{x}} > 0\,, \quad u^{\alpha_u} = \varphi^{\alpha_u}\left(t,x,u\right)$$

such that the transformed system, which has the same input output map like (2.12), (3.1), takes the form

$$\begin{aligned} \bar{x}_{1}^{\bar{\alpha}} &= \bar{f}^{\bar{\alpha}}\left(t,\bar{x},u\right) \\ y^{\alpha_y} &= \bar{c}^{\alpha_y}\left(t,\bar{x},u\right)\,, \quad \bar{c}^{\alpha_y} \circ \varphi = c^{\alpha_y} \end{aligned}$$

with $n_{\bar{x}} < n_x$. For the proof see (Isidori, 1995), (Nijmeijer and van der Schaft, 1990), (Sastry, 1999). The crucial observation for this case is that there exists at least one vertical field $v \in \mathcal{V}\left(\mathcal{E}\right)$, which meets

$$v\left(c^{\alpha_y}\right) = 0\,, \quad v\left(\varphi^{\bar{\alpha}}\right) = 0\,.$$

Let us introduce the codistribution $\Delta_0^* = \operatorname{span}\{\mathrm{d}c^{\alpha_v}, \mathrm{d}t, \mathrm{d}u^{\alpha_u}\}$, which is annihilated by v. Since this must hold along any trajectory emerging from x at t for any h, we consider the change of Δ_0^* along a trajectory given by f_e of (2.14). From

$$f_e(\mathrm{d}c^{\alpha_v}) = \mathrm{d}f_e(c^{\alpha_v}) \ , \quad f_e(\mathrm{d}t) = 0 \ , \quad f_e(\mathrm{d}u_I^{\alpha_u}) = \mathrm{d}u_{I+1}^{\alpha_u}$$

it follows that Δ_1^* is given by $\Delta_1^* = \operatorname{span}\{\mathrm{d}f_e(c^{\alpha_v}), \mathrm{d}u_1^{\alpha_u}\}$, which v must annihilate. Proceeding this way, we get a sequence of codistributions $\Delta_k^* = \operatorname{span}\left\{\mathrm{d}(f_e)^k(c^{\alpha_v}), \mathrm{d}u_k^{\alpha_u}\right\}$, where we used the abbreviation $(f_e)^0(c^{\alpha_v}) = c^{\alpha_v}$, $(f_e)^k(c^{\alpha_v}) = f_e\left((f_e)^{k-1}(c^{\alpha_v})\right)$. At first glance it seems that we have to construct an arbitrary number of codistributions Δ_k^*. Now, let us assume that the Δ_k^* are regular, and let $\bar{\Delta}^*(k)$ denote $\bar{\Delta}^*(k) = \sum_{i=0}^{k}\Delta_i^*$. We choose numbers k_{α_y} such that $\left\{\mathrm{d}t, \mathrm{d}(f_e)^{i_{\alpha_y}}(c^{\alpha_v}), \mathrm{d}u_i^{\alpha_u}\right\}$, $i_{\alpha_y} = 0, \ldots, k_{\alpha_y}$, $i = 0, \ldots, k$ is a basis of Δ^*, which is always possible. Let N be the smallest number such that $\bar{\Delta}^*(N+1) = \sum_{\bar{i}=0}^{N+1}\Delta_{\bar{i}}^*$ admits the basis $\left\{\mathrm{d}t, \mathrm{d}(f_e)^{i_{\alpha_y}}(c^{\alpha_v}), \mathrm{d}u_{\bar{i}}^{\alpha_u}\right\}$, $\bar{i} = 0, \ldots, N+1$. Then from

$$\mathrm{d}(f_e)^{k_{\beta_y}+1}(c^{\alpha_v}) = \mu^{\beta_v}\mathrm{d}t + \nu_{i_{\alpha_y},\alpha_y}^{\beta_y}\mathrm{d}(f_e)^{i_{\alpha_y}}(c^{\alpha_v}) + \xi_{\alpha_u}^{\beta_y,\bar{i}}\mathrm{d}u_{\bar{i}}^{\alpha_u}$$

and

$$\begin{aligned} f_e\left(\mathrm{d}(f_e)^{k_{\beta_y}+1}(c^{\alpha_v})\right) &= \mathrm{d}(f_e)^{k_{\beta_y}+2}(c^{\alpha_v}) \\ &= f_e(\mu^{\beta_v})\,\mathrm{d}t + f_e\left(\nu_{i_{\alpha_y},\alpha_y}^{\beta_y}\right)\mathrm{d}(f_e)^{i_{\alpha_y}}(c^{\alpha_v}) \\ &\quad + f_e\left(\xi_{\alpha_u}^{\beta_y,\bar{i}}\right)\mathrm{d}u_{\bar{i}}^{\alpha_u} + \nu_{i_{\alpha_y},\alpha_y}^{\beta_y}\mathrm{d}(f_e)^{i_{\alpha_y}+1}(c^{\alpha_v}) \\ &\quad + \xi_{\alpha_u}^{\beta_y,\bar{i}}\mathrm{d}u_{\bar{i}+1}^{\alpha_u} \end{aligned}$$

it follows that $\mathrm{d}(f_e)^{k_{\beta_y}+2}(c^{\alpha_v}) \in \operatorname{span}\left\{\mathrm{d}t, \mathrm{d}(f_e)^{i_{\alpha_y}}(c^{\alpha_v}), \mathrm{d}u_{\bar{i}}^{\alpha_u}, \mathrm{d}u_{N+2}^{\alpha_u}\right\}$ is met. In this case, the system is exhausted, and any further differentiation by f_e does not supply new information. Therefore, this procedure stops reliably after a finite number of steps. If $\dim\left(\bar{\Delta}^*(N)\right) = 1 + n_u(N+1) + n_x$, then the system (2.12), (3.1) is locally observable. Now, two facts are worth mentioning. First, one constructs the maximal, with respect to the dimension, involutive distribution $\bar{\Delta}$, such that $\bar{\Delta}$ is the annihilator of $\bar{\Delta}^*$. Second, the presented procedure can easily be implemented in any advanced computer algebra program.

4 Accessibility

The accessibility problem for the system (2.12) can be characterized in the following manner. Given a section $h \in \Gamma(\mathcal{Z})$ and a section $\sigma \in \Gamma(\mathcal{E})$ such that

$$\left(x_1^\alpha - f^\alpha \circ h\right) \circ j(\sigma) = 0 \tag{4.1}$$

is met on the interval $[t_1, t_2] \ni t$, $t_1, t_2 \in \mathbb{R}$, can we find a section $\bar{\sigma} \in \Gamma(\mathcal{E})$ of (2.12) by a suitable choice of $\bar{h} \in \Gamma(\mathcal{Z})$ such that $\bar{\sigma}$, $\bar{h}$ fulfill (4.1) together with $\sigma^\alpha(t_1) = \bar{\sigma}^\alpha(t_1)$, and furthermore $\bar{\sigma}^{\alpha}(t_2) = \bar{x}^{\alpha}$ is met for any $\bar{x}$ with $|\bar{x}^{\alpha} - \sigma^{\alpha}(t_2)| < \epsilon$ with a sufficiently small $\epsilon \in \mathbb{R}^{+}$? If this is possible, then we say, the system is accessible along $\sigma(T)$, $h \circ \sigma(T)$. We call the system locally accessible at (t, x), if this property holds for all trajectories sufficiently close to (t, x) with $\sigma^{\alpha}(t) = x^{\alpha}$. Now, one can show that the system (2.12) is locally not accessible, iff there exists a map φ,

$$t = \varphi^{1}(t) \ , \quad \bar{x}^{\bar{\alpha}} = \varphi^{\bar{\alpha}}(t, x) \ , \quad \bar{\alpha} = 1, \ldots, n_{\bar{x}} > 0$$

such that the transformed system takes the form

$$\bar{x}_{1}^{\bar{\alpha}} = \bar{f}^{\bar{\alpha}}(t, \bar{x}) \ . \tag{4.2}$$

For the proof, the interested reader is referred to (Isidori, 1995), (Nijmeijer and van der Schaft, 1990), (Sastry, 1999). Roughly speaking, the system (2.12) contains a subsystem (4.2) without input.

To check, whether φ exists, we rewrite (4.2) as a Pfaffian system

$$\bar{\omega}^{\bar{\alpha}}(t, \bar{x}) = \mathrm{d}\bar{x}^{\bar{\alpha}} - \bar{f}^{\bar{\alpha}}(t, \bar{x})\,\mathrm{d}t \ .$$

The crucial observation is that the forms $\bar{\omega}^{\bar{\alpha}}$ meet $\mathrm{d}\bar{\omega}^{\bar{\alpha}} \wedge \bar{\omega}^{1} \wedge \cdots \wedge \bar{\omega}^{n_{\bar{x}}} = 0$. Using the representation (2.15) for (2.12) we see that $\varphi^{*}(\bar{\omega}^{\bar{\alpha}}) \in \Delta_{0}^{*} = \operatorname{span}\{\omega^{\alpha}\}$ is met. Therefore, we have to check, whether Δ_{0}^{*} contains an integrable codistribution Δ_{N}^{*}. To construct Δ_{N}^{*} we calculate the derived system Δ_{1}^{*} of Δ_{0}^{*} according to

$$\Delta_{1}^{*} = \operatorname{span}\left\{\omega_{1}^{\alpha} \in \Delta_{0}^{*} | \mathrm{d}\omega_{1}^{\alpha} \wedge \omega_{0}^{1} \wedge \cdots \wedge \omega_{0}^{n_{x}} = 0\right\} \ , \quad \omega_{0}^{\alpha_{x}} = \omega^{\alpha_{x}} \ . \tag{4.3}$$

In a straightforward manner we can construct Δ_{2}^{*}, Δ_{3}^{*} and we derive the following filtration

$$\Delta_{0}^{*} \supset \Delta_{1}^{*} \supset \cdots \supset \Delta_{N}^{*} \ . \tag{4.4}$$

For the sake of simplicity, let us assume that all codistributions are regular at (t, x). Obviously, the last system Δ_{N}^{*} meets $\Delta_{N}^{*} = \Delta_{N+1}^{*}$ and the codistribution Δ_{N}^{*} is involutive. Therefore, the procedure stops reliably after a finite number of steps, and we conclude that the system is locally accessible near (t, x), iff $\Delta_{N}^{*} = \operatorname{span}\{0\}$ is met. Here, two facts are worth mentioning. First, one construct the with respect to the dimension maximal codistribution. Second, the presented procedure can easily be implemented in any advanced computer algebra program.

5 Input to Output Linearization

In the following we are going to investigate whether one can find for the system

$$\begin{array}{rcll} x_{1}^{\alpha} & = & f^{\alpha}(t, x, u) \ , & \alpha = 1, \ldots, n_{x} \\ y^{\alpha_{y}} & = & c^{\alpha_{y}}(t, x) \ , & \alpha_{y} = 1, \ldots, n_{y} \ , \end{array} \tag{5.1}$$

see also (2.1), with $n_{u} = n_{y}$ a static control law $u^{\alpha_{u}} = w^{\alpha_{u}}(t, x, v)$, $\alpha_{u} = 1, \ldots, n_{u}$ such that the input output map from the new input $v^{\alpha_{v}}$, $\alpha_{v} = 1, \ldots, n_{v} = n_{u}$ to the output

y^{α_y} is linear. According to (2.13) the system (5.1) can be seen as a section $\partial_1 + f^{\alpha}\partial_{\alpha} = f$ of $\pi^*_{\mathcal{Z}}(\mathcal{T}(\mathcal{E}))$, which is also a map $f : C^{\infty}(\mathcal{E}) \to C^{\infty}(\mathcal{Z})$.

Let in some neighborhood of a point (t, x, u) be met

$$(f)^{i_{\alpha_y}}(c^{\alpha_y}) = h^{\alpha_y, i_{\alpha_y}} \in C^{\infty}(\mathcal{E}) \ , \quad i_{\alpha_y} = 0, \ldots, r_{\alpha_y} - 1$$

and

$$(f)^{r_{\alpha_y}}(c^{\alpha_y}) = h^{\alpha_y, r_{\alpha_y}} \in C^{\infty}(\mathcal{Z}) \ , \quad h^{\alpha_y, r_{\alpha_y}} \notin C^{\infty}(\mathcal{E}) \ ,$$

as well as

$$\det\left(\left[\partial_{\alpha_u} h^{\alpha_y, r_{\alpha_y}}\right]\right) \neq 0 \ ,$$

then $\left[r_{\alpha_y}\right]$ is called the *vector relative degree* at (t, x, u) and

$$r = \sum_{\alpha_y=1}^{n_y} r_{\alpha_y} \tag{5.2}$$

the *relative degree* of the system (5.1). The reader can prove easily that the relative degree is independent of the choice of the coordinates. Due to the conditions stated above, see also the implicit function theorem, there exist functions w^{α_u} at least locally such that the relations

$$h^{\alpha_y, r_{\alpha_y}}(t, x, w(t, x, v)) = v^{\alpha_y} \tag{5.3}$$

are met. Now, we apply the diffeomorphic change of coordinates

$$t = \varphi^1(t) \ , \quad z^{\alpha} = \varphi^{\alpha}(t, x) \ , \quad v^{\alpha_v} = \varphi^{\alpha_v}(t, x, u) \tag{5.4}$$

yielding

$$\begin{aligned} z^{\alpha_y, i_{\alpha_y}} &= h^{\alpha_y, i_{\alpha_y}-1} \ , & i_{\alpha_y} &= 1, \ldots, r_{\alpha_y} \\ z^j &= \bar{h}^j \ , & j &= r+1, \ldots, n_x \\ v^{\alpha_y} &= h^{\alpha_y, r_{\alpha_y}}(t, x, u) \end{aligned} \tag{5.5}$$

with the functions $\bar{h}^j$ chosen such that φ is a diffeomorphism. Thus, in the new coordinates the system representation reads as

$$\begin{aligned} z_1^{\alpha_y, i_{\alpha_y}} &= f\left(h^{\alpha_y, i_{\alpha_y}-1}\right) \circ \varphi^{-1} = h^{\alpha_y, i_{\alpha_y}} \circ \varphi^{-1} \\ &= z^{\alpha_y, i_{\alpha_y}+1} \ , \qquad i_{\alpha_y} = 1, \ldots, r_{\alpha_y} - 1 \\ z_1^{\alpha_y, r_{\alpha_y}} &= f\left(h^{\alpha_y, r_{\alpha_y}-1}\right) \circ \varphi^{-1} = h^{\alpha_y, r_{\alpha_y}} \circ \varphi^{-1} \\ &= v^{\alpha_y} \\ z_1^j &= f\left(\bar{h}^j\right) \circ \varphi^{-1} \ , \qquad j = r+1, \ldots, n_x \ . \end{aligned} \tag{5.6}$$

Obviously, the map from the input v^{α_y} to the output $y^{\alpha_y} = c^{\alpha_y} \circ \varphi^{-1}$ of (5.6) is linear. Since this property is independent of the choice of the coordinates for the state, this is true also for the system (5.1) with the feedback law $u^{\alpha_u} = w^{\alpha_u}$, see (5.3).

The control law

$$v^{\alpha_y} = -\sum_{i_{\alpha_y}=1}^{r_{\alpha_y}} \mu^{\alpha_y}_{i_{\alpha_y}-1} z^{\alpha_y, i_{\alpha_y}} + \bar{v}^{\alpha_y} \tag{5.7}$$

with the Hurwitz polynomials

$$p^{\alpha_y}(s) = \sum_{i=0}^{r_{\alpha_y}} \mu_i^{\alpha_y} s^i \ , \quad \mu_{r_{\alpha_y}}^{\alpha_y} = 1$$

and the new inputs $\bar{v}^{\alpha_y}$ allows us to stabilize the input to output map of the system (5.6) with the new input $\bar{v}$. But the applicability of the control law (5.7) requires the stability of the subsystem

$$z_1^j = f\left(\bar{h}^j\right) \circ \varphi^{-1}\left(t, z, -\mu_{i_{\alpha_y}-1}^{\alpha_y} z^{\alpha_y, i_{\alpha_y}} + \bar{v}^{\alpha_y}\right) \tag{5.8}$$

with the inputs $\bar{v}^{\alpha_y}$ and $z^{\bar{\alpha}}$, $\bar{\alpha} = 1, \ldots, r$. In order to show that the stability of (5.8) is independent of the choice of (5.7), let us investigate the inverse system of

$$\begin{aligned} x_1^\alpha &= f^\alpha(t, x, u) \\ v^{\alpha_y} &= h^{\alpha_y, r_{\alpha_y}}(t, x, u) \ , \end{aligned}$$

which follows as

$$\begin{aligned} x_1^\alpha &= f^\alpha(t, x, w(t, x, v)) \\ u^{\alpha_u} &= w^{\alpha_u}(t, x, v) \ . \end{aligned} \tag{5.9}$$

The change of coordinates

$$t = \varphi^1(t) \ , \quad z^\alpha = \varphi^\alpha(t, x) \ , \quad v^{\alpha_v} = \varphi^{\alpha_v}(v)$$

with

$$\begin{aligned} z^{\alpha_y, i_{\alpha_y}} &= h^{\alpha_y, i_{\alpha_y}-1} \ , & i_{\alpha_y} &= 1, \ldots, r_{\alpha_y} \\ z^j &= \bar{h}^j \ , & j &= r+1, \ldots, n_x \end{aligned}$$

for the inverse system (5.9) yields

$$\begin{aligned} z_1^{\alpha_y, i_{\alpha_y}} &= f\left(h^{\alpha_y, i_{\alpha_y}-1}\right) \circ \varphi^{-1} = h^{\alpha_y, i_{\alpha_y}} \circ \varphi^{-1} \\ &= z^{\alpha_y, i_{\alpha_y}+1} \ , \qquad i_{\alpha_y} = 1, \ldots, r_{\alpha_y} - 1 \\ z_1^{\alpha_y, r_{\alpha_y}} &= f\left(h^{\alpha_y, r_{\alpha_y}-1}\right) \circ \varphi^{-1} = h^{\alpha_y, r_{\alpha_y}} \circ \varphi^{-1} = \\ &= v^{\alpha_y} \\ z_1^j &= f\left(\bar{h}^j\right) \circ \varphi^{-1} \ , \qquad j = r+1, \ldots, n_x \ . \end{aligned}$$

The subsystem

$$z_1^j = f\left(\bar{h}^j\right) \circ \varphi^{-1}$$

is called the zero-dynamics of the system (5.6). Obviously, the stability of the zero-dynamics is independent of the special choice for the control law (5.7).

6 Input to State Linearization

According to the previous section, it is straightforward to see, that for a relative degree $r = n_x$, see (5.2), no zero-dynamics exist. Therefore, the system is static feedback equivalent to a linear, time invariant one. The diffeomorphic coordinate transformation (5.4) and (5.5) yields the transformed system of the form

$$\begin{aligned} z_1^{\alpha_y,i_{\alpha_y}} &= z^{\alpha_y,i_{\alpha_y}+1} , \qquad i_{\alpha_y} = 1,\ldots,r_{\alpha_y}-1 \\ z_1^{\alpha_y,r_{\alpha_y}} &= v^{\alpha_v} , \end{aligned} \tag{6.1}$$

which is usually called the *Brunovsky normal form.* Obviously, the relative degree r depends on the output functions c^{α_v} of (5.1). The special choice

$$y^{\alpha_v} = z^{\alpha_y,1} \tag{6.2}$$

for the system (6.1) shows that any system (5.1), which is transformable to (6.1), must admit an output y with $r = n_x$.

Let us rewrite the system (6.1) as a Pfaffian system

$$\begin{aligned} &\mathrm{d}z^{\alpha_y,i_{\alpha_y}} - z^{\alpha_y,i_{\alpha_y}+1}\mathrm{d}t , \qquad i_{\alpha_y} = 1,\ldots,r_{\alpha_y}-1 \\ &\mathrm{d}z^{\alpha_y,r_{\alpha_y}} - v^{\alpha_v}\mathrm{d}t , \end{aligned}$$

then it is easy to see that the system (6.1) is locally accessible, since the filtration (4.4) of derived systems, see (4.3), meets $\Delta_N^* = \operatorname{span}\{0\}$. The crucial observation is that in addition the extended systems $\Delta_k^* \oplus \operatorname{span}\{dt\}$, $k = 0,\ldots,N-1$ are integrable, e.g. see (Sastry, 1999). Because the transformation (5.4) preserves this property, it must be met by the original system, too. From (5.4) with $z^{\alpha_y,i_{\alpha_y}} = \varphi^{\alpha_y,i_{\alpha_y}}(t,x)$, $i_{\alpha_y} = 1,\ldots,r_{\alpha_y}$ it follows that the output $y^{\alpha_v} = \varphi^{\alpha_y,1}(t,x)$ meets $r = n_x$. Furthermore, the functions meet the relations

$$\begin{aligned} \partial_{\alpha_u}\left((f)^{i_{\alpha_y}-1}\left(\varphi^{\alpha_y,1}\right)\right) &= 0 , \qquad i_{\alpha_y} = 1,\ldots,r_{\alpha_y} \\ \partial_{\alpha_u}\left((f)^{r_{\alpha_y}}\left(\varphi^{\alpha_y,1}\right)\right) &\neq 0 , \end{aligned}$$

or equivalently

$$\begin{aligned} \mathrm{ad}^{i_{\alpha_y}-1}(f,\partial_{\alpha_u})\left(\varphi^{\alpha_y,1}\right) &= 0 , \qquad i_{\alpha_y} = 1,\ldots,r_{\alpha_y} \\ \mathrm{ad}^{r_{\alpha_y}}(f,\partial_{\alpha_u})\left(\varphi^{\alpha_y,1}\right) &\neq 0 , \end{aligned} \tag{6.3}$$

where we used the abbreviation $\mathrm{ad}^k(f,\partial_{\alpha_u}) = \left[f, \mathrm{ad}^{k-1}(f,\partial_{\alpha_u})\right]$, $\mathrm{ad}^0(f,\partial_{\alpha_y}) = \partial_{\alpha_y}$. Now, the Theorem of Frobenius ensures that the set of first order partial differential equations (6.3) admits n_y functionally independent solutions, iff the condition from above is met. This proves that this condition is necessary and locally sufficient. One has to determine a solution of (6.3) to construct a linearizing control law. But often it is possible to guess the suitable output by some physical insight into the plant (5.1).

7 Conclusion

This contribution had two goals. The first one was to show, how one can identify dynamic systems with geometric objects defined on certain manifolds in a straightforward manner. The advantage of this approach is that one can exploit the geometric properties for the system analysis, as well as for control loop design. This way we were able to introduce and discuss the basic properties accessibility and observability for a quite general class of dynamic systems described by a set of explicit ordinary differential equations. To show the power of this approach for the control loop design, the two methods input to output and input to state linearization were presented. Both methods turned out to be very useful for many engineering applications. Although one has to find solutions of a set of partial differential equations in the latter case, often engineering insight helps to find the desired output functions. Finally, it is worth mentioning that the presented method can easily be implemented in any advanced computer algebra system. This explains, why also this method became quite popular.

Bibliography

A. Isidori. *Nonlinear Control Systems.* Springer, London, UK, 1995.

H. Nijmeijer and A.J. van der Schaft. *Nonlinear Dynamical Control Systems.* Springer, New York, 1990.

S. Sastry. *Nonlinear Systems Analysis, Stability and Control.* Springer, New York, 1999.

D.J. Saunders. *The Geometry of Jet Bundles.* Cambridge University Press, Cambridge, UK, 1989.

Some Applications of Differential Geometry in Mechanics

Kurt Schlacher, Gernot Grabmair, Helmut Ennsbrunner and Richard Stadlmayr
Institute of Automatic Control and Control Systems Technology,
Johannes Kepler University of Linz, Austria

Abstract Classical mechanics is not only one of the most successful scientific disciplines, it stands also at the beginning of modern physics. Furthermore, it demonstrates the deep connection between physics and geometry. Therefore, the development of differential geometry was pushed by ideas from mechanics. The goal of this contribution is to show the geometric interpretation of certain mechanical ideas like the time-space manifold or the metric on the spatial manifold and geodesics. The covariant derivative, derived from a special connection, allows us to represent the famous equations of point mechanics in a coordinate free manner. Based on this approach we extend these geometric ideas such that we are able to discuss concepts required for continuum mechanics. Here new geometric concepts like the covariant differential are introduced, which allow to transfer several ideas from point mechanics to continuum mechanics.

1 Introduction

Classical mechanics is a well established scientific discipline for centuries, whose importance, especially for application in advanced machinery, vibration control, robotics etc., is still, as ever increasing. Several advanced topics of this field are already presented in this book. Therefore, the reader will find here neither an introduction into mechanics nor a discussion of its advanced topics. But mechanics is rich on geometric ideas, even more one can say it represents in some way the geometry of the space, we are living in. E.g. one of the basic assumptions is that a point mass moves along straight lines, if no forces affect the point mass. Furthermore, the time appears in the equation as affine parameter. The mathematical formalization of these geometric ideas lead to the introduction of a connection such that fundamental mechanical assumptions become expressible in a pure geometric manner. Therefore, this contribution is devoted to the underlying geometric structures of the well known equations of point and continuum mechanics.

According to the remarks from above, this contribution is divided into two parts. The first one deals with point mechanics, where geometric objects like vector fields are introduced. A major part is devoted to the covariant derivative derived from a connection. In Riemannian geometry the covariant derivative is derived from the metric on a manifold. We present a more general approach, where the covariant derivative is derived from a connection, that is a splitting of a linear space in general. This general construction will be specialized to the metric case. Based on these ideas, we take a closer look at Newton's equations and derive their Hamiltonian counterpart. The second part deals with continuum mechanics, where we extend the previous approach to the case of field theories, where new geometric objects like forms and vector valued forms appear. Furthermore, a new geometric tool, the covariant differential, is introduced. Based on this approach, we discuss the Eulerian and Lagrangian picture of the equations of conservation of mass and linear momentum. Of course, all the results are well known, and the reader is asked to consult the rich literature. E.g. the books (Choquet-Bruhat and DeWitt Morette, 1982), (Frankel, 1998) deal with differential geometry, as well as with many applications in physics. Field theories are the topics of (Prástaro, 1996) and (Truesdell and Noll, 2004), where the second book is devoted to classical theories, whereas the first one presents also non classical field theories. The engineering point of view with many applications in mechanics can be found, e.g. in (Ziegler, 1991).

In the following, we assume that all mathematical objects are smooth, that is, all functions are sufficiently often differentiable. Although one can often weaken this assumption, it allows us to treat many problems in a significantly simpler way. Furthermore, one can use the presented methods also in the non smooth case, but one has to take more care, since some mathematical and physical subtleties may occur.

2 Point Mechanics

To start with, we choose an inertial frame and Cartesian coordinates. Let us consider the motion of a point with mass $m \in \mathbb{R}^+$, with the coordinates (x^α), $\alpha = 1, \ldots, 3$ for the position of the point and the coordinate (t) for the time. It is well known that the equations of motion are given by

$$m \frac{\mathrm{d}^2}{\mathrm{d}t^2} x^\alpha (t) = F^\alpha \left(t, x(t), \frac{\mathrm{d}}{\mathrm{d}t} x(t) \right) , \tag{2.1}$$

where F^α denotes the force which acts on the point mass. The velocity v and the acceleration a of the point with position x at t are given by $v^\alpha (t) = \frac{\mathrm{d}}{\mathrm{d}t} x^\alpha (t)$ and $a^\alpha (t) = \frac{\mathrm{d}^2}{\mathrm{d}t^2} x^\alpha (t)$. The representation of the velocity v and of the acceleration a depends on the choice of the coordinates. Roughly speaking, we derive the form of (2.1) in the case of flat metric, where the metric tensor is more or less given by the identity matrix. Since coordinate changes do not preserve the structure of the metric in general, the equations of motions take on a different form after such a change. Interestingly enough, we have to change the definition of the velocity v and the acceleration a only, because they are derived from certain geometric assumptions concerning the configuration manifold, that is the manifold which is used to describe the motion. Therefore, we present an approach, which is also valid in more general coordinate systems, in the following subsections.

2.1 Velocity and Vector Fields

Let us denote the time manifold by $\mathcal{B}$ with coordinate (t^0) and the spatial manifold by $\mathcal{M}$ with coordinates (x^α), $\alpha = 1, \ldots, n$. According to Newtonian mechanics, we introduce the trivial bundle $(\mathcal{E}, \pi, \mathcal{B})$ with $\mathcal{E} = \mathcal{B} \times \mathcal{M}$. This choice guarantees that for every point $p \in \mathcal{E}$ there exists a well defined time t^0. Furthermore, $\mathcal{M}$ is equipped with the invertible matrix $[g_{\alpha\beta}]$, $g_{\alpha\beta} \in C^\infty(\mathcal{M})$, $\alpha, \beta = 1, \ldots, n$, which defines the metric tensor $g = g_{\alpha\beta} \mathrm{d}x^\alpha \otimes \mathrm{d}x^\beta$. In an inertial frame with Cartesian coordinates the metric is given by $g = \delta_{\alpha\beta} \mathrm{d}x^\alpha \otimes \mathrm{d}x^\beta$ with the Kronecker symbol $\delta_{\alpha\beta}$. The special structure of the time space manifold $\mathcal{E}$ can be characterized by the tensor

$$\Lambda = \mathrm{d}t^0 \otimes \partial_0 \,, \quad \partial_0 = \frac{\partial}{\partial t} \tag{2.2}$$

as well as by the complementary tensor

$$\Lambda_c = \mathrm{d}x^\alpha \otimes \partial_\alpha \,. \tag{2.3}$$

The tensors Λ, Λ_c are related by $\Lambda \rfloor \Lambda_c = 0$, $\Lambda_c \rfloor \Lambda = 0$. From $\mathcal{B}$, $\mathcal{E}$ we derive the bundles $\mathcal{T}(\mathcal{B})$, $\mathcal{T}^*(\mathcal{B})$, $\mathcal{T}(\mathcal{E})$, $\mathcal{T}^*(\mathcal{E})$ with local coordinates $(t^0, \dot{t}^0)$, $(t^0, \dot{t}_0)$, $(t^0, x^\alpha, \dot{t}^0, \dot{x}^\alpha)$, $(t^0, x^\alpha, \dot{t}_0, \dot{x}_\alpha)$. Let $T(\pi) : \mathcal{T}(\mathcal{E}) \to \mathcal{T}(\mathcal{B})$ denote the tangent map of π. It is well known that $\ker(T(\pi))$, the vertical bundle, is a subbundle of $\mathcal{T}(\mathcal{E})$, which is denoted by $\mathcal{V}(\mathcal{E})$. But $\mathcal{V}(\mathcal{E})$ has no distinguished complement in absence of a connection. A connection is a linear map $\Gamma : \mathcal{T}(\mathcal{E}) \to \mathcal{T}(\mathcal{E})$, which is given in coordinates by the tensor

$$\Gamma = \mathrm{d}t^0 \otimes (\partial_0 + \Gamma_0^\alpha \partial_\alpha) \,, \quad \Gamma_0^\alpha \in C^\infty(\mathcal{E}) \,, \quad \partial_\alpha = \frac{\partial}{\partial x^\alpha} \,.$$

The canonical choice for the connection is $\Gamma = \Lambda$, see (2.2), and we obtain the splitting

$$\mathcal{T}(\mathcal{E}) = \mathcal{H}(\mathcal{E}) \oplus \mathcal{V}(\mathcal{E}) \,, \quad \mathcal{H}(\mathcal{E}) = \mathrm{span}\{\partial_0\} \,, \quad \mathcal{V}(\mathcal{E}) = \mathrm{span}\{\partial_\alpha\} \,.$$

A motion of a mass point is described by a section $s \in \Gamma(\mathcal{E})$. Let $j(s) \in \Gamma(J(\mathcal{E}))$ denote the first jet of s, then $j(s)$ can be identified with the differential $T(s) : \mathcal{T}(\mathcal{B}) \to \mathcal{T}(\mathcal{E})$ of the map $s : \mathcal{B} \to \mathcal{E}$. Introducing the tensor λ on $J(\mathcal{E})$,

$$\lambda = \mathrm{d}t^0 \otimes (\partial_0 + x_{1_0}^\alpha \partial_\alpha) \,, \tag{2.4}$$

we get $\lambda \circ j(s) = \mathrm{d}t^0 \otimes (\partial_0 + \partial_0(s^\alpha)\partial_\alpha) = T(s)$. General linear maps $\mathcal{T}(\mathcal{B}) \to \mathcal{T}(\mathcal{E})$ are given by $\lambda \circ r$ with $r \in \Gamma((J(\mathcal{E}), \pi_0^1, \mathcal{E}))$. A simple calculation shows that we get the vertical field $\partial_0 \rfloor (\lambda \circ r - \Lambda) \in \Gamma(\mathcal{V}(\mathcal{E}))$. Now, we are able to define the velocity $v \in \Gamma(\mathcal{V}(\mathcal{E}))$ of the moving mass by

$$v = \partial_0 \rfloor (\lambda \circ r - \Lambda) \,.$$

On the other hand, given a field $w \in \Gamma(\mathcal{V}(\mathcal{E}))$, we derive a system of explicit first order differential equations by (2.5)

$$\partial_0 \rfloor (\lambda \circ j(s) - \Lambda) = w \circ s \tag{2.5}$$

for $s \in \Gamma(\mathcal{E})$. Now a coordinate change of the type $\bar{t}^{\bar{0}} = \varphi^{\bar{0}}(t^0)$, $\bar{x}^{\bar{\alpha}} = \varphi^{\bar{\alpha}}(t, x)$, $\bar{\alpha} = 1, \ldots, n$, which preserve the bundle structure of $\mathcal{E}$, transforms Λ to

$$\bar{\Lambda} = \mathrm{d}\bar{t}^{\bar{0}} \otimes \left(\partial_{\bar{0}} + \left(\left(\partial_0 \varphi^{\bar{0}}\right)^{-1} \partial_0 \varphi^{\bar{\alpha}}\right) \circ \varphi^{-1} \partial_{\bar{\alpha}}\right) \,.$$

The reader is asked to check that (2.5) takes the form

$$\partial_{\bar{0}} \rfloor \left(\bar{\lambda} \circ j\left(\bar{s}\right) - \bar{\Lambda} \right) = \bar{w} \circ \bar{s}$$

according to

$$\begin{aligned} \dot{t}^{\bar{0}} &= \partial_0 \varphi^{\bar{0}} \dot{t}^0 \, , & \dot{t}_0 &= \partial_0 \varphi^{\bar{0}} \dot{t}_{\bar{0}} \, , \\ \dot{x}^{\bar{\alpha}} &= \partial_\alpha \varphi^{\bar{\alpha}} \dot{x}^\alpha + \partial_0 \varphi^{\bar{\alpha}} \dot{t}^0 \, , & \bar{x}^{\bar{\alpha}}_{1_{\bar{0}}} \partial_0 \varphi^{\bar{0}} &= \partial_\alpha \varphi^{\bar{\alpha}} x^\alpha_{1_0} + \partial_0 \varphi^{\bar{\alpha}} \, . \end{aligned}$$

Furthermore, the transformation rules for (2.5) are precisely the rules of the system

$$x^\alpha_{1_0} = f^\alpha\left(t, x\right) \, .$$

Since the equations (2.1) contain the acceleration of the mass point, we have to deal with second order derivatives. In mechanics this problem is solved by the introduction of another connection.

2.2 The Covariant Derivative

In general, the covariant derivative is obtained from a connection. In mechanics this connection is the Levi Civita connection, which is derived from a time independent metric in most cases. We restrict our considerations to the case of a time independent metric g, too. Again, let $\mathcal{E}$ denote the trivial bundle $(\mathcal{E}, \pi, \mathcal{B})$, $\mathcal{E} = \mathcal{B} \times \mathcal{M}$ with the one dimensional time manifold $\mathcal{B}$, the spatial manifold $\mathcal{M}$ with coordinate $\left(t^0\right)$ and coordinates $\left(x^\alpha\right)$, $\alpha = 1, \ldots, n$. Let Λ denote the special connection given by

$$\Lambda = \mathrm{d}t^0 \otimes \partial_0 \, , \quad \Lambda_c = \mathrm{d}x^\alpha \otimes \partial_\alpha \tag{2.6}$$

on $\mathcal{E}$, then we get the splitting

$$\mathcal{T}\left(\mathcal{E}\right) = \mathcal{H}\left(\mathcal{E}\right) \oplus \mathcal{V}\left(\mathcal{E}\right) \, , \quad \mathcal{H}\left(\mathcal{E}\right) = \mathrm{span}\left\{\partial_0\right\} \, .$$

Due to (2.6), adapted coordinates for $\mathcal{H}\left(\mathcal{E}\right)$, $\mathcal{V}\left(\mathcal{E}\right)$ are given by $\left(t^0, x^\alpha, \dot{t}^0\right)$, $\left(t^0, x^\alpha, \dot{x}^\alpha\right)$. From (2.6) we derive the splitting

$$\mathcal{T}^*\left(\mathcal{E}\right) = \mathcal{H}^*\left(\mathcal{E}\right) \oplus \mathcal{V}^*\left(\mathcal{E}\right) \, , \quad \mathcal{H}^*\left(\mathcal{E}\right) = \mathrm{span}\left\{\mathrm{d}t^0\right\} \, , \quad \mathcal{V}^*\left(\mathcal{E}\right) = \mathrm{span}\left\{\mathrm{d}x^\alpha\right\}$$

of $\mathcal{T}^*\left(\mathcal{E}\right)$. The adapted coordinates are $\left(t^0, x^\alpha, \dot{t}_0\right)$ for $\mathcal{H}^*\left(\mathcal{E}\right)$ and $\left(t^0, x^\alpha, \dot{x}_\alpha\right)$ for $\mathcal{V}^*\left(\mathcal{E}\right)$.

A vertical metric on $\mathcal{E}$ is a symmetric linear map $g : \mathcal{T}\left(\mathcal{E}\right) \to \mathcal{T}^*\left(\mathcal{E}\right)$ with $\ker\left(g\right) = \mathcal{H}\left(\mathcal{E}\right)$, $\mathrm{im}\left(g\right) = \mathcal{V}^*\left(\mathcal{E}\right)$. In adapted coordinates g is given by

$$\dot{x}_\alpha = g_{\alpha\beta} \dot{x}^\beta \, , \quad g_{\alpha\beta} = g_{\beta\alpha} \in C^\infty\left(\mathcal{M}\right) \tag{2.7}$$

or generally described by the tensor g,

$$g = g_{\alpha\beta} \mathrm{d}x^\alpha \otimes \mathrm{d}x^\beta$$

with an invertible matrix $\left[g_{\alpha\beta}\right]$. Since the mechanical connection is a linear connection derived from the metric (2.7), we discuss linear connections and the special case of a Riemannian manifold first.

Linear Connections on Vector Bundles Let $\mathcal{E} = (\mathcal{E}, \pi, \mathcal{B})$ denote a (vector) bundle with local coordinates (x^i, u^α), $i = 1, \ldots, p$, $\alpha = 1, \ldots, p$, (x^i) for $\mathcal{E}$ and $\mathcal{B}$. Analogously to (2.6), a connection on $\mathcal{E}$ defines a splitting of $\mathcal{T}(\mathcal{E})$. Let $(x^i, u^\alpha, \dot{x}^i, \dot{u}^\alpha)$, $(x^i, u^\alpha, u^\alpha_{1_i})$ be local coordinates of $\mathcal{T}(\mathcal{E})$, $J(\mathcal{E})$. A connection on $\mathcal{E}$ is derived from

$$\lambda = \mathrm{d}x^i \otimes \left(\partial_i + u^\alpha_{1_i}\partial_\alpha\right) , \quad \partial_i = \frac{\partial}{\partial x^i} , \quad \partial_\alpha = \frac{\partial}{\partial u^\alpha}$$

by $\Gamma = \mathrm{d}x^i \otimes (\partial_i + \Gamma^\alpha_i \partial_\alpha)$, $\Gamma \in \Gamma\left(\left(J(\mathcal{E}), \pi^1_0, \mathcal{E}\right)\right)$[1]. The *covariant derivative* $\tilde{\nabla}^\Gamma(s) : \mathcal{B} \to \mathcal{T}^*(\mathcal{B}) \otimes \mathcal{V}(\mathcal{E})$ of a section $s \in \Gamma(\mathcal{E})$ is defined as

$$\tilde{\nabla}^\Gamma(s) = \left(\mathrm{d}s^\alpha - (\Gamma^\alpha_i \circ s)\,\mathrm{d}x^i\right) \otimes \partial_\alpha . \tag{2.8}$$

If $\mathcal{E} = (\mathcal{E}, \pi, \mathcal{B})$ is a vector bundle, then let $\mathcal{E}^* = (\mathcal{E}^*, \pi, \mathcal{B})$ denote its dual bundle with local coordinates (x^i, u_α). Let $(x^i, u_\alpha, \dot{x}^i, \dot{u}_\alpha)$, $(x^i, u_\alpha, u_{\alpha 1_i})$ be local coordinates of $\mathcal{T}(\mathcal{E}^*)$, $J(\mathcal{E}^*)$, then a connection on $\mathcal{E}^*$ follows from

$$\lambda^* = \mathrm{d}x^i \otimes (\partial_i + u_{\alpha 1_i}\partial^\alpha) , \quad \partial^\alpha = \frac{\partial}{\partial u_\alpha}$$

together with a section $\Gamma^* \in \Gamma\left(\left(J(\mathcal{E}^*), \pi^1_0, \mathcal{E}^*\right)\right)$. The covariant derivative $\tilde{\nabla}^{\Gamma^*}(s) : \mathcal{B} \to \mathcal{T}^*(\mathcal{B}) \otimes \mathcal{V}(\mathcal{E}^*)$ of $s \in \Gamma(\mathcal{E}^*)$ is defined as

$$\tilde{\nabla}^{\Gamma^*}(s) = \left(\mathrm{d}s_\alpha - \Gamma^*_{i\alpha}\mathrm{d}x^i\right) \otimes \partial^\alpha . \tag{2.9}$$

A linear connection Γ on $\mathcal{E}$ meets

$$\Gamma^\alpha_i = \Gamma^\alpha_{i\beta} u^\beta , \quad \Gamma^\alpha_{i\beta} \in C^\infty(\mathcal{B}) \tag{2.10}$$

additionally. Let us consider the canonical map $\langle , \rangle : \mathcal{E} \times_\mathcal{B} \mathcal{E}^* \to C^\infty(\mathcal{B})$ given by $u^\alpha u_\alpha$ in coordinates and its lift to the map $j(\langle , \rangle) : J(\mathcal{E}) \times_\mathcal{B} J(\mathcal{E}^*) \to \mathcal{T}^*(\mathcal{B})$ given by $\left(u^\alpha_{1_i} u_\alpha + u^\alpha u_{\alpha 1_i}\right)\mathrm{d}x^i$. The linear connection Γ^*, $\Gamma^*_{i\alpha} = \Gamma^{*\beta}_{i\alpha} u_\beta$ on $\mathcal{E}^*$ is said to be dual to a linear connection Γ on $\mathcal{E}$, if it meets $j(\langle \Gamma, \Gamma^* \rangle) = 0$. From

$$\left(\Gamma^\alpha_{i\beta} u^\beta u_\alpha + u^\alpha \Gamma^{*\beta}_{i\alpha} u_\beta\right)\mathrm{d}x^i = \left(\Gamma^\alpha_{i\beta} + \Gamma^{*\alpha}_{i\beta}\right) u^\beta u_\alpha \mathrm{d}x^i$$

follows the coordinate expression

$$\Gamma^\alpha_{i\beta} = -\Gamma^{*\alpha}_{i\beta} . \tag{2.11}$$

From now on, we assume that $\mathcal{E}$ is a vector bundle, then the bundle $\mathcal{E} \times_\mathcal{B} \mathcal{E}$, $(x^i, u^\alpha, \bar{u}^\beta)$ and $\mathcal{V}(\mathcal{E})$ are isomorphic with the isomorphism $\bar{u}^\alpha = \dot{u}^\alpha$. This fact allows us to define a new covariant derivative by combination of this isomorphism with (2.8), (2.9). Let the vectors e_α, e^α correspond to the sections $u^\beta = \delta^\beta_\alpha$, $u_\beta = \delta^\alpha_\beta$ of $\mathcal{E}$, $\mathcal{E}^*$, then we write $v = v^\alpha e_\alpha$, $p_\alpha = p_\alpha e^\alpha$ for $v \in \Gamma(\mathcal{E})$, $p \in \Gamma(\mathcal{E}^*)$. The canonical map $\langle , \rangle : \mathcal{E} \times_\mathcal{B} \mathcal{E}^* \to C^\infty(\mathcal{B})$

[1]The symbol Γ is here used in different ways, but from the context no confusion may occur.

can be expressed as $v\rfloor p = v^\alpha p_\alpha$.. With this notation this special covariant derivative follows as

$$\begin{aligned}\nabla^\Gamma(v) &= \left(\mathrm{d}v^\alpha - (\Gamma_i^\alpha \circ v)\,\mathrm{d}x^i\right)\otimes e_\alpha \\ \nabla^\Gamma(p) &= \left(\mathrm{d}p_\alpha - (\Gamma_{i\alpha}^* \circ p)\,\mathrm{d}x^i\right)\otimes e^\alpha .\end{aligned}$$

Now we choose a linear connection Γ and its dual Γ^*. From

$$\begin{aligned}\nabla^\Gamma(v)\rfloor p + v\rfloor\nabla^\Gamma(p) &= \left(\partial_i v^\beta - \Gamma_{i\beta}^\alpha v^\beta\right)\mathrm{d}x^i p_\alpha + \left(\partial_i p_\alpha - \Gamma_{i\alpha}^{*\beta} p_\beta\right)\mathrm{d}x^i v^\alpha \\ &= \left(\mathrm{d}\left(v^\alpha p_\alpha\right) - \left(\Gamma_{i\beta}^\alpha + \Gamma_{i\beta}^{*\alpha}\right)p_\alpha v^\beta \mathrm{d}x^i\right)\end{aligned}$$

it follows that the product rule

$$\nabla^\Gamma(v\rfloor p) = \nabla^\Gamma(v)\rfloor p + v\rfloor\nabla^\Gamma(p) \tag{2.12}$$

is met, if we define

$$\nabla^\Gamma(f) = \mathrm{d}f\,,\quad f\in C^\infty(\mathcal{B})\,.$$

Finally, let $\mathcal{E}$, $\mathcal{E}^*$ be the tangent bundle $\mathcal{T}(\mathcal{M})$, cotangent bundle $\mathcal{T}^*(\mathcal{M})$ of a manifold $\mathcal{M}$ with coordinates (x^i), $i=1,\ldots,n$. Then we write

$$\nabla^\Gamma(v) = \left(\mathrm{d}v^i - (\Gamma_j^i \circ v)\,\mathrm{d}x^j\right)\otimes\partial_i \tag{2.13}$$

$$\nabla^\Gamma(p) = \left(\mathrm{d}p_i - (\Gamma_{ji}^* \circ p)\,\mathrm{d}x^j\right)\otimes\mathrm{d}x^i \tag{2.14}$$

with $v\in\Gamma(\mathcal{T}(\mathcal{M}))$, $p\in\Gamma(\mathcal{T}^*(\mathcal{M}))$. Now, the covariant derivative along a field $w\in\Gamma(\mathcal{T}(\mathcal{M}))$ is introduced as

$$\nabla_w^\Gamma(v) = w\rfloor\left(\mathrm{d}v^i - (\Gamma_j^i \circ v)\,\mathrm{d}x^j\right)\otimes\partial_i \tag{2.15}$$

$$\nabla_w^\Gamma(p) = w\rfloor\left(\mathrm{d}p_i - (\Gamma_{ji}^* \circ p)\,\mathrm{d}x^j\right)\otimes\mathrm{d}x^i \tag{2.16}$$

This definition allows us to extend the covariant derivative to higher order tensors by the requirement that the product rule with respect to the tensor product or

$$\nabla_w^\Gamma(a\otimes b) = \nabla_w^\Gamma(a)\otimes b + a\otimes\nabla_w^\Gamma(b)$$

is met for any tensor fields a, b defined on $\mathcal{M}$.

The Riemannian Connection Let us consider a manifold $\mathcal{M}$ with coordinates (x^i) together with its tangent and cotangent bundle $\mathcal{T}(\mathcal{M})$, $\mathcal{T}^*(\mathcal{M})$, where we use coordinates $(x^i,\dot{x}^i)$, $(x^i,\dot{x}_i)$. We assume the existence of an invertible map, the metric, $g:\mathcal{T}(\mathcal{M})\to\mathcal{T}^*(\mathcal{M})$ given by

$$\dot{x}_i = g_{ij}\dot{x}^j\,,\quad g_{ij}=g_{ji}\in C^\infty(\mathcal{M}) \tag{2.17}$$

or by the tensor

$$g = g_{ij}\mathrm{d}x^i\otimes\mathrm{d}x^j$$

such that $\mathcal{M}$ becomes a Riemannian manifold. The inverse $\bar{g}$ of g is written as

$$\bar{g} = \bar{g}^{ij}\partial_i\otimes\partial_j\,,\quad g_{ij}\bar{g}^{jk} = \delta_i^k\,. \tag{2.18}$$

If we extend g to a map $j(g) : J(\mathcal{T}(\mathcal{M})) \to J(\mathcal{T}^*(\mathcal{M}))$ by

$$\dot{x}_{j1_i} = g_{jk}\dot{x}^k_{1_i} + (\partial_i g_{jk})\,\dot{x}^k\,,$$

we see that $j(g)$ is affine. To make it linear, we choose a linear connection Γ and its dual Γ^*, such that

$$\left(\dot{x}_{j1_i} - \Gamma^*_{ij}\right) = g_{jk}\left(\dot{x}^k_{1_i} - \Gamma^k_i\right) \tag{2.19}$$

is met. Let Γ be symmetric, or $\Gamma^k_{ij} = \Gamma^k_{ji}$ is met, then from

$$\begin{aligned} \Gamma^*_{ij} &= g_{jk}\Gamma^k_i + (\partial_i g_{jk})\,\dot{x}^k \\ -\Gamma^k_{ij} g_{kl}\dot{x}^l &= g_{jk}\Gamma^k_{il}\dot{x}^l + (\partial_i g_{jl})\,\dot{x}^l \end{aligned}$$

it follows

$$g_{kl}\Gamma^k_{ij} + g_{jk}\Gamma^k_{il} + \partial_i g_{jl} = 0\,.$$

Furthermore, the sum of the relations

$$\begin{aligned} g_{kl}\Gamma^k_{ij} + g_{jk}\Gamma^k_{il} + \partial_i g_{jl} &= 0 \\ -g_{kj}\Gamma^k_{li} - g_{ik}\Gamma^k_{lj} - \partial_l g_{ij} &= 0 \\ g_{ki}\Gamma^k_{jl} + g_{lk}\Gamma^k_{ji} + \partial_j g_{li} &= 0 \end{aligned}$$

is

$$2g_{kl}\Gamma^k_{ij} = -\left(\partial_i g_{jl} + \partial_j g_{li} - \partial_l g_{ij}\right)\,,$$

or finally

$$-2\Gamma^k_{ij} = \bar{g}^{kl}\left(\partial_i g_{jl} + \partial_j g_{li} - \partial_l g_{ij}\right) \tag{2.20}$$

is met. Obviously, the functions Γ^k_{ij} are nothing else than the Christoffel symbols of the second kind, but most books use the opposite sign. Now it is straightforward to derive the Riemannian or Levi Civitta connection, which is nothing else then the linear connection (2.10) with the coefficients (2.20) applied to (2.13) or (2.14).

The Mechanical Covariant Derivative Again, let $\mathcal{E}$ denote the trivial bundle $(\mathcal{E}, \pi, \mathcal{B})$, $\mathcal{E} = \mathcal{B} \times \mathcal{M}$ with the one dimensional time manifold $\mathcal{B}$ the spatial manifold $\mathcal{M}$ with coordinate (t^0) and coordinates (x^α), $\alpha = 1, \ldots, n$. We define the mechanical covariant derivative ∇^Γ of $v \in \Gamma(\mathcal{T}(\mathcal{E}))$, $p \in \Gamma(\mathcal{T}^*(\mathcal{E}))$ as

$$\nabla^\Gamma(v) = \mathrm{d}v^0 \otimes \partial_0 + \left(\mathrm{d}v^\alpha - \Gamma^\alpha_{\beta\gamma} v^\gamma \mathrm{d}x^\beta\right) \otimes \partial_\alpha \tag{2.21}$$

$$\nabla^\Gamma(p) = \mathrm{d}p_0 \otimes \mathrm{d}t^0 + \left(\mathrm{d}p_\alpha - \Gamma^{*\gamma}_{\beta\alpha} p_\gamma \mathrm{d}x^\beta\right) \otimes \mathrm{d}x^\alpha \tag{2.22}$$

with the Christoffel symbols from (2.20). Furthermore, we have the relation

$$\nabla^\Gamma(v \rfloor g) = \nabla^\Gamma(v) \rfloor g \tag{2.23}$$

for any vertical field $v \in \Gamma(\mathcal{T}(\mathcal{E}))$. This relation is an immediate consequence of the construction used in (2.19).

Given the section $s \in \Gamma(\mathcal{E})$ one derives from (2.21) by

$$\nabla^{\Gamma}_{j(s)}(v) = 0$$

the first order differential equations

$$\partial_0 (v^\alpha \circ s)\,\partial_\alpha = \left(\Gamma^\alpha_{\beta\gamma} \partial_0 s^\beta v^\gamma \circ s\right)\partial_\alpha$$

for the field $v \in \Gamma(\mathcal{V}(\mathcal{E}))$. Here $j(s)$ is used as an abbreviation for the field $\partial_0 + \partial_0 s^\alpha \partial_\alpha = \partial_0 \rfloor \lambda \circ j(s)$. The flow of this equations is the linear map $\Sigma_{t^0,\bar{t}^0} : \mathcal{V}_{s(t^0)}(\mathcal{E}) \to \mathcal{V}_{s(\bar{t}^0)}(\mathcal{E})$. Now, the reader can easily convince oneself, that the relation

$$\nabla^{\Gamma}_{j(s)(t^0)}(v) = \lim_{\bar{t}^0 \to t^0} \frac{1}{\bar{t}^0 - t^0}\left(\Sigma^{-1}_{t^0,\bar{t}^0}\left(v \circ s\left(\bar{t}^0\right)\right) - v \circ s\left(t^0\right)\right)$$

is met. Although this equation resembles the definition of the Lie derivative, the main difference is, that the data of v are required on s, only. It is worth mentioning that $\Sigma_{t^0,\bar{t}^0}$ maps vertical vector fields to vertical vector fields. Similar results are easily obtained for vertical covector fields.

To give another interpretation of the covariant derivative, we consider a section $s \in \Gamma(\mathcal{E})$ together with its first jet $j(s)$. To keep the notation simple, we use for $\partial_0 \rfloor \lambda \circ j(s) \in \mathcal{T}(\mathcal{E})$ also the symbol $j(s)$ and derive from

$$\nabla^{\Gamma}_{j(s)}(j(s) \rfloor \Lambda_c) = 0$$

the system

$$\partial_0 (\partial_0 s^\alpha)\,\partial_\alpha = \left(\Gamma^\alpha_{\beta\gamma} \partial_0 s^\beta \partial_0 s^\alpha\right)\partial_\alpha \tag{2.24}$$

of second order ordinary differential equations. Now, we consider the variational problem

$$L(s) = \frac{1}{2}\int_{t_1}^{t_2} \left(g_{\beta\gamma} x_1^\beta x_1^\gamma\right) \circ j(s)\,\mathrm{d}t\,,$$

with the metric (2.17), where we look for a section $s \in \Gamma(\mathcal{E})$, which extremizes this integral. With the total time derivative $d_0 = \partial_0 + x_1^\alpha \partial_\alpha + x_2^\alpha \partial_\alpha^1$ the Euler-Lagrange equations of this problem are

$$\begin{aligned} d_1\left(\partial_\delta^1\left(\tfrac{1}{2} g_{\beta\gamma} x_1^\beta x_1^\gamma\right)\right) - \tfrac{1}{2}\partial_\delta g_{\beta\gamma} x_1^\beta x_1^\gamma &= 0 \\ d_1\left(\tfrac{1}{2} g_{\delta\gamma} x_1^\gamma + \tfrac{1}{2} g_{\beta\delta} x_1^\beta\right) - \tfrac{1}{2}\partial_\delta g_{\beta\gamma} x_1^\beta x_1^\gamma &= 0 \\ \tfrac{1}{2} g_{\delta\gamma} x_2^\gamma + \tfrac{1}{2} g_{\beta\delta} x_2^\beta + \tfrac{1}{2}\left(\partial_\beta g_{\delta\gamma} x_1^\beta x_1^\gamma + \partial_\gamma g_{\beta\delta} x_1^\beta x_1^\gamma\right) &= 0 \\ g_{\beta\delta} x_2^\beta + \tfrac{1}{2}\left(\partial_\beta g_{\delta\gamma} + \partial_\gamma g_{\beta\delta} - \partial_\delta g_{\beta\gamma}\right) x_1^\beta x_1^\gamma &= 0\,. \end{aligned}$$

With the inverse metric (2.18) we can rewrite these equations in the form

$$\bar{g}^{\alpha\delta}\left(g_{\beta\delta} x_2^\beta + \tfrac{1}{2}\left(\partial_\beta g_{\delta\gamma} + \partial_\gamma g_{\beta\delta} - \partial_\delta g_{\beta\gamma}\right) x_1^\beta x_1^\gamma\right) = 0$$

or

$$x_2^\alpha - \Gamma^\alpha_{\beta\gamma} x_1^\beta x_1^\gamma = 0$$

with the Christoffel symbols $\Gamma^\alpha_{\beta\gamma}$ from (2.20). Obviously, this is nothing else than (2.24). In this context a solution s of (2.24) is also called a *geodesic* curve.

2.3 A Mass Point

Now we are ready to give a geometric interpretation of the equations (2.1). Again, let $\mathcal{E}$ denote the trivial bundle $(\mathcal{E},\pi,\mathcal{B})$, $\mathcal{E}=\mathcal{B}\times\mathcal{M}$ with the one dimensional time manifold $\mathcal{B}$ the spatial manifold $\mathcal{M}$ with coordinate (t^0) and coordinates (x^α), $\alpha=1,\ldots,n$. We choose the connection (2.6). Furthermore, the manifold $\mathcal{M}$ is equipped with the metric g of (2.17).

Given a point mass with mass $m\in\mathbb{R}^+$, a motion is a section $s\in\Gamma(\mathcal{E})$. The velocity $v\in\mathcal{V}(\mathcal{E})$ is defined by

$$v=\partial_0\rfloor(\lambda\circ j(s)-\Lambda)=j(s)\rfloor\Lambda_c\,.$$

In the absence of external forces the mass point moves along geodesics, or

$$\nabla^{\Gamma}_{j(s)}(v)=0$$

with (2.21) is met. Again we use the symbol $j(s)$ also for $\partial_0\rfloor\lambda\circ j(s)\in\mathcal{T}(\mathcal{E})$, to keep the notation simple. Let $F=F_0\mathrm{d}t^0+F_\alpha\mathrm{d}x^\alpha$, $F_0,F_\alpha\in C^\infty(\mathcal{E})$ be a force field that acts on the mass, then we have the law

$$m\,\nabla^{\Gamma}_{j(s)}(v)=\bar{g}\rfloor F\circ s\,,\quad g_{\alpha\gamma}\bar{g}^{\gamma\beta}\mathrm{d}x^\alpha\otimes\partial_\beta=\Lambda_c \tag{2.25}$$

with the metric tensor (2.17) and its inverse (2.18). The equations (2.25) are the natural generalization of (2.1). One can generalize them even more like for gyroscopic forces. In this case the connection (2.6) must be replaced by a more complex one. But this will not be discussed here.

With the linear momentum $p=mv\rfloor g\in\Gamma(\mathcal{V}^*(\mathcal{E}))$ we can rewrite (2.25) as

$$m\,\nabla^{\Gamma}_{j(s)}(v)\rfloor g=\nabla^{\Gamma}_{j(s)}(p)=\Lambda_c\rfloor F\circ s \tag{2.26}$$

because of (2.23). Let us assume, there exists a function $V\in C^\infty(\mathcal{E})$ such that $\mathrm{d}V=-F$ is met, then we can rewrite (2.25), (2.26) as

$$\begin{aligned} m\,\nabla^{\Gamma}_{j(s)}(v) &= -\mathrm{grad}(V)\circ s\,,\quad \mathrm{grad}(V)=\bar{g}\rfloor\mathrm{d}V && (2.27)\\ \nabla^{\Gamma}_{j(s)}(p) &= -\Lambda_c\rfloor\mathrm{d}V\circ s\,. && (2.28)\end{aligned}$$

The relation (2.27) shows also the definition of the *gradient* of a function V. The equation (2.28) is nothing else than a compact form of the Euler Lagrange equations with Lagrangian $L=\frac{m}{2}v\rfloor v\rfloor g-V$. Furthermore, the relation

$$\begin{aligned} \frac{1}{2}\,\nabla^{\Gamma}_{j(s)}(v\rfloor p) &= \frac{1}{2}\,\nabla^{\Gamma}_{j(s)}(v)\rfloor p+\frac{1}{2}v\rfloor\,\nabla^{\Gamma}_{j(s)}(p)\\ \frac{m}{2}\,\nabla^{\Gamma}_{j(s)}(v\rfloor v\rfloor g) &= -\left(\frac{1}{2}(v\rfloor\Lambda_c\rfloor\mathrm{d}V)+\frac{1}{2}v\rfloor\Lambda_c\rfloor\mathrm{d}V\right)\circ s\\ \frac{m}{2}\partial_0(v\rfloor v\rfloor g\circ j(s)) &= -(v\rfloor\mathrm{d}V)\circ s\,,\end{aligned}$$

see (2.12), implies the relation

$$(d_0(E))\circ j^2(s)=\partial_0(E\circ j(s))\,,\qquad E=\frac{m}{2}v\rfloor v\rfloor g+V \tag{2.29}$$

along a solution s of (2.27). The function $E \in C^{\infty}(\mathcal{T}(\mathcal{E}))$ is nothing else than the total energy and the field d_0 denotes the total time derivative.

In order to derive the Hamilton equations, we start with

$$\begin{aligned} \nabla^{\Gamma}_{j(s)}(p) &= -\Lambda_c \rfloor \mathrm{d}V \circ s \\ j(s) \rfloor \left(\left(\mathrm{d}p_\alpha - \Gamma^{*\gamma}_{\beta\alpha} p_\gamma \mathrm{d}x^\beta \right) \otimes \mathrm{d}x^\alpha \right) &= -\left(\partial_\alpha V \mathrm{d}x^\alpha \right) \circ s \end{aligned}$$

and get from

$$\begin{aligned} j(s) \rfloor \mathrm{d}p_\alpha \otimes \mathrm{d}x^\alpha &= -\left(\left(\Gamma^{\gamma}_{\beta\alpha} p_\gamma v^\beta + \partial_\alpha V \right) \mathrm{d}x^\alpha \right) \circ s \\ &= -\left(\left(-\frac{1}{2m} p_\gamma \bar{g}^{\gamma\delta} \left(\partial_\beta g_{\alpha\delta} + \partial_\alpha g_{\delta\beta} - \partial_\delta g_{\beta\alpha} \right) \bar{g}^{\beta\zeta} p_\zeta + \partial_\alpha V \right) \mathrm{d}x^\alpha \right) \circ s \\ &= -\left(\left(-\frac{1}{2m} \bar{g}^{\gamma l} p_\gamma \left(\partial_\alpha g_{\delta\beta} \right) \bar{g}^{\beta\delta} p_\delta + \partial_\alpha V \right) \mathrm{d}x^\alpha \right) \circ s \\ &= -\left(\partial_\alpha \left(\frac{1}{2m} p_\gamma \bar{g}^{\beta\gamma} p_\beta + V \right) \mathrm{d}x^\alpha \right) \circ s \end{aligned}$$

the relation

$$j(s) \rfloor \mathrm{d}p_\alpha \otimes \mathrm{d}x^\alpha = -\left(\partial_\alpha H \mathrm{d}x^\alpha \right) \circ s \,, \quad H = \frac{1}{2m} p_\gamma \bar{g}^{\beta\gamma} p_\beta + V \,. \tag{2.30}$$

Furthermore, we have the equations

$$j(s) \rfloor \Lambda_c = \dot{\partial}^\alpha H \partial_\alpha \,, \quad \dot{\partial}^\alpha = \frac{\partial}{\partial p_\alpha} \,. \tag{2.31}$$

The reader should be aware of the fact that the equations (2.30), (2.31) are defined on $\pi^2_0(\mathcal{V}^*(\mathcal{E}))$, in particular $H \in \Gamma(J(\mathcal{E}))$ is met, since the p_α are considered as functions of $j(s)$. Now let $(t^0, x^\alpha, t_0, p_\alpha)$ be coordinates of $\mathcal{T}^*(\mathcal{E})$, where we use the coordinates ${p_\alpha}$[2] instead of $\dot{x}_\alpha$ to derive the results in standard notation. We see that the function H, now called the Hamiltonian, may also be considered as a function $H \in C^\infty(\mathcal{T}^*(\mathcal{E}))$. Let us introduce the special form $\omega_H \in \Gamma(\mathcal{T}^*(\mathcal{T}^*(\mathcal{E})))$,

$$\omega_H = p_\alpha \mathrm{d}x^\alpha - H \mathrm{d}t^0 \,. \tag{2.32}$$

The Hamilton vector field $v_H \in \Gamma(\mathcal{T}(\mathcal{T}^*(\mathcal{E})))$ follows from the relation as

$$v_H \rfloor \mathrm{d}\omega_H = 0$$

as

$$v_H = \partial_0 + \dot{\partial}^\alpha H \partial_\alpha - \partial_\alpha H \dot{\partial}^\alpha \,. \tag{2.33}$$

[2]Here, the symbol p_α is used as a coordinate for $\mathcal{T}^*(\mathcal{E})$ as well as a section $p_\alpha \mathrm{d}x^\alpha \in \Gamma(\mathcal{V}^*(\mathcal{E}))$, but no confusion should occur.

Obviously, the first order differential equations induced by v_H are the celebrated Hamilton equations. Additionally, we get from

$$v_H \rfloor v_H \rfloor \mathrm{d}\omega_H = 0$$

the relation

$$v_H \rfloor \mathrm{d}H = \partial_0 H$$

along a solution of the Hamilton equations. Obviously, the Hamilton equations (2.32), (2.33) live on $\mathcal{T}^*(\mathcal{E})$, whereas the equations (2.30), (2.31) are formulated on $\mathcal{V}^*(\mathcal{E})$. The reader should be aware of the fact that the dimensions of these vector bundles are different.

It is worth mentioning that the presented approach can easily be extended to more complex problems, especially in robotics. In this case the metric is derived from the mass matrix of a multibody system.

3 Continuum Mechanics

This section is devoted to the two balance equations of continuum mechanics, the conservation of mass and the conservation of linear momentum. We assume also the existence of the Cauchy stress tensor. Furthermore, we require that this tensor is symmetric, to avoid to deal with Cosserat continua. The most popular versions of these equations are their Eulerian and their Lagrangian formulation. The Eulerian version is more convenient for fluid dynamics, while the Lagrangian formulation meets better the requirements of elasticity. Of course, one can combine both approaches, or use other coordinate systems.

The transition from point mechanics to continuum mechanics requires the replacement of a vector by a new mathematical structure, the vector valued form. But the geometry, induced by the metric, remains, and allows us to construct new differential operators. Therefore, we start with the Eulerian picture, which is more or less the straightforward generalization from the point to the continuum case. The second part is devoted to the Lagrangian, which is derived here from the Eulerian one by a special coordinate transformation, the so called Piola transformation. This approach was chosen, to keep the mathematical effort as low as possible. But it is worth mentioning that both approaches are not equivalent in general.

3.1 The Eulerian Picture I

To start with, we consider the simple case of an intertial frame with Cartesian coordinates. Let $\mathcal{B}$ with coordinate (t^0) denote the time manifold, and let $\mathcal{M}$ be the spatial manifold with coordinates (x^α), $\alpha = 1, \ldots, n$, which is equipped with the trivial metric $g = \delta_{\alpha\beta}\mathrm{d}x^\alpha \otimes \mathrm{d}x^\beta$ in this case. The total manifold $\mathcal{B} \times \mathcal{M}$, also called the *configuration manifold*, is denoted by $\mathcal{E}$. Furthermore, we have the special spatial volume form $\mathrm{vol} = \mathrm{d}x^1 \wedge \cdots \wedge \mathrm{d}x^n$. A motion $\phi_\tau : \mathcal{E} \to \mathcal{E}$ is an isomorphism[3] that maps a configuration $\mathcal{C} \subset \mathcal{M}$ at t^0 to a configuration at $t^0 + \tau$, or equivalently, the position (x^α) of a material

[3]We do not discuss the case of non invertible maps ϕ_τ here.

point at t^0 is mapped to the position (ϕ_τ^α) at $\phi_\tau^0 = t^0 + \tau$. Roughly speaking, a configuration $\mathcal{C}$ is a set, whose elements are coordinates of the material points at a certain time. The infinitesimal generator of ϕ_τ is the field $v \in \Gamma(\mathcal{T}(\mathcal{E}))$,

$$v = \partial_0 + v^\alpha \partial_\alpha \,, \quad v^0 = \frac{\partial}{\partial \tau}\phi_\tau^0 = 1 \,, \quad v^\alpha = \frac{\partial}{\partial \tau}\phi_\tau^\alpha \,. \tag{3.1}$$

First we consider the conservation of mass. Let us assume there exists a function $\rho \in C^\infty(\mathcal{E})$, called the *mass density*. Furthermore, let $\mathcal{K} \subset \mathcal{C} \subset \mathcal{M}$ be a nice domain of integration, where $\mathcal{C}$ is a configuration of the elastic body. The principle of conservation of mass can be stated as

$$\frac{\mathrm{d}}{\mathrm{d}\tau} \int_{\phi_\tau(t^0,\mathcal{K})} \rho \operatorname{vol} \bigg|_{\tau=0} = 0 \tag{3.2}$$

in the simplest case. Since the equation (3.2) must hold for any $\mathcal{K}$ and any time t^0, we are able to derive its infinitesimal version as

$$\partial_0 \rho + \partial_\alpha (\rho v^\alpha) = 0 \,. \tag{3.3}$$

For the balance equation of linear momentum, we assume the existence and symmetry of the *Cauchy stress tensor*

$$\sigma = \sigma^{\alpha\beta} \partial_\alpha \otimes \partial_\beta \,, \quad \sigma^{\alpha\beta} = \sigma^{\beta\alpha} \in C^\infty(\mathcal{E}) \,.$$

Now, the equation of balance of *linear momentum* takes the form

$$\frac{\mathrm{d}}{\mathrm{d}\tau} \int_{\phi_\tau(t^0,\mathcal{K})} \rho \operatorname{vol} \otimes v^\alpha \partial_\alpha \bigg|_{\tau=0} = \int_{\mathcal{K}} \rho \operatorname{vol} \otimes b^\alpha \partial_\alpha + \int_{\partial\mathcal{K}} \sigma^{\alpha\beta} \partial_\alpha \rfloor \operatorname{vol} \otimes \partial_\beta \,. \tag{3.4}$$

The term $\rho\mathrm{vol} \otimes b^\alpha \partial_\alpha$ is the volume density of the body forces. The symbol $\partial\mathcal{K}$ denotes the boundary of $\mathcal{K} \subset \mathcal{C}$. Furthermore, the expression $\sigma^{\alpha\beta}\partial_\alpha \rfloor \mathrm{vol} \otimes \partial_\beta$ describes the surface density of the surface forces. Since the equation (3.4) must hold for any $\mathcal{K}$ and t^0, its differential version takes the form

$$(\partial_0 v^\alpha + v^\alpha \partial_\alpha v^\alpha)\, \rho = b^\alpha \rho + \partial_\beta \sigma^{\alpha\beta} \,, \tag{3.5}$$

where we already used the conservation of mass (3.2).

From a geometric point of view, the relation (3.2) has a clear interpretation, and the derivation of (3.3) is straightforward. Unfortunately, this is not true for (3.4), since special properties of the inertial frame with Cartesian coordinates are used. There are mainly two problems. The first one is that we sum up vectors with different footing. The second one is, that we have to transform a surface integral into a volume integral. Unfortunately, this can be done only for true forms but cannot be done for vector valued forms without additional concepts. To overcome these problems we will discuss the divergence and the covariant differential before we are able to derive the conservation of mass and linear momentum in curvilinear coordinates. Furthermore, we will clarify the connection between the integral and differential version of the balance relations.

Divergence and Covariant Differential Again, let $\mathcal{E}$ denote the trivial bundle $(\mathcal{E}, \pi, \mathcal{B})$, $\mathcal{E} = \mathcal{B} \times \mathcal{M}$ with the time manifold $\mathcal{B}$ and the spatial manifold $\mathcal{M}$ with coordinates (t^0) and (x^α), $\alpha = 1, \ldots, n$. Furthermore, the manifold $\mathcal{M}$ is a Riemannian manifold equipped with the metric g of (2.17). On $\mathcal{M}$ there exists the canonical volume element vol,

$$\mathrm{vol} = \sqrt{|\det([g_{\alpha\beta}])|}\mathrm{d}x^1 \wedge \cdots \wedge \mathrm{d}x^n \,. \tag{3.6}$$

Given a p–form $\omega \in \Gamma(\wedge^p \mathcal{T}^*(\mathcal{E}))$, we define the vertical part[4] $\mathrm{vert}(\omega) \in \Gamma(\wedge^p \mathcal{V}^*(\mathcal{E}))$ of ω by

$$\mathrm{d}t^0 \wedge \mathrm{vert}(\omega) = \mathrm{d}t^0 \wedge \omega \,. \tag{3.7}$$

Obviously, the canonical volume form vol is a vertical form.

Let us choose a flow $\phi_\tau : \mathcal{E} \to \mathcal{E}$, which meets $\phi_\tau^0 = t^0 + \tau$, then its infinitesimal generator is given by $v \in \Gamma(\mathcal{T}(\mathcal{E}))$. For a nice domain $\mathcal{K} \subset \pi^{-1}(t^0)$, we consider the integral

$$I = \frac{\mathrm{d}}{\mathrm{d}\tau} \int_{\phi_\tau(t,\mathcal{K})} \mathrm{vol} \bigg|_{\tau=0}$$

with the form vol of (3.6). From

$$\begin{aligned} I &= \int_{\mathcal{K}} \frac{\mathrm{d}}{\mathrm{d}\tau} \phi_\tau^*(\mathrm{vol}) \bigg|_{\tau=0} \\ &= \int_{\mathcal{K}} v(\mathrm{vol}) \\ &= \int_{\mathcal{K}} \mathrm{vert}(v(\mathrm{vol})) \end{aligned}$$

we derive the *divergence* of v as

$$\mathrm{vert}(v(\mathrm{vol})) = \mathrm{div}(v)\,\mathrm{vol} \tag{3.8}$$

with the coordinate expression

$$\mathrm{div}(v) = \frac{1}{\sqrt{|\det(g)|}} \partial_\alpha \left(v^\alpha \sqrt{|\det(g)|} \right) \,.$$

We need also a new type of differential the so called *covariant differential*, which acts on vector valued forms. In general the covariant differential assigns to a vector valued k-form a vector valued $(k+1)$-form. We will consider here only the special case of $\sigma \in \Gamma\left(\wedge^{n-1}(\mathcal{V}^*(\mathcal{E})) \otimes \mathcal{V}(\mathcal{E})\right)$ with

$$\sigma = \sigma^{\alpha\beta} \partial_\alpha \rfloor \mathrm{vol} \otimes \partial_\beta$$

and $\sigma^{\alpha\beta} \in C^\infty(\mathcal{E})$ and construct the map $\mathrm{d}^\Gamma : \wedge^{n-1}(\mathcal{V}^*(\mathcal{E})) \otimes \mathcal{V}(\mathcal{E}) \to \wedge^n(\mathcal{V}^*(\mathcal{E})) \otimes \mathcal{V}(\mathcal{E})$. Let us choose any section $p \in \mathcal{V}^*(\mathcal{E})$, then $\sigma \rfloor p$ is an $(n-1)$-form and we can apply the

[4]Of course, the definition of the vertical part depends on choice (2.6) for the connection.

d and vert operation in a straightforward manner. But the expression

$$\begin{aligned}\mathrm{vert}\left(\mathrm{d}\left(\sigma\rfloor p\right)\right) &= \mathrm{vert}\left(\mathrm{d}\left(p_\beta \sigma^{\alpha\beta}\partial_\alpha \rfloor \mathrm{vol}\right)\right)\\ &= \mathrm{vert}\left(\mathrm{d}p_\beta\right)\wedge \sigma^{\alpha\beta}\partial_\alpha\rfloor\mathrm{vol} + p_\beta \mathrm{vert}\left(\mathrm{d}\left(\sigma^{\alpha\beta}\partial_\alpha\rfloor\mathrm{vol}\right)\right)\end{aligned}$$

shows that the differential depends on p_β, as well as on its differentials $\mathrm{d}p_\beta$. We may rewrite this equation as

$$\begin{aligned}\mathrm{vert}\left(\mathrm{d}\left(\sigma\rfloor p\right)\right) &= \mathrm{vert}\left(\mathrm{d}p_\beta - \Gamma^{*\gamma}_{\delta\beta}p_\gamma \mathrm{d}x^\delta\right)\wedge \sigma^{\alpha\beta}\partial_\alpha\rfloor\mathrm{vol}\\ &\quad +\Gamma^{*\gamma}_{\delta\beta}p_\gamma \mathrm{d}x^\delta \wedge \sigma^{\alpha\beta}\partial_\alpha\rfloor\mathrm{vol} + p_\beta \mathrm{vert}\left(\mathrm{d}\left(\sigma^{\alpha\beta}\partial_\alpha\rfloor\mathrm{vol}\right)\right)\end{aligned}$$

with

$$\nabla^\Gamma(p) = \left(\mathrm{d}p_\beta - \Gamma^{*\gamma}_{\delta\beta}p_\gamma\mathrm{d}x^\delta\right)\otimes \mathrm{d}x^\beta$$

and the Christoffel symbols from (2.20). Now, we define the *covariant differential* of σ as

$$\mathrm{d}^\Gamma\sigma = \left(\Gamma^{*\beta}_{\alpha\gamma}\sigma^{\alpha\gamma}\mathrm{vol} + \mathrm{vert}\left(\mathrm{d}\left(\sigma^{\alpha\beta}\partial_\alpha\rfloor\mathrm{vol}\right)\right)\right)\otimes\partial_\beta\,. \tag{3.9}$$

It is straightforward to show that $\mathrm{d}^\Gamma\sigma$ meets the relation

$$\mathrm{vert}\left(\mathrm{d}\left(\sigma\rfloor p\right)\right) = \left(\mathrm{d}^\Gamma\sigma\right)\rfloor p + \hat{\otimes}\left(\tilde{\sigma}\rfloor\nabla^\Gamma(p)\right)\,, \tag{3.10}$$

where the symbol $\hat{\otimes}(e)$ denotes the replacement of $\otimes$ by $\wedge$ in the expression e and $\tilde{\sigma}$ is the tensor $\tilde{\sigma} = \sigma^{\beta\alpha}\partial_\alpha\rfloor\mathrm{vol}\otimes\partial_\beta$. A special case of (3.10) is given, if $\sigma^{\alpha\beta} = \sigma^{\beta\alpha}$ and $p = v\rfloor g$ is met. Let us introduce the tensor

$$d = \frac{1}{2}v(g)\,. \tag{3.11}$$

From

$$\begin{aligned}v(g) &= v\left(g_{\alpha\beta}\right)\mathrm{d}x^\alpha\otimes\mathrm{d}x^\beta + g_{\alpha\beta}\mathrm{d}v^\alpha\otimes\mathrm{d}x^\beta + g_{\alpha\beta}\mathrm{d}x^\alpha\otimes\mathrm{d}v^\beta\\ &= \left(v^\gamma\partial_\gamma g_{\alpha\beta}\right)\mathrm{d}x^\alpha\otimes\mathrm{d}x^\beta + 2g_{\alpha\beta}\mathrm{d}v^\alpha\otimes\mathrm{d}x^\beta\\ &= 2g_{\alpha\beta}\mathrm{d}v^\alpha\otimes\mathrm{d}x^\beta - \left(g_{\delta\beta}\Gamma^\delta_{\gamma\alpha} + g_{\alpha\delta}\Gamma^\delta_{\gamma\beta}\right)v^\gamma\mathrm{d}x^\alpha\otimes\mathrm{d}x^\beta\end{aligned}$$

with the Christoffel symbols $\Gamma^\delta_{\gamma\alpha}$ and

$$\begin{aligned}\hat{\otimes}\left(\sigma\rfloor\frac{1}{2}v(g)\right) &= \left(\sigma^{\alpha\beta}\partial_\alpha\rfloor\mathrm{vol}\right)\left(\partial_\beta\rfloor\left(g_{\gamma\delta}\mathrm{d}v^\gamma - \frac{1}{2}\left(g_{\xi\delta}\Gamma^\xi_{\chi\gamma} + g_{\gamma\xi}\Gamma^\xi_{\chi\delta}\right)v^\chi\mathrm{d}x^\gamma\right)\right)\wedge\mathrm{d}x^\delta\\ &= \left(\left(\sigma^{\alpha\beta}\partial_\alpha\rfloor\mathrm{vol}\right)\left(g_{\gamma\delta}\partial_\beta v^\gamma - \frac{1}{2}\left(g_{\xi\delta}\Gamma^\xi_{\chi\beta} + g_{\beta\xi}\Gamma^\xi_{\chi\delta}\right)v^\chi\right)\right)\wedge\mathrm{d}x^\delta\\ &= \left(\sigma^{\alpha\beta}\left(g_{\gamma\alpha}\partial_\beta v^\gamma - \frac{1}{2}\left(g_{\xi\alpha}\Gamma^\xi_{\chi\beta} + g_{\xi\beta}\Gamma^\xi_{\chi\alpha}\right)v^\chi\right)\right)\mathrm{vol}\\ &= \left(\sigma^{\alpha\beta}\left(g_{\gamma\alpha}\partial_\beta v^\gamma - g_{\xi\alpha}\Gamma^\xi_{\chi\beta}v^\chi\right)\right)\mathrm{vol}\end{aligned}$$

as well as

$$\begin{aligned}\stackrel{\wedge}{\otimes}\left(\sigma\rfloor\nabla^{\Gamma}\left(v\rfloor g\right)\right) &= \stackrel{\wedge}{\otimes}\left(\sigma\rfloor\left(\nabla^{\Gamma}\left(v\right)\rfloor g\right)\right)\\ &= \sigma^{\alpha\beta}g_{\beta\gamma}\partial_{\delta}v^{\gamma}\partial_{\alpha}\rfloor\mathrm{vol}\wedge\mathrm{d}x^{\delta}-\sigma^{\alpha\beta}g_{\beta\gamma}\Gamma^{\gamma}_{\xi\delta}v^{\delta}\partial_{\alpha}\rfloor\mathrm{vol}\wedge\mathrm{d}x^{\xi}\\ &= \sigma^{\alpha\beta}\left(g_{\beta\gamma}\partial_{\alpha}v^{\gamma}-g_{\beta\gamma}\Gamma^{\gamma}_{\alpha\delta}v^{\delta}\right)\mathrm{vol}\\ &= \sigma^{\alpha\beta}d_{\alpha\beta}\mathrm{vol}\end{aligned}$$

it follows

$$\stackrel{\wedge}{\otimes}\left(\sigma\rfloor\nabla^{\Gamma}\left(v\rfloor g\right)\right)=\stackrel{\wedge}{\otimes}\left(\sigma\rfloor d\right) . \tag{3.12}$$

3.2 The Eulerian Picture II

The main difference to the previous subsection *The Eulerian Picture I* is now, that the metric g on *configuration manifold* is not trivial any more. Therefore, we choose the canonical volume form vol of (3.6). Again, a motion $\phi_{\tau}:\mathcal{E}\to\mathcal{E}$ is an isomorphism that maps a configuration $\mathcal{C}\subset\mathcal{M}$ at t^0 to a configuration at $t^0+\tau$, whose infinitesimal generator of ϕ_τ is the field $v\in\Gamma\left(\mathcal{T}\left(\mathcal{E}\right)\right)$, see (3.1).

We consider here also the conservation of mass first and get

$$I=\frac{\mathrm{d}}{\mathrm{d}\tau}\int_{\phi_{\tau}(t^0,\mathcal{K})}\rho\mathrm{vol}\Big|_{\tau=0}=0\,,$$

with the mass density $\rho\in C^{\infty}\left(\mathcal{E}\right)$, the canonical volume form (3.6), and $\mathcal{K}\subset\mathcal{C}$ a nice domain of integration. From

$$\begin{aligned}I &= \int_{\mathcal{K}}\frac{\mathrm{d}}{\mathrm{d}\tau}\phi_{\tau}^{*}\left(\rho\mathrm{vol}\right)\Big|_{\tau=0}\\ &= \int_{\mathcal{K}}v\left(\rho\mathrm{vol}\right)\\ &= \int_{\mathcal{K}}\mathrm{vert}\left(v\left(\rho\mathrm{vol}\right)\right)\end{aligned}$$

we get

$$I=\int_{\mathcal{K}}\left(v\left(\rho\right)+\rho\mathrm{div}\left(v\right)\right)\mathrm{vol}=0 \tag{3.13}$$

with the divergence from (3.8). Since this relation must hold for any $\mathcal{K}$ and t^0, the differential version is given by

$$v\left(\rho\right)+\rho\mathrm{div}\left(v\right)=0\,. \tag{3.14}$$

To study the balance of linear momentum equation, we have to overcome the problem mentioned two subsections before. Therefore, we choose an arbitrary section $p\in\mathcal{V}^{*}\left(\mathcal{E}\right)$, which meets

$$\nabla^{\Gamma}\left(p\right)=0\,, \tag{3.15}$$

and consider the equation

$$\frac{\mathrm{d}}{\mathrm{d}\tau}\int_{\phi_\tau(t^0,\mathcal{K})}\rho\,\mathrm{vol}\,(\bar{v}\rfloor p)\bigg|_{\tau=0} = \int_{\mathcal{K}}\rho\mathrm{vol}\,(b\rfloor p) + \int_{\partial\mathcal{K}}\sigma\rfloor p = 0 \tag{3.16}$$

with the symmetric stress tensor $\sigma = \sigma^{\alpha\beta}\partial_\alpha\rfloor\mathrm{vol}\otimes\partial_\beta$, the volume density $\rho\mathrm{vol}\otimes b^\alpha\partial_\alpha$ of the body forces. Furthermore, $\bar{v} = \Lambda_c\rfloor v$ is the vertical part of v. Since only true forms appear in the integrals, we avoid the problems to integrate vector valued forms. The relations

$$\begin{aligned}
\int_{\mathcal{K}}\rho\mathrm{vol}\,\mathrm{vert}\,(v\,(\bar{v}\rfloor p)) &= \int_{\mathcal{K}}\rho\mathrm{vol}\,(b\rfloor p) + \int_{\mathcal{K}}\mathrm{vert}\,(\mathrm{d}\,(\sigma\rfloor p))\\
\int_{\mathcal{K}}\rho\mathrm{vol}\,((\nabla_v^\Gamma(\bar{v})\rfloor p) + (\bar{v}\rfloor\nabla_v^\Gamma(p))) &= \int_{\mathcal{K}}\rho\mathrm{vol}\,(b\rfloor p) + \int_{\mathcal{K}}\left((\mathrm{d}^\Gamma\sigma)\rfloor p + \overset{\wedge}{\otimes}(\sigma\rfloor\nabla^\Gamma(p))\right)\\
\int_{\mathcal{K}}\rho\mathrm{vol}\,(\nabla_v^\Gamma(\bar{v})\rfloor p) &= \int_{\mathcal{K}}\rho\mathrm{vol}\,(b\rfloor p) + \int_{\mathcal{K}}(\mathrm{d}^\Gamma\sigma)\rfloor p\,,
\end{aligned}$$

where we used (2.12), (3.10), (3.13), (3.15), show that

$$\int_{\mathcal{K}}\left(\rho\mathrm{vol}\otimes\nabla_v^\Gamma(\bar{v}) - \rho\mathrm{vol}\otimes b - \mathrm{d}^\Gamma\sigma\right)\rfloor p = 0 \tag{3.17}$$

must be met, if (3.16) is fulfilled for an admissible p.[5] Since equation (3.17) must hold for any $\mathcal{K}$, the infinitesimal version follows as

$$\rho\mathrm{vol}\otimes\nabla_v^\Gamma(\bar{v}) = \rho\mathrm{vol}\otimes b + \mathrm{d}^\Gamma\sigma\,. \tag{3.18}$$

To derive the energy relation analogous to (2.29) we consider the relations

$$\begin{aligned}
\rho\mathrm{vol}\otimes\nabla_v^\Gamma(\bar{v})\rfloor\bar{v}\rfloor g &= \rho\mathrm{vol}\otimes b\rfloor\bar{v}\rfloor g + (\mathrm{d}^\Gamma\sigma)\rfloor\bar{v}\rfloor g\\
\rho\mathrm{vol}\otimes\nabla_v^\Gamma(\bar{v})\rfloor\bar{v}\rfloor g + \overset{\wedge}{\otimes}(\sigma\rfloor d) &= (b\rfloor\bar{v}\rfloor g)\,\rho\mathrm{vol} + \mathrm{d}\,(\sigma\rfloor\bar{v}\rfloor g)\,,
\end{aligned}$$

where we used (3.12). Let us assume, there exists a function e_{el}, also called the stored energy function, which meets

$$\sigma^{\alpha\beta} = 2\rho\frac{\partial}{\partial g^{\alpha\beta}}e_{el}\,,$$

then we are able to rewrite the equation from above in the following form

$$\int_{\mathcal{K}} v\left(\left(\frac{1}{2}\bar{v}\rfloor\bar{v}\rfloor g + e_{el}(g)\right)\rho\mathrm{vol}\right) = \int_{\mathcal{K}}(b\rfloor\bar{v}\rfloor g)\,\rho\mathrm{vol} + \int_{\partial\mathcal{K}}\sigma\rfloor\bar{v}\rfloor g\,, \tag{3.19}$$

where again (3.13) is used. Obviously, the first integral on the right hand side of (3.19) expresses the flow of energy caused by the body forces and the second integral corresponds to the surface forces.

[5]The integral is understood here as the sum of the integrals of the components.

3.3 The Lagrangian Picture

The Lagrangian version of the equations of (3.13), (3.17) can be derived from a special coordinate transformation between the configuration manifold and a new manifold. Therefore, we introduce a new spatial manifold $\mathcal{M}_{\text{ref}}$, called the *reference manifold*, where we use coordinates (X^α), $\alpha = 1, \ldots, n$, which is equipped with a metric $G = G_{ij}\mathrm{d}X^i \otimes \mathrm{d}X^j$, $G_{ij} = G_{ji} \in C^\infty(\mathcal{M}_{\text{ref}})$, $i, j = 1, \ldots, n$. Since $\mathcal{M}_{\text{ref}}$ is a Riemannian manifold, we have the canonical volume form

$$\mathrm{VOL} = \sqrt{|\det[G_{ij}]|}\mathrm{d}X^1 \wedge \cdots \wedge \mathrm{d}X^n . \tag{3.20}$$

Again $\mathcal{B}$ with coordinate (t^0) is the time manifold such that we are able to introduce the bundles $\mathcal{E}_{\text{ref}} = (\mathcal{B} \times \mathcal{M}_{\text{ref}}, \pi_{\text{ref}}, \mathcal{B})$. A reference state of a body is a subset $\mathcal{R} \subset \mathcal{M}_{\text{ref}}$, such that an element of $\mathcal{R}$ corresponds to a material point of the body in a unique manner. We set $X^0 = t^0$ from now on to simplify the notation. The configuration manifold is the manifold $\mathcal{M}$, like in the Eulerian picture, with the coordinates (x^α), $\alpha = 1, \ldots, n$ and the metric $g = g_{\alpha\beta}\mathrm{d}x^\alpha \otimes \mathrm{d}x^\beta$. We discuss here only the case $\dim(\mathcal{M}_{\text{ref}}) = \dim(\mathcal{M})$, to simplify the following.

A motion is a section ψ of the bundle $\mathcal{E}_{\text{L}} = (\mathcal{E}_{\text{ref}} \times \mathcal{M}, \pi_{\text{L}}, \mathcal{E}_{\text{ref}})$. Roughly speaking ψ maps a reference state $\mathcal{R}$ to the configuration $\mathcal{C}$ at a certain time t^0. Let us consider its tangent map $T\psi : \mathcal{T}(\mathcal{E}_{\text{ref}}) \to \mathcal{T}(\mathcal{E}_{\text{L}})$ or the tensor

$$F_e = \mathrm{d}X^{\bar{i}} \otimes (\partial_{\bar{i}} + \partial_{\bar{i}}\psi^\alpha\partial_\alpha) , \quad \bar{i} = 0, \ldots, n , \tag{3.21}$$

where we used $X^0 = t^0$. It is worth mentioning that the time part of F_e,

$$V = (\partial_0 + V^\alpha\partial_\alpha) = \partial_0 \rfloor F_e \tag{3.22}$$

is not a field on $\mathcal{E} = (\mathcal{B} \times \mathcal{M}, \pi, \mathcal{B})$. Such a field v can only be constructed, iff ψ is invertible with respect to the coordinates (X^i) and one gets $v = V \circ \psi^{-1} \in \mathcal{T}(\mathcal{E})$. The vertical spatial part F of F_e,

$$F = F_i^\beta \mathrm{d}X^i \otimes \partial_\beta = \partial_i\psi^\beta \mathrm{d}X^i \otimes \partial_\beta , \quad i = 1, \ldots, n$$

is often called the *deformation gradient*. It is worth mentioning that F_e is nothing else than the first jet of ψ.

Conservation of Mass Before we are able to derive the formula for the conservation of mass, we need an analogous function to (3.7). Given a p–form $\omega \in \Gamma(\wedge^p\pi^*(\mathcal{T}^*(\mathcal{E}_{\text{ref}})))$, we define the vertical part $\mathrm{vert}(\omega) \in \Gamma(\wedge^p\pi^*(\mathcal{V}^*(\mathcal{E}_{\text{ref}})))$ of ω by

$$\mathrm{d}t^0 \wedge \mathrm{VERT}(\omega) = \mathrm{d}t^0 \wedge \omega . \tag{3.23}$$

Now we introduce the mass density ρ_{ref} by

$$\rho_{\text{ref}}\mathrm{VOL} = \mathrm{VERT}(\psi^*(\rho\mathrm{vol})) \tag{3.24}$$

From

$$\int_{\psi(t^0,\mathcal{S})} \rho \mathrm{vol} = \int_{\mathcal{S}} \rho_{\mathrm{ref}} \mathrm{VOL}$$

with a nice domain of integration and $\mathcal{S} \subset \mathcal{R}$, we derive the relation

$$\int_{\mathcal{S}} \partial_0 \left(\rho_{\mathrm{ref}}\right) \mathrm{VOL} = 0 \tag{3.25}$$

for the conservation of mass. Since the relation (3.25) must hold for any subset $\mathcal{S}$ and time t^0, we get the infinitesimal version as

$$\partial_0 \left(\rho_{\mathrm{ref}}\right) = 0 \,. \tag{3.26}$$

Before we can proceed with the balance of linear momentum principle, we need another mathematical tool, the Piola transformation.

The Piola Transformation Let us consider the stress tensor $\sigma = \sigma^{\alpha\beta}\partial_\alpha \rfloor \mathrm{vol} \otimes \partial_\beta$ of the Eulerian picture. We can use the map $\psi \in \mathcal{E}_{\mathrm{L}}$ to pull back the form part of σ from the configuration manifold $\mathcal{E}$ to the reference manifold. More precisely, the so called *Piola transformation* of the Cauchy stress tensor leads to the *first Piola Kirchhoff stress tensor* $P = P^{i\beta}\partial_i \rfloor \mathrm{VOL} \otimes \partial_\beta$, which meets the relation

$$\mathrm{VERT}\left(\psi^* \left(\sigma^{\alpha\beta}\partial_\alpha \rfloor \mathrm{vol}\right)\right) \otimes \partial_\beta = P^{i\beta}\partial_i \rfloor \mathrm{VOL} \otimes \partial_\beta \tag{3.27}$$

with $P^{i\alpha} \in C^\infty \left(J\left(\mathcal{E}_{\mathrm{L}}\right)\right)$.

In the subsection *Divergence and Covariant Differential* we introduced a new differential operator, which maps vector valued k-forms to vector valued $(k+1)$-forms. Therefore, we first consider a bundle $\mathcal{E}$ with local coordinates $\left(x^i\right)$, $i = 1, \ldots, p$ for the base manifold $\mathcal{B}$ and $\left(x^i, u^\alpha\right)$, $\alpha = 1, \ldots, q$ for $\mathcal{E}$, then the total derivatives d_i with respect to the independent coordinates are given by

$$d_i = \partial_i + x^\alpha_{I+1_i} \partial^I_\alpha \,.$$

The operator d_i meets

$$d_i \left(f\right) \circ j^{k+1} \left(s\right) = \partial_i \left(f \circ j^k (s)\right)$$

for $s \in \Gamma\left(\mathcal{E}\right)$ and $f \in C^\infty \left(J^k \left(\mathcal{E}\right)\right)$. Now, we consider a form $\omega \in \Gamma\left(\pi^{k,*} \wedge^p \left(\mathcal{T}^* \left(\mathcal{B}\right)\right)\right)$ with $\left(J^k \left(\mathcal{E}\right), \pi^k, \mathcal{B}\right)$ and define the new operator $\mathrm{d}_H : \pi^{k,*} \wedge^p \left(\mathcal{T}^* \left(\mathcal{B}\right)\right) \to \pi^{k+1,*} \wedge^{p+1} \left(\mathcal{T}^* \left(\mathcal{B}\right)\right)$ by

$$\mathrm{d}_H \left(\omega\right) \circ j^{k+1} \left(s\right) = \mathrm{d}\left(\omega \circ j^k \left(s\right)\right) \,.$$

In coordinates the operator d_H is given by

$$\mathrm{d}_H \left(\omega\right) = \mathrm{d}x^i \wedge d_i \left(\omega\right) \,. \tag{3.28}$$

Now, we are ready to define a new covariant differential $\mathrm{d}^\Gamma_H : \wedge^{n-1} \left(\mathcal{V}^* \left(\mathcal{E}_{\mathrm{ref}}\right)\right) \otimes \mathcal{V}\left(\mathcal{E}_{\mathrm{L}}\right) \to \wedge^n \left(\mathcal{V}^* \left(\mathcal{E}_{\mathrm{ref}}\right)\right) \otimes \mathcal{V}\left(\mathcal{E}_{\mathrm{L}}\right)$ and choose any section $p \in \mathcal{V}^* \left(\mathcal{E}_{\mathrm{L}}\right)$. Since $P \rfloor p$ is an $(n-1)$-form, we can apply the d_H and VERT operation in a straightforward manner.

Roughly speaking, we copy now the derivation of the d^Γoperator of the subsection *Divergence and Covariant differential*, but we replace the d operator by the d_H operator. From

$$\begin{aligned}\mathrm{VERT}\left(\mathrm{d}_H\left(P\rfloor p\right)\right) &= \mathrm{VERT}\left(\mathrm{d}_H\left(p_\beta P^{i\beta}\partial_i\rfloor\mathrm{VOL}\right)\right)\\ &= \mathrm{VERT}\left(\mathrm{d}_H p_\beta\right)\wedge P^{i\beta}\partial_i\rfloor\mathrm{VOL}+p_\beta\mathrm{VERT}\left(\mathrm{d}_H\left(P^{i\beta}\partial_i\rfloor\mathrm{VOL}\right)\right)\\ &= \mathrm{VERT}\left(\mathrm{d}_H p_\beta-\Gamma^{*\gamma}_{\delta\beta}p_\gamma F^\delta_i\mathrm{d}X^i\right)\wedge P^{j\beta}\partial_j\rfloor\mathrm{VOL}\\ &\quad+\mathrm{VERT}\left(\Gamma^{*\gamma}_{\delta\beta}p_\gamma F^\delta_i P^{i\beta}\mathrm{VOL}+p_\beta\mathrm{d}_H\left(P^{i\beta}\partial_i\rfloor\mathrm{VOL}\right)\right)\end{aligned}$$

we derive the horizontal *covariant differential* of P as

$$\mathrm{d}^\Gamma_H P=\mathrm{VERT}\left(\Gamma^{*\beta}_{\delta\gamma}F^\delta_i P^{i\gamma}\mathrm{VOL}+\mathrm{d}_H\left(P^{i\beta}\partial_i\rfloor\mathrm{VOL}\right)\right)\partial_\beta\ . \tag{3.29}$$

If the matrix $[F^\alpha_i]$, $i=1,\ldots,n$, see (3.21), is invertible, then we are able to introduce the *second Piola Kirchhoff stress tensor* S,

$$S=S^{ij}\partial_i\rfloor\mathrm{VOL}\otimes\partial_j=P\rfloor\bar{F}\ ,\quad \bar{F}=\bar{F}^i_\alpha\mathrm{d}x^\alpha\otimes\partial_i\ , \tag{3.30}$$

where $\left[\bar{F}^\alpha_i\right]$ is the inverse matrix of the deformation gradient. If the Cauchy stress tensor is symmetric, then the tensor S is also symmetric. Furthermore, the operator d^Γ_H meets the relation

$$\mathrm{VERT}\left(\mathrm{d}_H\left(P\rfloor p\right)\right)=\mathrm{d}^\Gamma_H P\rfloor p+\overset{\wedge}{\otimes}\left(\tilde{S}\rfloor h\left(\nabla^\Gamma\left(p\right)\right)\right) \tag{3.31}$$

with $\tilde{S}=S^{ji}\partial_i\rfloor\mathrm{VOL}\otimes\partial_j$,

$$h\left(\nabla^\Gamma\left(p\right)\right)=\left(\mathrm{d}_H p_\beta-\Gamma^{*\gamma}_{\delta\beta}p_\gamma F^\delta_i\mathrm{d}X^i\right)\otimes\mathrm{d}_H x^\beta$$

and the Christoffel symbols from (2.20).

Let us introduce the tensor C, and its Lie derivative D along V,

$$C=g_{\alpha\beta}\mathrm{d}_H x^\alpha\otimes\mathrm{d}_H x^\beta\ ,\quad D=\frac{1}{2}\partial_0\left(C\right)\ . \tag{3.32}$$

Then the Lagrangian counterpart to formula (3.12) for symmetric S is given by

$$\overset{\wedge}{\otimes}\left(S\rfloor h\left(\nabla^\Gamma\left(V\rfloor g\right)\right)\right)=\overset{\wedge}{\otimes}\left(S\rfloor D\right) \tag{3.33}$$

with V from (3.22) and symmetric S. Since one can get this relation in a manner similar to (3.12), its derivation is omitted here.

3.4 The Balance of Linear Momentum

Again let ψ denote a motion with velocity V, see (3.22). For a nice domain of integration $\mathcal{S}\subset\mathcal{R}$, we get the Lagrangian counterpart to (3.17) as

$$\int_{\mathcal{S}} \left(\rho_{\text{ref}}\text{VOL} \otimes \nabla^{\Gamma}_{V}\left(\bar{V}\right) - \rho_{\text{ref}}\text{VOL} \otimes B - \text{d}^{\Gamma}_{H}P\right) \rfloor p = 0\,,$$

where P denotes the first Piola Kirchhoff stress tensor of (3.27), $\bar{V} = \text{VERT}\,(V)$ is the vertical part of V, and B is given by $B = b^{\alpha}\partial_{\alpha}$. Here we used the balance of mass (3.25) and the horizontal covariant differential from (3.29). Furthermore, this relation must hold for any p, which meets (3.15). Since this relation must hold for every nice domain $\mathcal{S} \subset \mathcal{R}$ we derive the infinitesimal version as

$$\rho_{\text{ref}}\text{VOL} \otimes \nabla^{\Gamma}_{V}\left(\bar{V}\right) = \rho_{\text{ref}}\text{VOL} \otimes B + \text{d}^{\Gamma}_{H}P \tag{3.34}$$

with $\nabla^{\Gamma}_{V}\left(\bar{V}\right) = \left(\partial_0 V^{\alpha} - \Gamma^{\alpha}_{\beta\gamma}V^{\beta}V^{\gamma}\right)\partial_{\alpha}$.

To derive the energy relation corresponding to (3.19), we assume that S is symmetric and consider the relation

$$\begin{aligned} \rho_{\text{ref}}\text{VOL} \otimes \nabla^{\Gamma}_{V}\left(\bar{V}\right)\rfloor\bar{V}\rfloor g &= \rho_{\text{ref}}\text{VOL} \otimes B\rfloor\bar{V}\rfloor g + \left(\text{d}^{\Gamma}_{H}P\right)\rfloor\bar{V}\rfloor g \\ \rho_{\text{ref}}\text{VOL} \otimes \nabla^{\Gamma}_{V}\left(\bar{V}\right)\rfloor\bar{V}\rfloor g + \overset{\wedge}{\otimes}\left(S\rfloor D\right) &= \rho_{\text{ref}}\text{VOL} \otimes B\rfloor\bar{V}\rfloor g + \text{d}_H\left(P\rfloor\bar{V}\rfloor g\right) \end{aligned}$$

with the second Piola Kirchhoff tensor S from (3.30) and the tensors C, D from (3.32). Furthermore, the relations (3.31) and (3.33) are used here. Let us assume, there exists a function E_{el}, which meets

$$S^{ij} = 2\rho_{\text{ref}}\frac{\partial}{\partial C_{\alpha\beta}}E_{el}\,,$$

then we are able to rewrite the equation from above in the following form

$$\int_{\mathcal{S}} \partial_0\left(\left(\frac{1}{2}\bar{V}\rfloor\bar{V}\rfloor g + E_{el}\left(C\right)\right)\rho_{\text{ref}}\text{VOL}\right) = \int_{\mathcal{S}}\left(B\rfloor\bar{V}\rfloor g\right)\rho_{\text{ref}}\text{VOL} + \int_{\partial\mathcal{S}} P\rfloor\bar{V}\rfloor g\,.$$

4 Conclusion

The goal of this contribution was to give a geometric interpretation of the well known relations of point and continuum mechanics. The standard mathematical formalization of these geometric ideas leads to the introduction of a metric and the corresponding connection, as well as to several new differential operators derived from the metric. We present even a more general approach, where the covariant derivative was derived from a connection. Based on these ideas, we gave a geometric interpretation of Newton's equations. At first glance the mathematical effort seems too high for these results. Yet it could be well justified in continuum mechanics. Following this approach, we were able to develop the balance equations in their Eulerian or Lagrangian formulation in a straightforward manner. From a geometric point of view, we had only to replace points by volume forms and vectors by vector valued forms.

Bibliography

Y. Choquet-Bruhat and C. DeWitt Morette. *Analysis, Manifolds and Physics.* Elsevier Science B.V., Amsterdam, 1982.

T. Frankel. *The Geometry of Physics, An Introduction.* Cambridge University Press, Cambridge, 1998.

A. Prástaro. *Geometry of PDEs and Mechanics.* World Scientific, Singapore, 1996.

C. Truesdell and W. Noll. *The Non-Linear Field Theories of Mechanics, 3rd ed.* Springer-Verlag, New York, 2004.

F. Ziegler. *Mechanics of Solids and Fluids.* Springer-Verlag, New York, 1991.

MIX
Papier aus verantwortungsvollen Quellen
Paper from responsible sources
FSC® C105338

If you have any concerns about our products,
you can contact us on
ProductSafety@springernature.com

In case Publisher is established outside the EU,
the EU authorized representative is:
Springer Nature Customer Service Center GmbH
Europaplatz 3, 69115 Heidelberg, Germany

Printed by Libri Plureos GmbH
in Hamburg, Germany